Lecture Notes in Computer Science   14889

Founding Editors

Gerhard Goos
Juris Hartmanis

AF173703

The series Lecture Notes in Computer Science (LNCS), including its subseries Lecture Notes in Artificial Intelligence (LNAI) and Lecture Notes in Bioinformatics (LNBI), has established itself as a medium for the publication of new developments in computer science and information technology research, teaching, and education.

LNCS enjoys close cooperation with the computer science R & D community, the series counts many renowned academics among its volume editors and paper authors, and collaborates with prestigious societies. Its mission is to serve this international community by providing an invaluable service, mainly focused on the publication of conference and workshop proceedings and postproceedings. LNCS commenced publication in 1973.

Matthias Galster · Patrizia Scandurra ·
Tommi Mikkonen · Pablo Oliveira Antonino ·
Elisa Yumi Nakagawa · Elena Navarro
Editors

Software Architecture

18th European Conference, ECSA 2024
Luxembourg City, Luxembourg, September 3–6, 2024
Proceedings

 Springer

Editors
Matthias Galster 🆔
University of Canterbury
Christchurch, New Zealand

Patrizia Scandurra 🆔
University of Bergamo
Bergamo, Italy

Tommi Mikkonen 🆔
University of Helsinki
Helsinki, Finland

Pablo Oliveira Antonino 🆔
Fraunhofer IESE
Kaiserslautern, Germany

Elisa Yumi Nakagawa 🆔
University of São Paulo
São Carlos, São Paulo, Brazil

Elena Navarro 🆔
University of Castilla-La Mancha
Albacete, Spain

ISSN 0302-9743 ISSN 1611-3349 (electronic)
Lecture Notes in Computer Science
ISBN 978-3-031-70796-4 ISBN 978-3-031-70797-1 (eBook)
https://doi.org/10.1007/978-3-031-70797-1

Welcome to ECSA 2024

The European Conference on Software Architecture (ECSA) is the premier European software architecture conference, providing researchers, practitioners, and educators with a platform to present and discuss the most recent, innovative, and significant findings and experiences in the field of software architecture research and practice.

The theme of the 18th edition of ECSA was "Software Architectures for Trustworthy Software". With the increasing complexity of software and the black box nature of software components, it becomes more and more difficult - but at the same time also more important - to demonstrate that we can trust software. At the same time, legislators introduce certification and regulations of software-based technology (e.g., the Digital Services Act and the AI Act in the European Union, the Software Bill of Materials mandated by the US government, or domain-specific compliance and data privacy requirements in sectors like banking). Also, end users demand explainable and understandable software. ECSA 2024 aimed to explore how software architecture can help to build trustworthy systems that are maintainable and evolve over time. This includes socio-technical aspects of trust (e.g., what trust means for different types of stakeholders), how we define and measure trust as a quality attribute, how software architecture can support transparency in software to increase the trust of technical and non-technical stakeholders, etc. In detail, topics of interest included provenance at the architecture level, data sovereignty and data ownership and control (e.g., in systems that utilize Large Language Models), explainable software and explainable AI from a software architecture perspective, software supply chains (e.g., Google's SLSA) and architectural implications such as dependencies and security, and how software architecture can support certification, as well as governance and regulatory compliance. ECSA 2024 allowed participants to learn about software architecture principles and practices, assurance methods from and for industry to architect trustworthy software, and methods to show the suitability of the software architecture for its intended application, including risk-driven architectures, and governance tactics to mitigate security threats in design decisions. Furthermore, ECSA 2024 covered research and practical experiences in particular domains of trustworthy systems (e.g., large-scale national digital infrastructures and national digital ID systems).

This edition of ECSA was held during September 3–6, 2024, as an in-person/physical conference in the beautiful and historic city of Luxembourg. The core technical program included sessions that blended contributions from the research, industry, journal-first, and tools & demonstration tracks, plus two keynote talks. Moreover, ECSA 2024 offered a doctoral symposium track. ECSA 2024 also encompassed one tutorial on Virtual Engineering Techniques for CI/CD Pipelines and two workshops, the 3rd International workshop on Quality in Software Architecture (QUALIFIER), and the 7th Context-Aware, Autonomous and Smart Architectures International Workshop (CASA).

For ECSA 2024, we received 80 submissions for the Research track and 9 for the Industry track. For the third time, this year's ECSA Research track followed a double-blind review process. After desk-rejecting 3 papers, each paper submitted to the Research track received three reviews. Based on the discussions and recommendations of the Program Committee, we accepted 24 papers to the Research track (14 papers as full research papers, 3 papers as experience report papers, 7 papers as short papers). The Industry track accepted 3 full papers. The conference attracted papers (co-)authored by researchers, practitioners, and academics from 34 countries (Argentina, Australia, Austria, Belgium, Brazil, Canada, Chile, China, Croatia, Czechia, Denmark, Ecuador, Egypt, Estonia, Finland, France, Germany, India, Italy, Japan, Netherlands, New Zealand, Pakistan, Poland, Portugal, Russia, Singapore, Spain, Sweden, Switzerland, Tunisia, Turkey, the UK, and the USA).

The Research track of ECSA 2024 for the first time explicitly encouraged authors to support Open Science. We encouraged all contributing authors to disclose (anonymized and curated) data/artifacts to increase reproducibility. Authors were encouraged to share artifacts in a Zenodo community for the ECSA conference series (https://zenodo.org/communities/ecsa/). However, sharing research artifacts was not mandatory for submission or acceptance. To support the ECSA Open Science initiative, we also introduced a new role on the Organizing Committee.

The main ECSA program featured two keynotes:

1. The title of the first keynote, by Ralf Reussner, Full Professor at the Karlsruhe Institute of Technology in Germany, was "Software Engineering – Quo Vadis?"
2. The second keynote speaker was Jürgen Hamm, Solutions Architect SAP EMEA at NetApp Twin Solution in Germany, who spoke about "Exploring Collaborative Innovations: Digital Twins and Architectural Challenges".

We are grateful to the members of the Program Committee for their valuable and timely reviews which contributed to a high-quality technical program for ECSA 2024. Furthermore, we would like to thank the members of the Organizing Committee for successfully organizing the event with several tracks, as well as the workshop organizers, who made significant contributions to this year's event.

We thank our sponsor Springer, which funded the best paper award of ECSA 2024 and supported us by publishing the proceedings in the Lecture Notes in Computer Science series. Finally, we thank the authors of all the ECSA 2024 submissions and the attendees of the conference for their participation.

July 2024

Pablo Oliveira Antonino
Matthias Galster
Tommi Mikkonen
Elena Navarro
Elisa Yumi Nakagawa
Patrizia Scandurra

Organization

Steering Committee

Paris Avgeriou (Chair)	University of Groningen, The Netherlands
Thais Batista	Federal University of Rio Grande do Norte, Brazil
Stefan Biffl	Vienna University of Technology, Austria
Tomas Bures	Charles University, Czechia
Carlos E. Cuesta	Rey Juan Carlos University, Spain
Laurence Duchien	University of Lille, France
Matthias Galster	University of Canterbury, New Zealand
Ilias Gerostathopoulos	Vrije Universiteit Amsterdam, The Netherlands
Paola Inverardi	University of L'Aquila, Italy
Patricia Lago	Vrije Universiteit Amsterdam, The Netherlands
Grace Lewis	Carnegie Mellon Software Engineering Institute, USA
Ivano Malavolta	Vrije Universiteit Amsterdam, The Netherlands
Raffaela Mirandola	Politecnico di Milano, Italy
Henry Muccini	University of L'Aquila, Italy
Elisa Yumi Nakagawa	University of São Paulo, Brazil
Elena Navarro	University of Castilla-La Mancha, Spain
Flavio Oquendo	IRISA, University of South Brittany, France
Ipek Ozkaya	Carnegie Mellon University, USA
Patrizia Scandurra	University of Bergamo, Italy
Bedir Tekinerdogan	Wageningen University, The Netherlands
Chouki Tibermacine	University of Montpellier, France
Catia Trubiani	Gran Sasso Science Institute, Italy
Danny Weyns	KU Leuven, Belgium

Research Track

Program Co-chairs

Patrizia Scandurra	University of Bergamo, Italy
Matthias Galster	University of Canterbury, New Zealand

Program Committee

Muhammad Ali Babar	University of Adelaide, Australia
Jesper Andersson	Linnaeus University, Sweden
Paolo Arcaini	National Institute of Informatics, Japan
Chetan Arora	Monash University, Australia
Hernan Astudillo	Universidad Andrés Bello, Chile
Paris Avgeriou	University of Groningen, The Netherlands
Luciano Baresi	Politecnico di Milano, Italy
Thais Batista	Federal University of Rio Grande do Norte, Brazil
Steffen Becker	University of Stuttgart, Germany
Justus Bogner	Vrije Universiteit Amsterdam, The Netherlands
Radu Calinescu	University of York, UK
Matteo Camilli	Politecnico di Milano, Italy
Jan Carlson	Mälardalen University, Sweden
Urtado Christelle	IMT Mines Alès, France
Filipe Correia	University of Porto, Portugal
Martina De Sanctis	Gran Sasso Science Institute, Italy
Khalil Drira	LAAS-CNRS, France
David Garlan	Carnegie Mellon University, USA
Ilias Gerostathopoulos	Vrije Universiteit Amsterdam, The Netherlands
Fabian Gilson	University of Canterbury, New Zealand
Robert Heinrich	Karlsruhe Institute of Technology, Germany
Sebastian Herold	Karlstad University, Sweden
Paola Inverardi	University of L'Aquila, Italy
Anne Koziolek	Karlsruhe Institute of Technology, Germany
Patricia Lago	Vrije Universiteit Amsterdam, The Netherlands
Grace Lewis	Carnegie Mellon Software Engineering Institute, USA
Ivano Malavolta	Vrije Universiteit Amsterdam, The Netherlands
Tommi Mikkonen	University of Helsinki, Finland
Raffaela Mirandola	Karlsruhe Institute of Technology, Germany
Henry Muccini	University of L'Aquila, Italy
Angelika Musil	Technische Universität Wien, Austria
Juergen Musil	Vienna University of Technology, Austria
Evangelos Ntentos	University of Vienna, Austria
Pablo Oliveira Antonino	Fraunhofer IESE, Germany
Flavio Oquendo	IRISA (UMR CNRS), Université de Bretagne-Sud, France
Ipek Ozkaya	Carnegie Mellon University, USA
Patrizio Pelliccione	Gran Sasso Science Institute, L'Aquila, Italy
Jennifer Perez	Universidad Politécnica de Madrid, Spain

Diego Perez-Palacin   Linnaeus University, Sweden
Ana Petrovska   Technical University of Munich, Germany
Maryam Razavian   Eindhoven University of Technology,
                                   The Netherlands
Elvinia Riccobene   University of Milan, Italy
António Rito Silva   University of Lisbon, Portugal
Salah Sadou   IRISA, University of South Brittany
Riccardo Scandariato   University of Bergamo, Italy
Bradley Schmerl   Carnegie Mellon University, USA
Lionel Seinturier   University of Lille, France
Ali Shokri   Virginia Tech, USA
Jacopo Soldani   University of Pisa, Italy
Romina Spalazzese   Malmö University, Sweden
Chouki Tibermacine   University of Montpellier, France
Catia Trubiani   Gran Sasso Science Institute, Italy
Christos Tsigkanos   University of Athens, Greece
Katja Tuma   Vrije Universiteit Amsterdam, The Netherlands
André van Hoorn (†)   University of Hamburg, Germany
Dimitri Van Landuyt   KU Leuven, Belgium
Rainer Weinreich   Johannes Kepler University Linz, Austria
Anna Wingkvist   Linnaeus University, Sweden
Andrzej Zalewski   Warsaw University of Technology, Poland
Uwe Zdun   University of Vienna, Austria

Additional Reviewers

Cesar A.P. Batista   Francisco Ponce
João Biazotto   Victor Prokhorenko
Everton Cavalcante   Alexandre Sabbadin
Abdel Kader Chabi Sika Boni   Simon Schneider
Aufeef Chauhan   Ioannis Stefanakos
Raziyeh Or Razieh Dehghani   Emna Taktak
Xinwei Fang   Sylvain Vauttier
Iffat Fatima   Gricel Vazquez
Markus Funke   Jeisson Andrès Vergara Vargas
Calum Imrie   Rui Yang
Floriment Klinaku   Karim Zarour
Etienne Lemonnier   Limeng Zhang

Industrial Track

Program Committee Co-chairs

Pablo Oliveira Antonino   Fraunhofer IESE, Germany
Tommi Mikkonen   University of Helsinki, Finland

Program Committee

Klara Borowa   Warsaw University of Technology, Poland
Mario Benitez   Software Engineering Institute, Carnegie Mellon
                               University, USA
Stefan Malich   Architectural Change Management, Germany
Federico Ciccozzi   Mälardalen University, Sweden
Ian Gorton   Northeastern University Seattle, USA
Heiko Koziolek   ABB Corporate Research, Germany
Henry Muccini   University of L'Aquila, Italy
Elisa Yumi Nakagawa   University of São Paulo, Brazil
Miroslaw Staron   University of Gothenburg, Sweden
Marion Wiese   Universität Hamburg, Germany
Eoin Woods   Endava, UK
Remco de Boer   ArchiXL & Vrije Universiteit Amsterdam,
                               The Netherlands
Xabier Larrucea   Straumann Group, Spain
Olaf Zimmermann   University of Applied Sciences of Eastern
                               Switzerland, Switzerland
Uwe Zdun   University of Vienna, Austria
Anton Jansen   Philips, The Netherlands
Fabio Moretti   ENEA, Italy
Mirko D'Angelo   Ericsson Research, Sweden
Antonino Sabetta   SAP Labs, France
Damian Andrew Tamburri   TU/e, The Netherlands
Juha Savolainen   Danfoss, Finland
Semih Çetin   .
Deniz Akdur   ASELSAN, Turkey
Antero Taivalsaari   Nokia Technologies, Finland
Carlos Solis   ION Analytics, UK
Geylani Kardas   Ege University, Turkey
Carlos E. Cuesta   Rey Juan Carlos University, Spain
Andreas Morgenstern   Mercedes-Benz, Germany

Organizing Committee

General Co-chairs

Elena Navarro	University of Castilla-La Mancha, Spain
Elisa Yumi Nakagawa	University of São Paulo, Brazil

Program Co-chairs

Patrizia Scandurra	University of Bergamo, Italy
Matthias Galster	University of Canterbury, New Zealand

Industry Co-chairs

Tommi Mikkonen	University of Helsinki, Sweden
Pablo Oliveira Antonino	Fraunhofer IESE, Germany

Workshops Co-chairs

Colin C. Venter	University of Huddersfield, UK
Uwe Zdun	University of Vienna, Austria

Tools and Demos Co-chairs

Apostolos Ampatzoglou	University of Macedonia, Greece
Pablo Oliveira Antonino	Universidad Politécnica de Madrid, Spain

Doctoral Symposium Co-chairs

Barbora Buhnova	Masaryk University, Czechia
Valentina Lenarduzzi	University of Oulu, Finland

Journal First Co-chairs

Jesper Andersson	Linnaeus University, Sweden
Robert Heinrich	Karlsruhe Institute of Technology, Germany

MIP Co-chairs

Paris Avgeriou   University of Groningen, The Netherlands
Eoin Woods   Endava, UK

Open Science Chair

Justus Bogner   Vrije Universiteit Amsterdam, The Netherlands

Proceedings Co-chairs

Vasilios Andrikopoulos   University of Groningen, The Netherlands
Jamal El Hachem   University of South Brittany, IRISA Laboratory,
                                        France

Publicity and Social Media Co-chairs

Matteo Camilli   Politecnico di Milano, Italy
Joanna C. S. Santos   University of Notre Dame, USA

Student Volunteer Chair

Aurora Macias   University of Castilla-La Mancha, Spain

Web Chair

Wallace Manzano   University of São Paulo, Brazil

Local Chair

Magali Martin   University of Luxembourg, Luxembourg

Keynotes

Software Engineering – Quo Vadis?

Ralf Reussner

Karlsruhe Institute of Technology, Germany
ralf.reussner@kit.edu

Abstract. The increasing role of software in technical innovations offers new opportunities for software engineering as a discipline. In this keynote Ralf Reussner presents a SWOT analysis of our field. Strengths have usually been created through importing concepts from other disciplines, like the term architecture. In the keynote software architecture-based analysis is treated as an example of an engineering approach in software design. Opportunities are given through software engineering methods helping in challenges of systems engineering, from view consistency and model analysis to the handling of versions and variants. This means that software engineering can also play a central role in system development in terms of providing new software engineering originated systems engineering methods. However, this is threatened by the increasing trend to primarily optimize the acceptance of the next conference paper without any further scientific vision. This focus on the next published paper may seem to help us optimize our own careers, but – as a substitute for a research strategy and vision – misses the opportunity to further develop our discipline.

Keywords: software engineering · software architecture · software architecture analysis

Biography

After studying computer science at the University of Karlsruhe (T.H.) and receiving his doctorate there in 2001 Ralf Reussner worked as a senior research scientist at DSTC Pty Ltd in Melbourne. From 2003 he was awarded with a junior research group "Palladio" within German Science Foundation Emmy Noether Excellence Programme while being one of the scientific directors at the OFFIS-Institute for technology transfer and Junior Professor at University of Oldenburg. At the age of 33, he accepted an offer for professorship (W3) in software engineering at the University of Karlsruhe (T.H.), later declining offers on full professorships from Universities Osnabrueck, Hamburg and TU Munich. In summer 2006 he was appointed as the youngest director of the FZI Research Centre for Information Technology in Karlsruhe. 2011-2017 he was member of the Executive Board of the FZI. Ralf was co-founder of the "International Conference on the Quality of Software Architecture" series in 2005, one of the pillars of today's "International Conference on Software Architecture". Since 2023, Ralf Reussner has been the spokesperson for the newly established special research area Convide - Consistency in the View-based Development of Cyber-Physical Systems, where he researches on consistency between views in the development of software-intensive systems together with over 70 scientists from TU Munich, TU Dresden and U Mannheim.

Exploring Collaborative Innovations: Digital Twins and Architectural Challenges

Jürgen Hamm

NetApp Twin Solution, Germany

Abstract. Digital Twins (DTs) offer significant potential for optimizing, analyzing, and adapting complex engineered systems, particularly post-deployment. They leverage historical data and real-time streaming data from sensors to achieve these objectives. As the adoption of DTs grows, ecosystems of digital twins emerge. Multiple DTs may serve different goals, represent various components in an architecture, embody different levels of detail, abstraction, or fidelity, and be applied at both type/aggregate and instance levels. This talk will present how NetApp has been addressing these challenges with its partners, emphasizing the key motivations behind NetApp's sustained collaboration with Fraunhofer IESE, which focuses on utilizing architecture robust DTs for various architecture-centric engineering objectives, including anomaly detection, predictive maintenance, and optimization to address broader business drivers such as financial considerations, time-to-market pressures, and competitive dynamics that shape industry strategies. Attendees will gain insights into how industry-research partnerships can drive technological innovation and address critical quality requirements, including security, scalability, portability, automation, and sustainability.

Keywords: software architecture · digital twins · industry-research partnerships

Biography

Since March 2012, Jürgen Hamm has been holding the position Solutions Architect SAP EMEA at NetApp Germany. In this role, Hamm focuses on consulting customers and partners on IT-infrastructures, network technologies, SAP technologies and virtualization. Hamm builds cross-functional teams to secure the successful execution of SAP-related customer projects in EMEA. Since 2018, Jürgen Hamm is also Lead Architect NetApp Twin Solution. In this role he is promoting the development of NetApp's value offering in the Internet of Things (IoT), thus expanding opportunities with new groups of customers and broadening NetApp's go-to-market. The IoT solution "NetApp Twin Solution" is just one example of multiple solutions and demos that Hamm set up to showcase and prove NetApp's role in the IoT market. Before joining NetApp, Jürgen Hamm worked as a technical consultant at IT consultancy GOPA. He started his career as SAP technology expert at Novasoft in 1998. Jürgen is a state-certified technician in the field of automation and production engineering.

Contents

Architecture Modeling and Design

Integrating Data Quality in Industrial Big Data Architectures: An Action
Design Research Study 3
   Ipek Ustunboyacioglu, Indika Kumara, Dario Di Nucci,
   Damian Andrew Tamburri, and Willem-Jan van den Heuvel

Case Study: Applying Automated Optimization Tooling to Microservice
Environments that Scale Safely at Ancestry.com and the Learnings 20
   Darek Gajewski, Muhammad Ashfakur Rahman Arju,
   Amr S. Abdelfattah, and Tomas Cerny

The Nature of Questions that Arise During Software Architecture Design 37
   Neil B. Harrison and Ademar Aguiar

Automated Architecture Recovery for Embedded Software Systems:
An Industrial Case Study 53
   Domenico Amalfitano, Marco De Luca,
   Domenico Francesco De Angelis, and Anna Rita Fasolino

An Analysis of MLOps Architectures: A Systematic Mapping Study 69
   Faezeh Amou Najafabadi, Justus Bogner, Ilias Gerostathopoulos,
   and Patricia Lago

Attention-Based Method for Design Pattern Detection 86
   Rania Mzid, Ilyes Rezgui, and Tewfik Ziadi

Architecture Evaluation

Cause-Effect Chain-Based Diagnosis of Automotive Onboard Energy
Systems 105
   Stefan Kugele, Lorenz Schreyer, and Martin Lamprecht

Architecture-Based Issue Propagation Analysis 121
   Sandro Speth, Niklas Krieger, Robert Heinrich, and Steffen Becker

MDEPT: Microservices Design Evaluator and Performance Tester 138
   Raghad Matar and Jasmin Jahic

Microservices Architecture

Exploring Architectural Evolution in Microservice Systems Using
Repository Mining Techniques and Static Code Analysis   157
    Patric Genfer and Uwe Zdun

Temporal Community Detection in Developer Collaboration Networks
of Microservice Projects   174
    Alexander Bakhtin, Xiaozhou Li, and Davide Taibi

Uncertainty Calculation-as-a-Service: Microservice-Based Metrology
Applications   183
    Anil Cetinkaya, M. Cagri Kaya, Teklie Belay Bzuneh,
    and Halit Oguztuzun

Performance Impact of Microservice Granularity Decisions: An Empirical
Evaluation Using the Service Weaver Framework   191
    Ricardo César Mendonça Filho and Nabor C. Mendonça

Improving Comprehensibility of Event-Driven Microservice Architectures
by Graph-Based Visualizations   207
    Sven Schoop, Erik Hebisch, and Thomas Franz

Sustainability

Energy Consumption of IoT Monitoring Software Architectures in the Edge . . .   217
    Juan Sebastian Ochoa, Jennifer Pérez, Javier García,
    Daniel Guamán, Norberto Cañas, and Vanessa Rodriguez-Horcajo

Software Architecture Assessment for Sustainability: A Case Study   233
    Iffat Fatima and Patricia Lago

Trustworthiness

Modeling and Analyzing Zero Trust Architectures Regarding Performance
and Security   253
    Nicolas Boltz, Larissa Schmid, Bahareh Taghavi,
    Christopher Gerking, and Robert Heinrich

Towards Secure Management of Edge-Cloud IoT Microservices Using
Policy as Code   270
    Samodha Pallewatta and Muhammad Ali Babar

Electric Vehicle Fast-Charging Software: Architectural Considerations
Towards Trustworthiness 288
    Vick Dini, Damian Andrew Tamburri, and Elisabetta Di Nitto

Architecture Decision Making

Exploring Architectural Design Decisions in Mailing Lists and Their
Traceability to Issue Trackers 307
    Mohamed Soliman

Helping Novice Architects to Make Quality Design Decisions Using
an LLM-Based Assistant 324
    J. Andrés Díaz-Pace, Antonela Tommasel, and Rafael Capilla

Architecture Decision Records in Practice: An Action Research Study 333
    Bardha Ahmeti, Maja Linder, Raffaela Groner, and Rebekka Wohlrab

Towards Teamwise Informed Decisions On Microservice Security Smells 350
    Francisco Ponce, Jacopo Soldani, Hernán Astudillo, and Antonio Brogi

Automated Quality Concerns Extraction from User Stories and Acceptance
Criteria for Early Architectural Decisions 359
    Khubaib Amjad Alam, Hira Asif, Irum Inayat,
    and Saif-Ur-Rehman Khan

Architecture Documentation

SCATS Framework for Software Integration in Software-Defined Vehicle
with Cross-Organizational Agile Teams 371
    Jasmin Jahić

The Execution Perspective in Software Architecture Descriptions:
A Systematic Mapping 379
    Tales Viglioni, Thais Batista, Everton Cavalcante, and Flavio Oquendo

Architectural Views: The State of Practice in Open-Source Software
Projects 396
    Sofia Migliorini, Roberto Verdecchia, Ivano Malavolta,
    Patricia Lago, and Enrico Vicario

Author Index 417

Architecture Modeling and Design

Integrating Data Quality in Industrial Big Data Architectures: An Action Design Research Study

Ipek Ustunboyacioglu[1], Indika Kumara[1(✉)] (iD), Dario Di Nucci[2],
Damian Andrew Tamburri[3], and Willem-Jan van den Heuvel[1]

[1] Jheronimus Academy of Data Science, Tilburg University, 's-Hertogenbosch, The Netherlands
{i.ustunboyacioglu,i.p.k.weerasinghadewage,
w.j.a.m.v.d.heuvel}@uvt.nl
[2] University of Salerno, Salerno, Italy
d.a.tamburri@tue.nl
[3] Eindhoven University of Technology, Eindhoven, The Netherlands
ddinucci@unisa.it

Abstract. In today's data-driven business environments, organizations heavily rely on high-quality data to make informed decisions and gain a competitive advantage. Organizations typically use big data architectures to store, process, and manage their exponentially growing enterprise data. However, ensuring data quality in such scenarios remains a significant challenge for many organizations. Despite the vast number of data quality tools available, integrating such tools into big data architectures has not been fully explored. In this study, we aim to formulate design principles to support systematically incorporating data quality testing into big data architectures. For this purpose, we performed an action design research study at a large organization in the Netherlands. Finally, we employed the architecture trade-off analysis method (ATAM) to evaluate our solution design.

Keywords: Action Design Research · Data Quality · Big Data Architecture · Architecture Trade-off Analysis · ATAM

1 Introduction

Many organizations generate a large volume of data that can be used to improve their business operations and decisions [28]. The quality of the data used for decision-making is critical to make those decisions reliable, accurate, and complete [2]. As a result, data quality assessment/testing has become a prevalent topic among researchers and practitioners [5, 12]. In particular, the adoption and development of data quality tools in different organizational settings have received increased attention from researchers [1, 17, 20, 28, 32]

Organizations generally use specific data architectures (e.g., data lake, data warehouse, and Lambda [9]) to ingest, store, and process their data. Hence, to ensure the quality of organizational data, it is essential to integrate the capabilities for detecting and correcting data quality issues into existing data architectures. Many data quality tools exist [12], and the importance of ensuring big data quality has already been acknowledged in the literature [2]. However, to our knowledge, integrating data quality tools into

M. Galster et al. (Eds.): ECSA 2024, LNCS 14889, pp. 3–19, 2024.
https://doi.org/10.1007/978-3-031-70797-1_1

big data architectures has not yet received sufficient attention from researchers. Therefore, we aim to develop prescriptive design knowledge on incorporating data quality checks into big data pipelines. The following research question guides it:

What are the general design principles for a big data architecture that incorporates data quality assessment into data pipelines?

The design principles are *"fundamental propositions that aid designers in achieving a successful transfer of requirements to design"* [22]. To answer our research question, we conducted an Action Design Research (ADR) study within an organizational setting. ADR draws on design science research (DSR) and action research (AR) and provides a method for building design knowledge through developing and evaluating artifacts through researcher-practitioner collaboration within an organizational context [27]. The participating organization is a leading manufacturing company headquartered in the Netherlands. Due to confidentiality, we cannot disclose its name and details. We carried out our ADR study in collaboration with a team responsible for managing the company's data platforms and providing high-quality data to its consumers. We identified five design principles and developed a general data architecture integrating data quality checks into big data pipelines. Moreover, we used the Architecture Trade-off Analysis Method (ATAM) to evaluate the proposed architecture.

The paper is organized as follows. Section 2 presents the ADR-based research design. Section 3 describes the problem formation phase in our ADR process, whereas Sect. 4 presents the artifact's building, intervention, and evaluation cycles, including the ATAM process. Section 6 reviews the academia-industry studies on data quality assessment. Section 7 discusses the threats to validity. Finally, Sect. 8 concludes the paper and suggests future work.

2 Methodology

As we aimed to identify the design principles for integrating data quality checks into a big data architecture by developing and evaluating the relevant artifact at an organization through a researcher-practitioner collaboration, we followed the Action Design Research (ADR) methodology [27]. ADR brings Action Research (AR) practices to Design Science Research (DSR) to address its lack of consideration of the organizational settings in creating IT artifacts [8,27]. ADR puts the organizational phenomenon first place and incrementally builds, intervenes, and evaluates the artifact through continuous researcher-practitioner interactions. This iterative nature of ADR also aligns with the agile software development practices adopted by the participating company.

Figure 1 shows our research design using the ADR method, which is based on previous studies [1,10,11,27]. The ADR team conducts research and development activities. In our project, the ADR team consists of the first two authors of this paper (researchers), all members from the R&D data analytics team (practitioners), and two participants (end-users) from the two other teams that use the data platform managed by the R&D team. The ADR process consists of four main stages: 1) problem formulation, 2) cycles of building, intervention, and evaluation (BIE), 3) reflection and learning, and 4) formulation of learning.

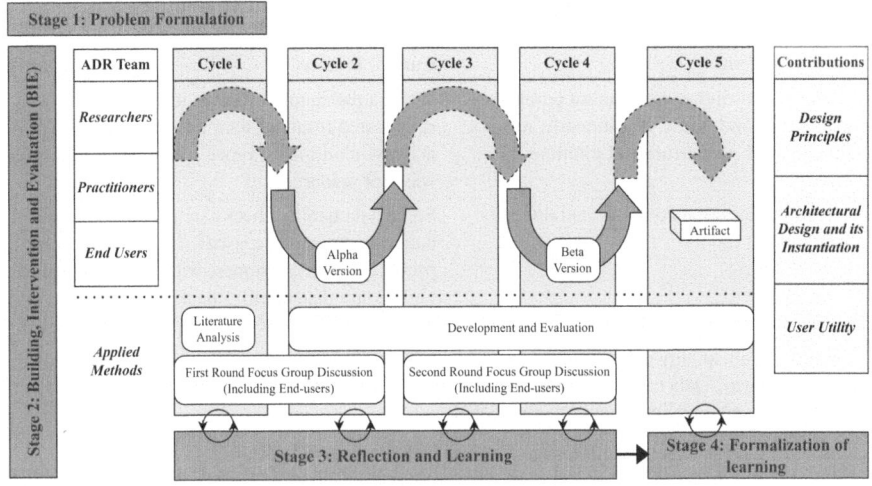

Fig. 1. Research design based on ADR

Problem Formulation. This stage defines the problem in detail based on the inputs from practitioners and researchers. A problem can be an issue/challenge the practitioners observe in their organizational environment or researchers and practitioners foresee. Section 3 describes the data quality problems we identified at the company.

Building, Intervention, and Evaluation. This stage consists of developing and evaluating the artifact with contributions from researchers, practitioners, and end-users. The initial design of the artifact is created to address the problems identified in Stage 1. The literature primarily provides the necessary design knowledge. The artifact and its design principles are refined and improved through its organizational use in multiple design cycles. In our study, the artifact was evaluated leveraging the Architecture Trade-off Analysis Method (ATAM) [14, 18], given that we develop the software architecture for a data platform. This process systematically analyzes software architectures, looking at their support for software quality attributes. It can help identify the architecture's risk areas and sensitivity points and enables practitioners to design and evaluate the architecture in parallel [19]. Section 4 presents the BIE cycles and the ATAM process we followed.

Refection and Learning. This stage aims to identify general knowledge that can be applied to a broader class of problems (instead of a solution to a specific problem instance). The reflection process is continuous and occurs together with the first two stages [27]. It consciously reflects on the problem framing, the theories chosen, and the emerging artifact in the organization context [27].

Formulation of Learning. The last stage of ADR focuses on generalizing and formalizing the learning achieved in the previous stage. According to Sein et al. [27], this stage needs to consider the generalization of problem and solution instances and the formalization of design principles that govern the artifact. Section 5 presents the design principles we formulated based on our learning from the ADR project at the company.

Table 1. Limitations of current data quality process at the company

Identified Problem	Current Process
Detecting and resolving data quality problems is labor-intensive and costly. Additionally, it is not possible to track past errors and solutions	When a data quality issue is identified, data is re-ingested from the source to the data lake. This is a time-consuming process for employees and a waste of resources
Data quality rules are incomplete and differ for each dataset	Some data quality checks are performed when data consumers raise issues. The team wants to provide data to consumers with high confidence, i.e., certain data quality checks have been completed and passed
The results of data quality checks are not stored in a structured way. Data consumers are not aware of which data quality checks are completed	Data consumers are informed about data quality check results in person or by email

3   Problem Formulation

The data analytics team under the R&D IT team of the company builds and manages a data lake to store data from various sources. The company was aware of the importance of data-driven decision-making and, hence, the importance of data quality. When this project started, ensuring data quality was the team's highest priority, as data consumers were continuously reporting problems related to data quality, such as missing data, duplicate data, and incorrect field types in data records.

The researcher organized several informal discussion sessions with some ADR team members to define the company's current data quality management processes and identify the problems related to ensuring data quality through the data lake managed by the data analytics team. Besides informal discussions with the team members, we conducted a focus group discussion session with all members of the ADR team to identify the problems in the company's current data quality assessment process. Table 1 lists the identified problems.

The first problem concerned a lack of automated data quality checks in data pipelines. A team member in a discussion session also quoted this as *"the detection of arriving data growth problems is not automated in the current process"*. The data quality checks were performed only when a consumer reported a problem with the data tables in the data lake. In this case, two sub-problems arose: how to solve the identified problem and what happens if the data consumer does not recognize the problem. To solve the problem, the data was re-ingested from the source since it was impossible to trace where the error occurred. Re-ingestion of the dataset means, at a minimum, that an employee has to spend additional time on a task they have already completed. Second, they have to use additional computing resources in their public cloud account.

The second problem was the low confidence in the data. Since the quality checks were executed upon the consumer's request, the team did not know whether a data set met specific data quality criteria, with each data set having different characteristics and requiring different data quality checks.

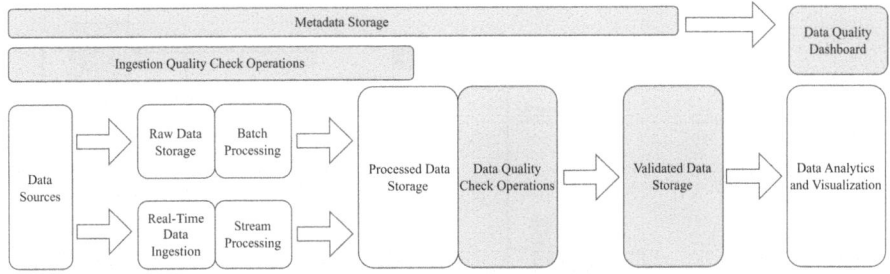

Fig. 2. Data quality assessment integrated in the big data architecture

The third problem emerged from the sub-problems of the first one. Because the results of data quality checks were not maintained anywhere, there is currently no source of information for data consumers to know when and what data quality checks have been performed.

4 Building, Intervention, and Evaluation

4.1 Cycle 1 Cycle 2: Architectural Design of Data Quality Check Integration

To address the identified problems, the ADR team built a big data architecture that allows the data analytics team and data consumers to monitor data quality in the data lake continuously. We used the existing literature and the focus group sessions with the data analytics team to design and evaluate the solution architecture. Regarding literature, we primarily used the surveys or reviews on big data architectures and big data quality, for example, [2,9,15,24,25,29].

The first round of focus group discussion was organized with the data analytics team to discuss the findings from the literature on integrating data quality tools into big data architectures. The meeting outcome was sufficient to define general functionalities and verify the initial architectural design. The discussion session was organized as a semi-structured interview via Microsoft Teams. Four data engineers from the data analytics team with different experience levels participated in the session. Three of them have been working primarily on the data lake. In contrast, the other data engineer was involved in Splunk projects, focusing on collecting and analyzing logs from the data platforms at the participating company.

Based on the output from the focus group session and the literature, the researchers in the ADR team designed the initial solution architecture, integrating data quality assessments to different phases in a big data pipeline. The team used the ATAM approach to evaluate and improve the initial architecture. Section 4.3 discusses the results of the ATAM process.

Figures 2 and 3 show the designed architecture and its instantiation at the company using the AWS cloud, which was selected as the company had already been using it. The data architecture consists of general components of a big data platform: data ingestion, data processing (batch and streaming), data storage, and serving layer (analysis and

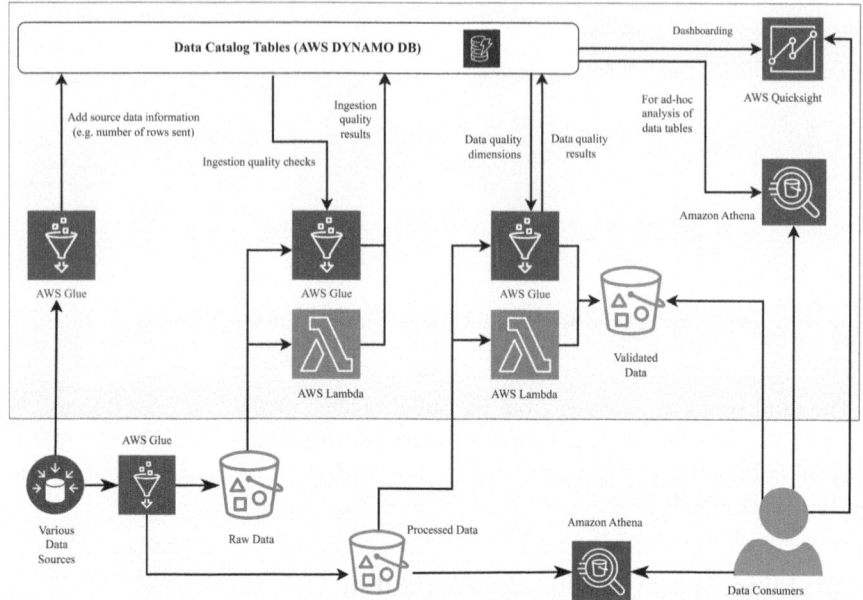

Fig. 3. Implementation of the solution architecture in Fig. 2 in AWS.

visualization) [9, 25]. In addition, to support data quality assessments, we introduced the components for running different quality checks and storing the relevant data and metadata, such as data set schemas, data quality constraints, and results.

This architecture decomposes quality checks into two sections: ingestion quality and data quality. Ingestion quality is a simple yet critical component of an ETL (Extract, Transform, and Load) pipeline. It refers to the complete data extraction from various sources without producing duplicates [28]. The first step to check the ingestion quality was calculating and storing each table's row count in the source system. In our instantiation in AWS, we queried the batch data sources and counted the records using AWS Glue data processing jobs. We used the record counting metric provided by the AWS Kinesis service for streaming data sources. AWS Glue is used for ETL-based batch data processing, and AWS Kinesis is used for ingesting and processing streaming data. An AWS lambda function triggers the Glue job whenever new data arrives. The ingestion quality check operation is designed to be run immediately after the data extraction operation. By doing so, the data analytics team can detect the errors as early as possible. It can also continuously provide data to data consumers with high confidence if no problem is identified when checking ingestion quality. One data engineer in the first focus group meeting commented on this part of the architecture as follows:

"Let's say we have all of these checks in place on the ingestion side. Then we can say, for example, automatically go back and refresh them, which reduces our workload a lot. So, ..., you get notified instead of spending your minutes on it.

... So in effect, the automation plus monitoring do reduce the time-consuming process".

Additionally, the focus group highlighted that the result of each check should be stored in a structured way. The results tables could be used as a source for dashboards, which would help identify the patterns in data errors and determine the root causes of problems related to ingestion quality.

The second part of the data quality checks focused on data quality dimensions other than ingestion quality, specifically defined for each data table. Our architecture encourages data quality constraints to be determined by data consumers. Data consumers who analyze the data to add value to the business are the end-users of the data produced by the data analytics team. Providing high-quality data is critical to the success of their work. Hence, data consumers are the main point of contact when obtaining data quality constraints. A team member also justified this decision with the following comment:

"Data quality dimensions should be provided by the data owners. So, for example, we do not understand the data, we're not the owners of that, so we should go to the people in the source databases and say what checks are required.".

For the metadata storage, we first considered AWS Glue Catalog, a data catalog tool from AWS. However, it does not support adding custom fields and calculations or cover all data quality constraints we wanted to implement. Thus, we implemented a custom metadata store using the AWS DynamoDB key-value database. A web user interface was designed and connected to the *constraints* table in DynamoDB to collect the data quality constraints from different users in a single predefined format. Data consumers are asked to enter both generic dimensions such as *completeness, uniqueness, consistency* and specific dimensions such as *max value is 5 ($\leq$ 5), mean is 10 (mean = 10)*. The system converts those defined constraints into tool-specific constraints following the approach by Schelter et al. [26]. Table 2 shows the translation of user-defined constraints into data quality tool functions. This implementation increases the users' confidence in datasets and ensures that all necessary data quality checks are automatically done for each table without manual intervention. The data quality checks were implemented using Pydeequ[1]. We used it because the company had already chosen it for data quality projects.

Regarding data quality monitoring, our solution architecture enabled the data analytics team to create data quality dashboards using the data tables that store the results of data quality checks. These dashboards helped address the third limitation identified in the company's current data quality checking process because data consumers could quickly assess the quality of data tables and act according to their needs based on their findings.

4.2   Cycle 3 - Cycle 4 - Cycle 5: Continuous Improvement

The alpha prototype is the initial artifact created in Cycles 1 and 2 of the BIE. At the beginning of Cycle 3, the researchers organized another focus group meeting with the

---

[1] https://github.com/awslabs/python-deequ.

Table 2. Data quality constraints: translations and definitions based on [26]

User Defined Constraints	Translated Constraint	Explanation
Completeness	isComplete()	Checks if there is any missing value in the given column
Uniqueness	isUnique()	Checks if there are any duplicates in the given column
≥ 1000	hasMin(1000)	Checks if the minimum value of the given column is 1000
≤ 1000	hasMax(1000)	Checks if the maximum value of the given column is 1000
$mean = 1000$	hasMean()	Checks if the mean value of the given column is 1000
string	hasDataType(String)	Checks if the column type is String
numeric	hasDataType(Numeric)	Checks if the column type is Numeric.

data analytics team to get feedback on the artifact and discuss necessary improvements for the beta prototype. Cycles 3, 4, and 5 were dedicated to improving the prototype to reach a fully functional end product. In this phase, weekly meetings were organized with a senior data engineer from the data analytics team, primarily responsible for designing and maintaining the pipelines in the data lake.

At the end of Cycle 3, the second round of focus group meetings was conducted with seven participants from the ADR team. In this meeting, the three data engineers working on the data pipelines of the data lake were present, together with two other data engineers and the product manager. The meeting started with a presentation, including a brief overview of the project and a live demonstration of the data pipeline utilizing selected data sources. The live demo was intended to demonstrate the designed pipeline's working principles and initiate the discussion of the solution architecture among the participants. The feedback given in this session was very positive and encouraging for the future improvement phase.

The architecture in the alpha prototype was evaluated in small parts, and for each part, possible areas of improvement were identified during the second round of focus group discussions. During the meeting, the participants emphasized that the structure and design of the data catalog tables play an essential role in the successful implementation of the architecture. Thus, all DynamoDB tables were carefully evaluated at this stage, and necessary improvements were made. For example, the quality check results table obtained from Pydeequ did not provide results in a user-friendly way by clearly outputting the name of the constraint, the name of the column, and the other constraint parameters. Therefore, we adopted PySpark[2] to perform some transformations on the results data frame objects accordingly, allowing us to improve the visualization of the results. Figure 4 shows a screenshot of the data quality dashboards.

At the end of the BIE cycles, the project manager, the researchers, and the senior data engineer evaluated the results. They stated that the detailed analysis of the architecture using the ATAM and the initial implementation in AWS was very informative for the company in deciding how to improve its big data architecture further.

4.3 Evaluation of the Solution Architecture Using ATAM

The ADR team employed the ATAM approach to assess the quality attributes of our solution architecture and to find and mitigate any associated risks. The ATAM process

---

[2] https://spark.apache.org/.

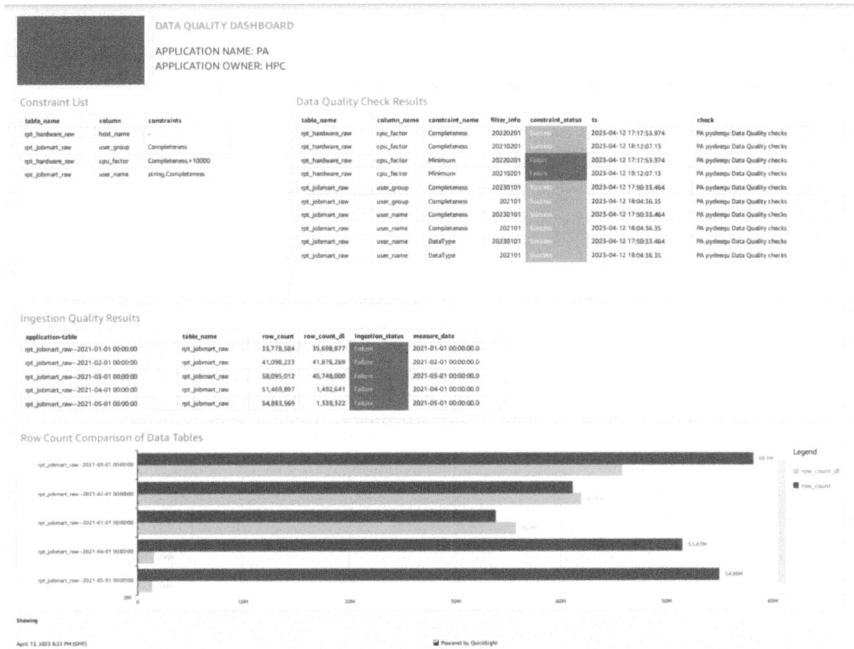

Fig. 4. AWS Quicksight visualization of data quality

includes nine steps [4], and the rest of this section briefly presents how we followed those steps.

The first three steps are introducing ATAM, discussing business drivers for the system, identifying desired quality attributes, and presenting the system's architecture. In our ADR study, the problem formulation covers the central part of the presentation phase of ATAM, where we presented the business drivers and the architecture. The stakeholders were the evaluation team, the project decision-makers, and the customers of the architecture. The evaluation team consisted of three engineers from the data analytics team who are not primarily involved with the data lake and can provide objective evaluations. The project decision-makers were a project manager, a data engineer, and a researcher. While the project manager determined the scope of the problem, the data engineer and the researcher worked together on the design and implementation of the architecture.

The next three steps of the ATAM process aim to identify the key architectural decisions and options, elicit a utility tree, and analyze the identified options based on the scenarios in the utility tree. The project decision-makers decided to consider the design options for three key architectural decisions as architectural approaches/options: 1) how to trigger data quality tests (scheduled-based and threshold-based options), 2) how to support new data sources and types (configurable pipelines for all sources and a specific pipeline per source options), and 3) how to prevent data duplication in the data

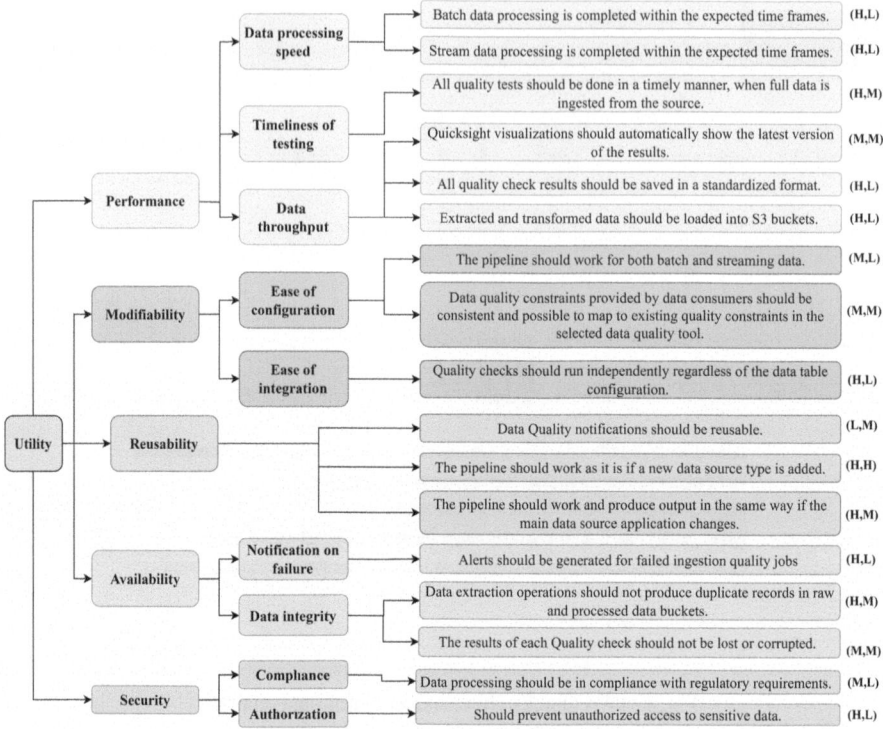

Fig. 5. The utility tree

lake (proactively prevent ingesting redundant data, executing deduplication operations, and informing consumers about the level of duplication options).

A utility tree organizes quality attribute requirements top-down, from high-level goals to fine-grained quality scenarios. Figure 5 shows our utility tree. We identified scenarios related to *performance, modifiability, reusability, availability,* and *security* of the system. Furthermore, each architectural goal, i.e., the leaf nodes, is ranked according to its importance for the success of the architecture and the difficulty/risk of achieving the goal (architect's assessment). Following the case study by Kazman et al. [18], we used relative rankings such as *high, medium, low* as they are easier to interpret and less prone to error by stakeholders. For instance, in our utility tree, $\{H, M\}$ means that this functionality has *high* importance for the success of the architecture and *medium* complexity level.

Although each quality attribute and its associated sub-factors in the utility tree need to be evaluated individually, due to time constraints (for practitioners), we only considered the ones that are ranked $\{H, H\}$ and $\{H, M\}$ to analyze further to detect risks, sensitivity points, and trade-off points. We considered all chosen scenarios. However, we only present the analysis for a single scenario due to the limited space. Our online

Table 3. An analysis example. R-risks, S-sensitive, T-trade-off, AA-architectural approaches

Scenario		All quality tests should be done promptly when full data is ingested from the source
Quality Attribute		Performance
Environment		Cloud environment (AWS) with enough capacity
Stimulus		Quality tests are initialized in ETL pipeline
Response		Ingestion and data quality Glue jobs are run for a given period.
Architectural Approach		Analysis of Architectural Approach
ID	Approach	R/T./S ID	Analysis
AA01	Quality tests will be performed according to a daily or monthly schedule	R01	There could be a delay in error detection in the data ingestion part of the pipeline.
		S01	Sensitivity point for streaming data
		T01	A risk factor for the reliability of the results in streaming data but a non-risk factor for automation of the quality checks. A risk factor for detecting ingestion quality problems on time
AA02	Quality tests will be executed once the ingested data amount reaches a threshold value	R02	Data gaps can be found at the time of the check, and consequently, low data consumer satisfaction could happen
		S02	Sensitivity point for data completeness
		T02	A risk factor for high accuracy of the results. A non-risk factor for detecting errors on time

Table 4. System requirement scenarios elicited by the stakeholders

No	Category	Scenarios
1	Use Case Scenario	AWS Glue job should be able to get row count from the sources and write the results to a DynamoDB table
2	Use Case Scenario	All quality tests should be done timely when full data is ingested from the source
3	Use Case Scenario	The data extraction operation should not produce duplicate records in raw and processed data buckets
4	Use Case Scenario	The ingestion quality job should create a notification in the system in case of failure
5	Use Case Scenario	The output of quality checks should not be overwritten. Specific keys should be defined for each table
6	Use Case Scenario	The quality checks should work independently. (No need for onboarding data from source)
7	Use Case Scenario	The data consumers' input on data quality constraints should be consistent. They should map to the quality constraints of the chosen tool
8	Use Case Scenario	Quicksight visualizations should automatically show the latest version of the results
9	Use Case Scenario	The data transformations in the pipeline should follow company standards
10	Growth Scenario	The pipeline should work as it is if a new data source is added
11	Growth Scenario	The pipeline should produce output the same way if the main data source application changes (with small to no change needed)
12	Growth Scenario	The pipeline should work for batch and streaming data
13	Exploratory Scenario	The same architecture should be usable if the main data source changes
14	Exploratory Scenario	The pipeline should prevent unauthorized access to sensitive data

appendix includes analyses of the other selected scenarios[3]. As shown in Table 3, we considered two options for executing data ingestion quality tests: schedule-based and threshold-based. Each architectural design option has its risks, sensitivity points, and trade-offs. Hence, we conducted a brainstorming session with all stakeholders to determine the best choice and decide on the risk mitigation strategies for recognized risks. The schedule-based triggering of data quality checks may fail to detect errors on time, especially for streaming data, leading to processing and storing of data with quality issues. When quality checks are conducted based on a predefined threshold, there is a risk of overlooking data gaps or missing data that falls below the threshold. For example, if the threshold for triggering a quality check is set based on the volume of ingested data or the frequency of data arrival, it is possible that smaller or intermittent data arrivals may not meet this threshold and consequently go unchecked. Moreover, if the

---

[3] https://doi.org/10.5281/zenodo.11058489.

ingestion quality check is done after a certain data amount, the results of those checks will be stored and shared with data consumers; however, since the data may not be fully ingested by that time, there is a risk for data completeness for the remaining data.

The final three steps in the ATAM process involve eliciting quality attribute scenarios using a broad group of stakeholders, prioritizing them, analyzing the high-priority scenarios and architectural approaches, and reporting the results. Table 4 presents the 14 scenarios the ADR team identified through the focus group meetings with the stakeholders, categorized into use case, growth, and exploratory scenarios. Use case scenarios consider the stakeholders' intended actions, and growth scenarios represent future improvements/changes to the system. Exploratory scenarios focus on the limits and the boundary conditions of the system. Scenarios 2, 3, and 10 in Table 4 were selected as the high-priority scenarios. As they were also in the utility tree and the analysis of the relevant architectural approaches for them had already been done in the previous phase, the ADR team decided to stop the evaluation process and produce a report summarizing the results of the ATAM process.

5   Reflection, Learning, and Formalization

By completing the BIE stage, we designed and implemented an architecture to integrate data quality assessment into big data pipelines. Incorporating data quality checks into different phases of a data pipeline is a common challenge for many organizations [2,28]. Moreover, our solution architecture in Fig. 2 is not specific to the company and can be used as a reference architecture by other organizations. To further generalize our findings, we derive five design principles for integrating data quality checks into big data pipelines by reflecting on the knowledge and experience gained from our ADR project at the company. When formulating our design principles, we followed previous guidelines [7,22,27].

Design Principle 1: Separate Ingestion Quality Checking and Data Quality Checking. To identify and correct the errors in a data pipeline before they cause problems to the downstream data consumers, it is essential to incorporate data validation checks into the correct place in the pipeline and schedule them promptly. To successfully check the data quality after the data is transformed and loaded into a data lake, the system must first ensure that the data is wholly ingested without duplicates (i.e., ingestion quality). More granular control over the data quality checks can be achieved by separating these two aspects of data quality. For example, ingestion quality checks can focus on ensuring that data is complete, accurate, and correctly formatted at the point of ingestion. In contrast, data quality checks after the transformation phase can focus on ensuring that the transformed data is consistent, valid, and meets specific business rules. Moreover, the practitioners at the company stated that by having this separation in the architecture, they could quickly identify the places in the pipeline where data quality issues occurred.

Design Principle 2: Use the Appropriate Techniques for Accurately Counting Streaming Data. When conducting quality checks, it is crucial to consider the nature and characteristics of streaming data, such as its large volume and speed [28]. Our solution architecture entails an artifact generalizable to many data applications. However, in the BIE cycle, we faced the problem of checking the record count of streaming data accurately and timely. Real-time streaming data is a collection of items that appear sequentially and may only be seen once [16]. Hence, streaming data pipelines must respond quickly, and queries must be answered promptly [21]. Several methods for advanced data stream mining have been proposed [13]. Adding a buffer or interval to the expected record count is a solution to anticipate the record count to compensate for any potential delays or lags in the data. The domain experts at the company found this solution helpful and practical.

Design Principle 3: Enable Users to Define Data Quality Constraints Consistently Based on Their Business Rules. The consistency of the definitions of the quality constraints plays a crucial role in successfully automating data quality in the pipeline [28]. For example, the absence of standardized and consistent rules for assessing data quality impeded the company's ability to ensure uniform quality across various data sources. A tool (a component in the architecture) allowing users to specify and enforce data quality constraints in a standardized way can fulfill this task. The requirements for data quality constraints could change for the given dataset. However, data quality checking jobs in the data pipelines should be robust and capable of handling all scenarios. Furthermore, the tool should support collecting business rules from data consumers. Translating business rules into data quality rules is critical for enhancing data quality in big data environments [3]. Hence, data consumers also play a crucial role in successfully integrating data quality tools into big data pipelines.

Design Principle 4: Structure the Design and Storage of Quality Checks Appropriately, as they are Invaluable Assets. As a tool to track and monitor the quality of data along the data processing pipeline, quality check tables are a crucial part of data quality management. A data catalog may benefit significantly from the systematic design and storage of quality check tables. A data catalog can inventory all data sources in an organization by storing their metadata and providing capabilities such as searching data and governing data access [23]. It is efficient to incorporate quality check tables into a data catalog by constructing and storing them systematically. The catalog can give consumers a summary of the data quality checks carried out on the data, along with any problems found and how they were fixed. Users can use such information to identify quality issues or restrictions affecting their analysis or decision-making. The proper organization of quality check tables makes the results easily accessible. When working with massive and complicated datasets, where it could be challenging to maintain and check data quality manually, this might be very crucial. Users may find accessing and comprehending the data simpler with the catalog's structured and ordered view of the quality checks.

Design Principle 5: Support Continuous Quality Monitoring and Intervention in Data Pipelines. The ability to continuously monitor the quality of the data being processed is of great value in ensuring the reliability and effectiveness of the entire data integration process [2]. By implementing structured and automated data quality checks, practitioners can continuously monitor data quality, identify and address issues promptly, validate new data sources, and contribute to overall data quality improvement. As a result, data quality issues can be recognized and resolved early before they reach data consumers, minimizing the risk of making decisions based on unreliable or erroneous information.

6 Related Work

Several studies reported industrial experience on data quality assessment [17,28,32]. Gualo et al. [17] conducted a case study research on data quality certification based on the ISO/IEC 25012 standard and reported the best practices and lessons learned. They considered the data stored in the relational databases of three organizations and developed SQL-based rules to detect data quality issues. Zhang et al. [32] developed a set of tests to validate various versions of a dataset produced by data preparation pipelines used in data science projects. Swami et al. [28] presented Data Sentinel, the data validation platform at LinkedIn. It allows users to define data quality checks declaratively and visualize data validation results.

Several AR (action research) and ADR studies exist on data quality assessment [1,20,30]. Lwakatare et al. [20] conducted an AR study at a telecommunication organization to identify the best practices for integrating data validation into machine learning (ML) projects. Their findings indicate that data validation checks should be done at various data levels (e.g., raw data sets, feature sets, and across data sets) and that tools should offer actionable feedback to the developers (e.g., showing data errors and suggestions to improve them). Altendeitering and Guggenberger [1] performed an ADR study at a pharmaceutical company to identify the design principles of a data quality tool that can detect quality issues using ML algorithms and suggest data validation rules. Westin and Sein [30] applied the ADR approach to develop and evaluate a tool for assessing and proving data quality in construction engineering projects. They primarily focus on the data sources used for engineering drawings.

Compared with the above studies, we aimed to identify the design principles for integrating data quality assessment into big data architectures through an ADR study at a large manufacturing organization in the Netherlands. We discovered five design principles, created a generic solution architecture, and instantiated it using the AWS cloud. Moreover, we applied the ATAM process to evaluate the architecture.

7 Threats to Validity

The internal validity concerns scrutinizing interventions in BIE cycles [10] and the validity of the methods employed to collect and interpret data [31]. A potential threat to this validity is the maturity and stability of the ADR team [20]. Throughout the ADR project, the team was unchanged. The primary practitioners in the team (i.e., data engineers and the product manager) had more than two years of experience in their roles. We used two formal focus group sessions for data collection. We performed the thematic analysis on the transcripts from those sessions [6]. Besides the two focus groups, we conducted many informal discussions within the ADR team. While we wrote down the key takeaways from those meetings, we did not record them as the practitioners objected to recording informal meetings.

The external validity can threaten the generalizability of our findings [31]. A potential threat to this validity is the design knowledge is derived from the ADR study at a single organization, which might not apply to other organizations. To minimize this threat, we carefully followed the guidelines proposed by the ADR literature on formalizing design principles [22,27]. More ADR studies in different organizational settings using multiple big data platforms and architectures can help validate and refine the discovered design principles and identify new ones.

8   Conclusion

We investigated the design principles and architectural support for integrating data quality assessment capabilities into big data architectures by conducting an Action Design Research (ADR) study at an organization in the Netherlands. We analyzed the designed architecture using the ATAM process, revealing its risks and sensitivity points. Our solution architecture and the prescriptive design knowledge developed throughout the ADR process can help data-driven organizations improve their big data architectures and pipelines to provide high-quality data to their data consumers. In future research, we aim to extend our architecture to support machine learning pipelines. In particular, we plan to support detecting data and concept drifts and validating ML models. Finally, we will consider automating the extraction of data quality constraints from the natural language descriptions of business rules.

References

1. Altendeitering, M., Guggenberger, T.M.: Designing data quality tools: findings from an action design research project at boehringer ingelheim. In: ECIS (2021)
2. Baldassarre, M.T., Caballero, I., Caivano, D., Rivas Garcia, B., Piattini, M.: From big data to smart data: a data quality perspective. In: Proceedings of the 1st ACM SIGSOFT International Workshop on Ensemble-Based Software Engineering, EnSEmble 2018, pp. 19–24. Association for Computing Machinery, New York (2018)
3. Baldassarre, M.T., Caballero, I., Caivano, D., Rivas Garcia, B., Piattini, M.: From big data to smart data: a data quality perspective. In: Proceedings of the 1st ACM SIGSOFT International Workshop on Ensemble-Based Software Engineering, pp. 19–24 (2018)
4. Bass, L., Clements, P., Kazman, R.: Software Architecture in Practice. Addison-Wesley Professional (2003)
5. Batini, C., Cappiello, C., Francalanci, C., Maurino, A.: Methodologies for data quality assessment and improvement. ACM Comput. Surv. (CSUR) 41(3), 1–52 (2009)
6. Braun, V., Clarke, V.: Thematic Analysis. American Psychological Association (2012)
7. Chandra, L., Seidel, S., Gregor, S.: Prescriptive knowledge in is research: conceptualizing design principles in terms of materiality, action, and boundary conditions. In: 2015 48th Hawaii International Conference on System Sciences, pp. 4039–4048. IEEE (2015)
8. Cole, R., Purao, S., Rossi, M., Sein, M.: Being proactive: where action research meets design research. In: ICIS 2005 Proceedings, p. 27 (2005)
9. Davoudian, A., Liu, M.: Big data systems: a software engineering perspective. ACM Comput. Surv. 53(5) (2020)

10. Díaz, O., Montalvillo, L., Medeiros, R., Azanza, M., Fogdal, T.: Visualizing the customization endeavor in product-based-evolving software product lines: a case of action design research. Empir. Softw. Eng. **27**(3), 75 (2022)
11. Dreyer, S., Olivotti, D., Lebek, B., Breitner, M.H.: Towards a smart services enabling information architecture for installed base management in manufacturing. Wirtschaftsinformatik und Angewandte Informatik (2017)
12. Ehrlinger, L., Wöß, W.: A survey of data quality measurement and monitoring tools. Front. Big Data **28** (2022)
13. Gaber, M.M., Zaslavsky, A., Krishnaswamy, S.: Mining data streams: a review. ACM SIGMOD Rec. **34**(2), 18–26 (2005)
14. Gallagher, B.P.: Using the architecture tradeoff analysis method to evaluate a reference architecture: a case study. Carnegie-Mellon Univ Pittsburgh PA Software Engineering Inst, Technical report (2000)
15. Gao, J., Xie, C., Tao, C.: Big data validation and quality assurance – issuses, challenges, and needs. In: 2016 IEEE Symposium on Service-Oriented System Engineering (SOSE), pp. 433–441 (2016)
16. Golab, L., Özsu, M.T.: Issues in data stream management. ACM SIGMOD Rec. **32**(2), 5–14 (2003)
17. Gualo, F., Rodriguez, M., Verdugo, J., Caballero, I., Piattini, M.: Data quality certification using ISO/IEC 25012: industrial experiences. J. Syst. Softw. **176**, 110938 (2021)
18. Kazman, R., Klein, M., Barbacci, M., Longstaff, T., Lipson, H., Carriere, J.: The architecture tradeoff analysis method. In: Proceedings. Fourth IEEE International Conference on Engineering of Complex Computer Systems (Cat. No. 98EX193), pp. 68–78. IEEE (1998)
19. Lee, J., Kang, S., Chun, H., Park, B., Lim, C.: Analysis of van-core system architecture-a case study of applying the ATAM. In: 2009 10th ACIS International Conference on Software Engineering, Artificial Intelligences, Networking and Parallel/Distributed Computing, pp. 358–363. IEEE (2009)
20. Lwakatare, L.E., Rånge, E., Crnkovic, I., Bosch, J.: On the experiences of adopting automated data validation in an industrial machine learning project. In: 2021 IEEE/ACM 43rd International Conference on Software Engineering: Software Engineering in Practice (ICSE-SEIP), pp. 248–257. IEEE (2021)
21. Manku, G.S., Motwani, R.: Approximate frequency counts over data streams. In: VLDB 2002: Proceedings of the 28th International Conference on Very Large Databases, pp. 346–357. Elsevier (2002)
22. Möller, F., Guggenberger, T.M., Otto, B.: Towards a method for design principle development in information systems. In: Hofmann, S., Müller, O., Rossi, M. (eds.) DESRIST 2020. LNCS, vol. 12388, pp. 208–220. Springer, Cham (2020). https://doi.org/10.1007/978-3-030-64823-7_20
23. Olesen-Bagneux, O.: The Enterprise Data Catalog. O'Reilly Media, Inc. (2023)
24. Pääkkönen, P., Pakkala, D.: Reference architecture and classification of technologies, products and services for big data systems. Big Data Res. **2**(4), 166–186 (2015)
25. Reis, J., Housley, M.: Fundamentals of Data Engineering. O'Reilly Media, Inc. (2022)
26. Schelter, S., Lange, D., Schmidt, P., Celikel, M., Biessmann, F., Grafberger, A.: Automating large-scale data quality verification. Proc. VLDB Endow. **11**(12), 1781–1794 (2018)
27. Sein, M.K., Henfridsson, O., Purao, S., Rossi, M., Lindgren, R.: Action Design Research. MIS Q. 37–56 (2011)
28. Swami, A., Vasudevan, S., Huyn, J.: Data sentinel: a declarative production-scale data validation platform. In: 2020 IEEE 36th International Conference on Data Engineering (ICDE), pp. 1579–1590 (2020)
29. Taleb, I., Serhani, M.A., Bouhaddioui, C., Dssouli, R.: Big data quality framework: a holistic approach to continuous quality management. J. Big Data **8**(1), 76 (2021)

30. Westin, S., Sein, M.K.: Improving data quality in construction engineering projects: an action design research approach. J. Manag. Eng. **30**(3), 05014003 (2014)
31. Wohlin, C., Runeson, P., Höst, M., Ohlsson, M.C., Regnell, B., Wesslén, A.: Experimentation in Software Engineering. Springer, Heidelberg (2012). https://doi.org/10.1007/978-3-642-29044-2
32. Zhang, L., Howard, S., Montpool, T., Moore, J., Mahajan, K., Miranskyy, A.: Automated data validation: an industrial experience report. J. Syst. Softw. **197**, 111573 (2023)

Case Study: Applying Automated Optimization Tooling to Microservice Environments that Scale Safely at Ancestry.com and the Learnings

Darek Gajewski[2], Muhammad Ashfakur Rahman Arju[1,2],
Amr S. Abdelfattah[2], and Tomas Cerny[2](✉)

[1] Computer Science, Montana University, Missoula, MT, USA
`muhammadashfaku.arju@student.montana.edu`
[2] Systems and Industrial Engineering, University of Arizona, Tucson, AZ, USA
`{dgajewski,tcerny}@arizona.edu, amr_elsayed1@baylor.edu`

Abstract. In 2018, Ancestry migrated into Amazon's public cloud. Hundreds of applications that were hosted on physical hardware were migrated to the cloud onto internally supported Kubernetes environments. This established a microservice architecture for many applications at Ancestry. In 2020, Ancestry's cloud expenses goals were set at $12M in Operating expenses savings over 2 years. To overcome the problem of scale, an automated right-sizing effort needed to take place. Ancestry purchased a tool called Opsani. Opsani allowed the team to use dynamic canary testing in combination with a CanaryAdvisor to act as the AI agent. Opsani was integrated into Ancestry's continuous integration and continuous delivery platform to automate right-sizing and help reach cost optimization targets with limited resources. After a successful proof of concept was developed by the *DNA Matches* agile team, there was a company-wide integration of Opsani into Ancestry's CI/CD platform. The tool would need to scale out continuous optimization efforts across many of the teams utilizing Kubernetes and virtual machine infrastructure. This case study details the experience and knowledge gained from scaling automation optimizations using an AI agent with more than 200 individual applications across 30 agile teams. The challenges introduced by the rate of changes in the system and monitoring feedback delays began to affect the customer experience with the product. As teams adopted optimization changes at higher rates, new guard rails were introduced to control the release cycles and ensure stability for paying customers.

Keywords: Case Study · Cloud Systems · Canary Testing · Enterprise Architecture · Container Right-sizing Optimization

1 Introduction

Microservice architecture (MSA) is the current industry standard for building scalable systems [4,6]. However, there is a challenge for larger systems and

M. Galster et al. (Eds.): ECSA 2024, LNCS 14889, pp. 20–36, 2024.
https://doi.org/10.1007/978-3-031-70797-1_2

microservices, as it creates an order of magnitude more resources to maintain and polynomial growth on artifacts to manage (network configs, JVM configs, Terraform scripts, auto-scaling groups, etc.). The most cost-effective way to reach Ancestry's 2-year saving goal of $12M in operational expenses (OpEx) savings was to automate some of the tasks to resize containers and virtual machines (VMs). The secondary effect was to reduce the development efforts and reach smaller optimizations with automation. As a business, Ancestry only spent 25-30% of its development time maintaining the code; this included the work hours needed each year to reach its optimization targets. High-level analysis showed Ancestry Kubernetes and VM environments were over-provisioned on central processing units (CPUs) and high speed memory (MEM) resources, resulting in $millions in unused physical resources Ancestry was paying for and not utilizing.

Much of Ancestry's infrastructure was maintained directly on AWS Linux and Windows-based VM infrastructure. Larger VMs make scaling harder, as more apps are bound to a larger set of resources. There are far fewer artifacts that development teams have to support. What was 100 s of VMs with established redundant models were now 1000 s of containers with many of the same practices inherited from previous development practices. When Ancestry introduced microservices applications wrapped in containers, resources for one application were fractured into many pieces. Teams now needed to maintain resources for different parts of their applications and services instead of sharing those resources. Microservice meant micro-problems and a lot of them. MSA can increase agility of the infrastructure, however, it creates more code to maintain an applications deployment templates and settings. Microservices create management overhead, and agile teams do not scale the same way. Ancestry introduced Opsani as a tool to support teams. Opsani is a tool that uses canary testing methods and a CanaryAgent AI agent, as summarized well by Tarvo et al. [15], to automate right-sizing efforts on containerized applications.

This paper shares the automation efforts and challenges experienced by Ancestry with right-sizing microservices containers at scale. The learnings of the right-sizing optimization to over 200 applications, while involving hundreds of developers, trying to reach optimization goals set by the business of $12M over 2 years. The attempted roll-out of an automated continuous optimization effort using a CanaryAdvisor tool, like Opsani, to reach the thousands of containers Ancestry now had and was planning to double by 2024 ($n_{future_{k8s}} = 20,000+$).

This paper describes the canary testing and CanaryAdvisor tool processes that enabled Ancestry to automate right-sizing efforts and gain cost-savings. It also describes the adverse effects of automation and the multiple challenges raised through implementation and scaling into a production environment. As Tarvo et al. [15] illustrates, creating baselines requires multiple data agents, data pre-processing settings, and metric comparison, adding to the overhead. The intervals needed to attain high confidence rates in those baselines needed adjustments based on the type of traffic being covered.

This paper includes these sections: Sect. 2 outlines related work about right-sizing, canary testing, and CanaryAdvisor as an AI agent. Section 3 details background information about Ancestry and the problem at scale. Section 4 elaborates on the methodologies that were applied to the problem. Section 5 details Ancestry's perspective on implementing automation from one to many applications. How one team's learnings helped scale the process to customers safely. Section 6 discusses the effects of metric delays and other business costs on large-scale infrastructure optimization. Section 7 discusses additional avenues for this approach, safeguards, and further research needed. Finally, Sect. 8 provides a conclusion.

2   Background and Related Work

An MSA involves building applications as a group of smaller sets of applications. Each part of the applications runs and scales independently while they talk to each other using simple methods like HTTP APIs [9]. MSAs pose some unique challenges. In the MSA paradigm, each service communicates over the network. This adds potential delays and makes secure interactions tricky. Managing data across different services can lead to inconsistencies, and each service might need its own security measures. Testing and deploying these services is also challenging because you need to check them individually and together. Additionally, monitoring multiple services to ensure they are working correctly requires advanced tools and skills. MSA demands changes in how teams operate and the technologies they use, which can be a big shift for any organization [7].

With MSAs, codebases change by nature and have more frequent but smaller changes as compared to monolith systems. This is to write more code just to cover tests needing to be performed in each phase; then the results must be analyzed to determine if the new version can move to the next phase. Automating this decision-making process by monitoring and analyzing the metrics saves a lot of time because this needs to be done often [16]. From this need came different types of testing like Shadow testing, Canary testing, A/B testing, etc. [16].

Canary testing is a strategy where the new version of a product is released to a small group of users first. If this new version does not perform well with this smaller group, it can be automatically switched back to the earlier stable version. This method allows for quick rollbacks, no downtime, and makes it easier to monitor errors and performance [16]. There is also a requirement to carefully observe and monitor, which often results in a slower roll-out. Typically, monitoring is done manually or with rule-based methods. Google and Netflix released open-source tools for automated canary analysis named Spinnaker [2] and Kayenta [1]. Spinnaker helps with canary deployments, and then Kayenta validates the deployments and acts as a judge to assess their success. Tarvo et al. [15] introduced CanaryAdvisor, a tool designed for automatic canary testing of cloud-based applications. Similar to Spinnaker, CanaryAdvisor uses statistical models to assess the correctness of the newly deployed version.

Another difficulty the MSA challenges involve is allocating optimal resources at smaller scales, such as CPU and MEM. Each microservice application needs unique tuning to perform efficiently without overutilization or wasteful under-utilization. Paganini et al. [12] developed a model for a cloud computing infrastructure that spreads across various locations, each with multiple servers. Using a fluid queue model, the authors initially defined the system's capacity with a fixed number of servers per location and confirmed the effectiveness of known policies. Then, they changed the number of servers and tried to minimize costs while maintaining system stability. Many stochastic models were used to determine what can be the optimal factors for identifying the "right size." Shen et al. [14] developed a stochastic model using Queueing Theory to understand data center dynamics, discovered a trade-off between energy use and performance, and then created a Broyden-Fletcher-Goldfarb-Shanno-based algorithm to find the best balance. They tested their Stochastic Right-sizing Model (SRM) in a real cloud data center, where it significantly cut energy use while keeping performance high.

3   Background on Ancestry's Container Right-Sizing

MSA brings many benefits, but it also brings new challenges [4,6], especially when scaled to large systems $n_{k8s} = 10,000+$. This case study discusses Ancestry's experience of integrating continuous right-sizing optimization using Opsani acting as a CanaryAdvisor (acting as an AI agent) [15], allowing Ancestry to scale to the 200 applications. It describes the cost-saving benefits of automating right-sizing efforts and the challenges that occur when automation is applied to production and customer-facing applications.

Current technical culture demonstrates the benefits of creating an MSA by leveraging virtualization technologies, such as Kubernetes (k8s) [5]. In many cases, agile methodologies improve agility and product development. Having achieved a microservice architecture, Ancestry proceeded to implement agile processes using Rally as a workforce management system for product and development. Ancestry Operations LLC[1] gained many of the benefits of fracturing a monolithic data-center-hosted infrastructure into the cloud. Infrastructure costs decreased, a well-formed CI/CD automated release cycle, and more frequent and smaller changes resulted in a more stable system. Ancestry successfully adopted an agile development model through microservice architecture. This allowed Ancestry's infrastructure (in part) to scale quickly based on customer demand, product feature growth, or marketing events. Microservices allowed for a more active auto-scaling application and decreased costs. Paying micro-dollars a day for the microservices and individual team efforts yielded substantial savings on individual applications; however, it did not scale to the tens of applications teams now supported.

At the time, Ancestry maintained 200+ applications in internally supported *k8s'* solution. Managing *k8s* clusters with $n_{k8s} = 10,000+$ number of containers of varying sizes, the problem of scale becomes pronounced. The *k8s* capacity at

---

[1] Presently, none of the authors are affiliated with Ancestry Operations LLC, Opsani, or Cisco AppDynamics.

the time totaled over \$5 M each year in AWS costs[2]. An optimization goal of 20% (given each year for two years), in theory, would reduce resource usage and save \$1.2 M each year in Kubernetes expenses.

Previous resource optimization efforts yielded large savings because resources were larger and there were fewer of them. At Ancestry, teams used established templates to launch new applications. After the initial launch, teams adjusted the size of containers, JAVA environment variables, and other metrics to create a more optimal service on a cost-per-transaction level. As a result of building using pre-defined templates, the entire environment had a homogeneous nature. Ancestry's infrastructure also shared many common practices. This allowed the CanaryAdvisor approach to be applied to over 200 applications within the k8s clusters and scale continuous optimization efforts using machine learning and new guardrails to safely automate a time-consuming task. While humans are slow at analyzing patterns, they are also able to analyze data streams from multiple sources and identify dependencies. The CanaryAdvisor agent, similar to what Opsani and Ancestry were integrating, needs to be able to make decisions like a human and ensure the applications do not suffer from the changes. With many applications undergoing changes to the size of their physical resources, additional downstream changes needed to be made to coincide with previous changes. For example, a container that is downsized from 2GB of memory to 1GB of memory needs to have some of its operating system or JVM settings changed, like `heapsize` and `garbage collection`. Beyond the need to restrict logical settings to coincide with physical resources, tuning JAVA environment variables has other benefits. As a known example, a small but serious problem that can affect large-scale infrastructures is the intervals at which garbage collection processes run. Large-scale systems can have major impacts on performance and downstream dependencies if these variables stay static or in default states. Moreover, safety and caution are needed when using JAVA in critical applications [13]. Java garbage collection takes time and resources to perform. Those activities can have queuing effects on downstream applications, which cascades to further queuing and potential cascading failures throughout the system. Precautions need to be made to ensure stability in the system as it goes through its own cleanup cycles.

By 2021, right-sizing efforts using Opsani were optimizing applications for over 30 agile teams due to the homogeneous nature of the team's CI/CD and monitoring pipelines, but applications for another 50 agile teams were not using the same technology stacks and could not utilize an auto-optimizing tool such as Opsani. While only covering 10% of Ancestry's MSA applications, daily changes to the size and capacity of each application ramped up to a significant scale. Changes to the k8s environment were made multiple times a day. More frequent changes meant the possibility of configurations that affected customer-facing applications in adverse ways and needed a rollback. Opsani's tool allowed the team to move faster than before, while the scale of the optimization efforts would cause unintended consequences to the system when multiple changes to the

---

[2] Not including support/licensing costs.

system were made frequently, and propagation issues arose. At peak, Ancestry's k8s environment had more than $n_{apps} = 200+$. The entire cluster also managed more overhead capacity than needed. The cost of k8s in 2021 was \$5M+ per year. Top-level analysis revealed that the k8s clusters still had too many cores provisioned for the daily peaks. While the apps enjoyed an agile and consistent environment, there were potential savings in the system. The goal was to reduce this overhead. A small architecture team could not scale all of Ancestry's agile teams and support all the applications in a constantly evolving environment; however, more than half the teams were using k8s, and more were scheduled to migrate. Scaling to more than one hundred engineers to yield as little as \$2.00 in savings each day in their k8s infrastructure needed tooling to overcome the problems of a high rate of change and propagation and ensure the customer experience was not affected. While the tool was only optimizing k8s loads, plans were in place to start optimizing the large VM-run applications. This would enable Ancestry to go after an environment worth another \$10M in OpEx costs each year.

Opsani was designed to automate right-sizing efforts through canary testing methodologies, and the use of a CanaryAgent as described by Tarvo et al. [15]. Coupled with usage metrics and cost data, Opsani would analyze the result of the changes it made to a canary and provide optimal sizing recommendations for release to production. See Fig. 1 that describes the optimization cycle. By using canary testing methodologies, Opsani would make small changes to one of the containers in the cluster, analyze the results, and then roll that change forward or back. If it went forward, it would also need to define new JVM settings, like a new heap-size setting to coincide with the new container size. Opsani fit the objective of helping remove the cognitive load on engineering teams, who had to tediously try new container configurations to find an optimal size.

4   Methodology Applied

Hundreds of microservice applications needed optimization. There were two physical capacity settings to try: CPU and MEM, along with multiple logical JAVA environment variables to tune. As a side effect of optimizing one constraint, another one will immediately constrain the efforts for optimizations. In Ancestry's case, physically constraining the system more meant that the CanaryAdvisor would need to be aware of logical settings in the application. As previously mentioned, scaling memory on a container requires heap-size adjustments so the application does not misbehave when a conflict occurs. For example, when there is less physical memory than is allocated by JVM settings, Ancestry needed a system that could overcome the large number of applications that were sizing changes with compounding variables across multiple dimensions. Physical settings and JVM settings would not be where the CanaryAdvisor would stop tuning. The Ancestry infrastructure team had plans to create a third dimension by which the CanaryAdvisor could optimize on. Variables to auto-scale groups in VMs (or HPAs adjustments in k8s) to maintain scale and replicas at different

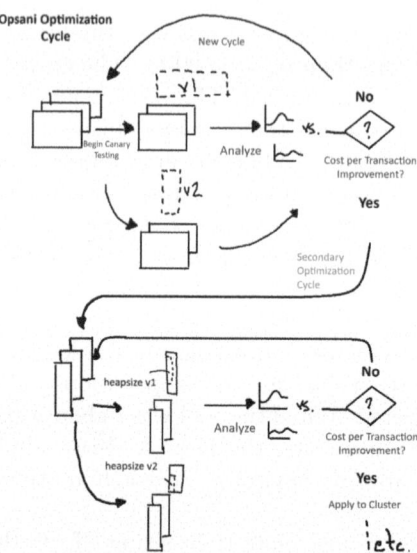

Fig. 1. Two Stage Canary Optimization Cycle; First Cycle tunes size of the container, second cycle tunes the heap size settings.

rates. In addition to making CI/CD adjustments, the CanaryAdvisor needed to control the flow of traffic to a canary it had launched for testing. Ancestry's front-end network tool that was capable of routing traffic internally within the environment based on normal round-robin or load-based distribution, it also allowed for the CanaryAdvisor to route external and internal traffic based on Ancestry's established testing methodologies. Consistent data collection pipelines needed to be established so that Opsani could consume and understand how the data was presented. Ancestry's observability platform at the time used NewRelic for a majority of its infrastructure and application monitoring. As an example, Opsani maintains a public GitHub repository that outlines the NewRelic Servo (data agent) [11] and how it reads in metrics. For example, CPU and MEM metrics the following were being consumed to establish an application's used cores and memory;

```
SELECT
    average(cpuUsedCores) AS cpuUsedCores,
    average(cpuRequestedCores) AS cpuRequestedCores,
    average(memoryUsedBytes) AS memoryUsedBytes,
    average(memoryRequestedBytes) AS memoryRequestedBytes
```

These metrics, as an example from Opsani's Servo Code, request from NewRelic, basic CPU and MEM used metrics. The data agent established resource usage patterns from a baseline and compared that to a Canary, which is allowed to launch. Other metrics included timeout responses and network indicators that could tell Opsani the capacity or health of an application. Machine Learning algorithms were applied by Opsani at this stage to remove anomalous data and create an accurate baseline for the system. Further learning of the tool would be applied to allow it to reach these metrics. At the core, one of Opsani's responsibilities was to baseline the cluster using Ancestry-supplied metrics, clean the data up, and recreate those baselines as necessary to compare it to new size iterations when optimizing.

The tool also maintained a direct interface to the team's CI/CD pipeline, allowing it to make changes directly to the containers and redeploy canaries needed. Opsani's CanaryAdvisors, were only to right-size the container, and see how the application performed. Usually, it would cut the resources to notice some change in performance. When performance dropped, it would revert the change and stop optimizing. Small right-sizing efforts were automated by the tool and Ancestry's well-architected CI/CD. As more teams adopted the tool, sizing of the containers alone was no longer just the issue. Beyond sizing changes to the containers, Ancestry, and Opsani would have to introduce changes to JAVA environment variables, like garbage-collection and heap-size variables. While the CanaryAdvisor could react to adverse changes it had made to a canary it deployed, the data collection feedback loop was found to be delayed. The delay was due to Ancestry's overall monitoring platform ingestion and processing of GB per sec of metrics. Metric propagation across the system can take seconds if not minutes. The data agent used by Opsani would need to sanitize data sets across multiple data streams (NewRelic, CloudWatch, Logs). The overall health of the system would be at risk if the CanaryAgent could not react fast enough $MTTR_{seconds} = 60$ (mean time to recovery). This negative effect of frequent changes meant the system as a whole needed to be regulated and controlled with more consistent data collection streams and improved pre-processing intervals to ensure the health of the applications was maintained as canary testing methodologies were scaled out.

5 From One to Many

At the time, Ancestry had already implemented Opsani on a single application within one of Ancestry's services, called DNA Matches. DNA Matches is one of many products for Ancestry and makes up for a considerable expense cost. It consists of many applications and costs millions a year in cloud expenses. The team managed to create a proof of concept at Ancestry that showed Opsani could ingest monitoring metrics and produce right-sizing recommendations that the team could push out and test through their integration of Opsani's AI agent directly to the teams' CI/CD pipelines.

Opsani's objective was to optimize the flow of performing application performance testing through canary testing methodologies and a CanaryAgent that allowed iteration of changes and analysis of the outcomes as positive or negative. While the POC only addressed a small amount of savings for the effort. The scale of the cluster and subsequent growth would return the costs to Ancestry and provide new, safer approaches to continuous optimization efforts at scale.

5.1 DNA Match Service as a POC

The DNA Match service comprises multiple containerized applications interacting with each other horizontally and vertically across the network. Customers interact with an HTML front-end service; those applications intersect other applications that serve a customer with cousin-matching data the product offers. The service supports Ancestry's 3 million subscribers and over 25 million people in the world's largest consumer DNA network [3]. It stores subscriber matches and their potential cousin relationships using the DNA material provided. The service consumes new subscriber DNA patterns on a schedule and re-processes that new data set against the existing set. The service, as a whole, used many different technologies and was a great candidate for applying automation. It served both internal business customer (applications), and external applications (customers).

DNA Matches as a full product allowed Ancestry to learn how many applications would be affected when automation took over more applications at once.

5.2 Optimization Process and Added Dimensions

By adopting an automated optimization process, the DNA Match team was able to produce more frequent changes to the size of the chosen container sizes across more of their applications. Going out of the prescribed sizes, the tool would suggest different containers of different sizes. Before any optimization changes take place, the tool ingests Ancestry's monitoring metrics (see NewRelic Servo metrics as an example). Having an established Data collection method. Automation took place as Opsani AI Agent would be integrated into the Ancestry's CI/CD platform. It was able to baseline the cluster through custom data collection and processing pipelines. Opsani was allowed to read and change Infrastructure as code (IaC), deploy a canary version of the application, test it against the baseline, and make the output a performance/cost delta. The final goal was to find a configuration that satisfies the service's Service Layer Agreements (SLA) while becoming more cost-effective.

5.3 Baselines and Artificial vs. Organic Loads

The DNA Match Service was used as it had the ability to scale its applications to serve millions of customers. The application consisted of two types of loads: artificial loads, generated by the business, and organic loads, those created by customer traffic $customers_{DAU} = 1,000,000+$[3].

---

[3] DAU = Daily Active Users.

The first applications tested by the DNA Matches team were subject to artificial loads only. The time intervals that the data agent would need to consume to create a correct baseline were based on the system's time to process the entire data set one time. For example, DNA re-processing needed to process trillions of connections and consume hours of CPU resources. The interval to measure and establish a baseline is well-defined by the business process. Organic loads, like customers, which can be predicted, are rarely consistent throughout the day or even during the hour. Also, any application that undergoes a load should produce measurable metrics within the observability platform of the specific infrastructure. The initial loads that the DNA matching application underwent were long-lived processes, and the baselines against canaries were only verified after the full process would run. Allowing the baselines to stay consistent $basetime_{hours} = 2+$, spanning many hours, to be consumed for verification. DNA Matches, as a service, needed to expand on their application set, to also focus on optimizing their containerized applications on applications subject to organic loads. Creating and measuring baselines for organic loads was the first challenge. It should also be understood that the applications underwent auto-scaling thanks to k8s. This scaled the application based on traffic, which meant that containers were constantly formed and torn down. Creating a consistent baseline for future analysis meant that organic loads needed more time to reduce the noise and create a consistent baseline metric.

5.4 Scaling Safely for Customer Loads

For artificial loads, the process worked well, as the system was subject to a schedule, and not effected by release timing and health-checks. However, DNA Matches front end applications had organic loads. Data pre-processing and baseline intervals of the application was subject to more noise in an ever changing system, and the load was not consistent throughout the day.

While customer load is never the same day-to-day, patterns can be formed mainly for forecasting purposes. Using daily or weekly usage patterns to create a baseline of the performance of an application that is undergoing automated scaling changes. If a canary had the wrong configuration when testing with customer traffic, some customers would experience issues. Those customers, unlike the back-end services, can not be rescheduled for later, and they might never come back. That is lost revenue. Scaling safely meant that the system as a whole needed to react when changes were affecting customers' experience.

Questions first needed to be answered before a solution could be invented.

- How long does it take to release one change?
- How much time do NewRelic metrics take to propagate?
- Can we add multiple health checks?
- When do we trigger a rollback?

Fig. 2. User Load Example

As an example from Ancestry, Fig. 2 shows an example of how customer traffic would look. A containerized application that runs 10–12 containers during peak load will scale down to 2 for nighttime loads. Two issues arise when making optimization efforts for container servicing customers. The first is that the release at night does not have enough load to know if the changes were beneficial. Two containers is the minimum high-available deployment pattern chosen by Ancestry. The traffic in the scenario is split 50/50. Introducing a canary has the effect of moving 33% of the traffic (third one in the pool); the system is not at load to scale to three containers and now has an artificially deflated load. Secondly, nighttime scaling means that one small change has the potential to cause fail-over routines in the system. When half of the containers see traffic drop off, monitoring tools will panic and cause alerts and false positives for operational support teams. Downstream dependencies could misbehave and initiate fail-over routines. Causing instability in the system.

Making changes during the day is a high-risk activity, but the load on the system is the most pronounced and accurate at those peaks. If the tool was to be used during the day on customer traffic, it would need to reduce the risk introduced by making more frequent changes to the application. While a canary will limit the blast-radius to the percentage of traffic it takes (usually less than 10% at peak for most applications within Ancestry).

When developers release code and make small changes, the added benefit of small changes reduces the overall risk of the entire feature set. The same benefits apply with a Canary methodology. Make a small change to one container and observe the change in behavior. However, developers make a few changes a day at most to production code. Opsani was capable of making iterations to the canary all day long. When the frequency of changes increases, the number of customers affecting changes also increases. A typical release cycle has humans as part of the validation steps to ensure safe integration and release cycles of code to customers. While health checks are automated and rollbacks are initiated automatically. Monitoring metrics from other sources will indicate issues. These additional metrics would be the basis of minimizing the blast radius to 1% of the canary traffic. Monitoring the change to the one canary, the tool would monitor the health of the container and maintain health checks throughout the release of additional traffic to the canary until it was at capacity and healthy. By introducing more health checks, the CanaryAdvisor would stage traffic to the Canary at intervals of:

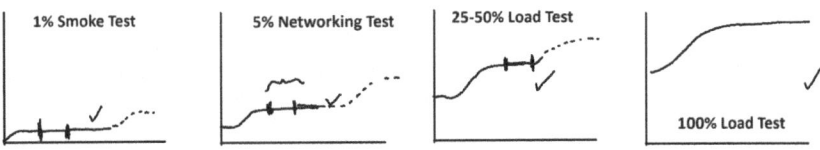

Fig. 3. Four Graphs to represent the additional time and tests needed to reduce the effect on customer traffic.

- 1% of traffic for one container, a simple smoke test.
- 5%–10% of traffic for one container, simple load test.
- 25%–50% of traffic for one container, test a significant load ramp.
- 100% of traffic for one container, full load test.

This additional mechanism, introduced by Ancestry's architecture team, could control risk profile of the more frequent changes taking place. Slowing down the release of traffic, would protect the customers, but it would also slow down optimization tactics. The rate of changes, speed of release, and post-change baselining all needed additional time to establish confidence.

Using BigIP's F5 network tools, Ancestry controls the ramp-up of traffic directly at the IP level and specifically to the canary within a given application's pre-defined load-balancing methods. See Fig. 3 for an example of a four-staged health check the Opsani CanaryAgent would initiate to ensure smaller effects on customers when testing new configurations.

The CanaryAgent controlled the release of customer traffic through the CI/CD pipeline controls. It monitored the health of the application at each stage in the ramp up of traffic. The Enterprise Architecture team established the ramp-up times based on test cases to be covered. Test case 1, at 1%, to ensure smoke tests pass and traffic flows from end to end. Test case 2, 5–10% to ensure load went up gradually and noting panics. Test case 3 goes to 25–50% of traffic to ensure stability across the networking for the rest of the service as traffic offloads from other containers. Finally, to 100%, where the system would do full health checks and confirm release, and begin baseline the canary performance to compare to the cluster's performance. This allows the system to take in more health checks than before, allowing for safer release cycles with reduced customer effects.

6   Issues at Scale

Monitoring metrics come in many flavors. The most important metrics are typically related to response times. The system is healthy so long as it processes requests within the desired response times. Monitoring this metric is important, to derive issues in the system, the tool needs to be able to consume metrics from different data agents. At Ancestry, the issue that presented itself was that multiple monitoring tools were used at the same time. With plans to retire or add new suites of tools, Opsani's data agents needed to be continuously developed to support Ancestry's evolving infrastructure software.

This section elaborates on combining data streams and more variables by which to optimize. The issues presented when considering metrics and log data streams that output at GBs per sec in metrics are not only that data agents need to ingest more metrics, but agents also need continual support for changes to data streams.

6.1   Metrics Gaps and Delays

One issue that a data agent needs to overcome is to produce a baseline and comparison with confidence. While changes to live applications were taking place, the propagation of monitoring metrics made these changes riskier by extending the mean time to response for rollback. Opsani would stand up a canary, the CanaryAgent would shift 1% of traffic to the canary and establish good health by reading monitoring metrics downstream, in another platform (NewRelic). The metrics took seconds to appear and minutes to establish confidence with enough consistency to give the tool confidence to move forward to the next traffic ramp-up health test. This feedback loop is not perceived by humans as much, but when an AI agent is waiting to check the health of a system before it moves to 5% or 50% traffic, the noise in the system is very pronounced, and a lot of effort is needed to teach the system how to ignore delays caused by the tools themselves. Figure 3 shows how a four-stage approach would be delayed by minutes when data metrics needed to catch up. With direct results to revenue if customers' applications were not rolled back quickly. The delay in getting to those metrics meant the CanaryAgent could not react fast enough without affecting some customers.

To further complicate monitoring delays, a heterogeneous monitoring tool suite meant Ancestry had data metrics and logging sets in multiple languages. Not all monitoring was covered by NewRelic Application Performance monitoring agents, CloudWatch, and other tools used at Ancestry to cover the totality of observable metrics. Many integration points in a large-scale enterprise architecture cause undue stress for the business and tool providers when maintaining its overall architectural degradation. By automating a change cycle, Ancestry quickly learned that real-time metrics were not as real-time as advertised. Sometimes, taking minutes to fully trust the intervals, the CanaryAgent would artificially cause delays, extending potential time to affected customers. While the number of affected customers in this situation will never be zero, the time to react based on metrics could be reduced.

Many monitoring data streams were underutilized for this process to work for all of Ancestry's infrastructure. However, as long as the application is maintained and deployed using standard Git and CI/CD procedures at Ancestry, Ancestry could fully automate right-sizing optimization using this method for more of its infrastructure.

Constant changes and targets to technology migrations at Ancestry meant Opsani, as a small company, could not support the needed development changes for multiple data streams and added dimensions to tune at Ancestry. This created gaps for certain teams within Ancestry. Opsani's tool set could not ingest different types of monitoring and logging streams at Ancestry and needed more development to consume data streams from different sources.

6.2   Validation Other Business Metrics

A cycle that needed to be validated at each step by human eyes was now being automated; transactional costs could drop even further. However, other compounding variables were affected by the constant changes and small customer issues that would occur before a health check kicked in. The constant changes to the system also exert a load that costs CPU cycles, licenses, and resources in other places in the enterprise. Smaller containers that scale out horizontally need more interactions with supporting infrastructure. Many of which are billed based on usage. More containers meant more money for some other tool providers, which was not as easy to calculate.

- API calls for deployments, security scans
- Logging and monitoring resources
- Service agreements to the customers

While teams were able to reduce the AWS portion of their expenses, other supporting costs that were not fully understood became affected. Monitoring costs for services across the enterprise's ecosystem, for example, is adversely affected by automation efforts at the application layer. The system is polled more frequently, and data streams need to be kept active for longer. This has the adverse effect of costing more money to support an observability platform. Potentially offsetting further costs and optimizations.

Not all business metrics were being recognized by the tool. Decomposing logging or monitoring usage based on one tiny container application in a monitoring tool that costs $2M each year was not possible and would have to become an estimate in the overall process of optimizing applications at Ancestry.

Licensing costs across tools were not consistent either. The growth in one tool did not scale the same way Ancestry's resources grew. For example, the CPU cost for each VM is linear. However, when fracturing the number of systems, tools will need more agents. So, although additional CPUs are not required, you will need ten times the number of agents or licenses you originally had to manage. Tool costs were not scaling linearly with the slope of the growth of the infrastructure. Ancestry needs to analyze the value and subsequent costs of maintaining an MSA, licensing issues, and development support costs as just a few larger cost variables that are much harder to measure in real time.

7   Discussion

Having hardened releases to customer production traffic, teams who adopted the tool saved on infrastructure costs and time by removing some of the cognitive load of mundane configuration variations and change exploration. Leading to other downstream tasks that could be automated, like deeper health-checks. This allowed agile development teams to impact other projects.

Not all teams were able to adopt the tool due to the nature of their release cycles and workload schedules. Teams that adopted the tool were able to successfully continue optimization tactics while removing the load from the team, allowing the team to get to their optimization targets sooner in the year. However, the tool could not cover all teams and environments and needed extensively more development to support Windows/.NET environments because they were built using Jenkins pipelines and monitored differently, needing new data agents to be added to the tool.

Many of the metrics along the business axis, such as upfront contractual commitments, labor costs, licensing costs, and so on. Those costs were still outside of the scope of the tool and team when calculating cost efficiency. Many factors can be introduced in later revisions to include variables in the system to allow for better constraints and solutions that fit the business. While the machine learning aspects of Opsani were around ingesting monitoring metrics and fitting them to those curves, the tool needed development to support more data agents and be sufficiently valuable across the entire enterprise.

The other major issue to overcome, as seen by the author, was the perceived "real-time" nature of metrics. Metrics like CPU and MEM metrics took minutes to update, rolling up metric interval time to minutes or hours, causing many downstream delays to ensure accuracy in data. As a result, each step had to be artificially slowed down. AI Optimizations were allowing teams to go faster, exposing the problem with monitoring solutions and the scale of data that causes delays in metric propagation. We can also hypothesize other issues related to monitoring platforms and their delays; as an example, the mean time to recovery (MTTR) goes up. With more changes, alerting faster and backing out was more important than ever.

There is an opportunity to apply Large Language Models (LLMs) [8,10] to integrate and improve the speed of detecting issues. Tailing logs could be a much faster way to detect issues in the health of an application. It can take minutes for issues to propagate through a company's observability platform. By tailing logs and ingesting those logs through LLMs, triggers could be created during periods when monitoring is still propagating the metrics. An alert can trigger an LLM to ingest the logs and read the logs to sanitize the potential fault from the application. Logging standards are often malformed or not consistent. DevOps teams are very good at looking through logs to find issues, and LLMs could supplement those efforts. Operational support staff teach the LLM how to digest the logs and correct false positives.

8 Conclusions

The optimization process of container right-sizing is difficult to establish and requires great efforts. The presented automated process undertaken by Ancestry yielded significant cost reductions, covering more than 50 of the largest 200 applications within its k8s. Ancestry saved more than half of its target ($0.6M) on the effort. The process was able to work when the environment was sufficiently homogeneous to enable the collection of specified metrics and allow for tuning of similar variables. This paper shares processes for such optimizations and the best practices that Ancestry has developed and proved in a production environment.

This case study provides practical enterprise-setting guidelines for the community to use if an AI CanaryAgent-based solution is adopted for optimization or testing efforts. It shares practical observations from large systems in operation and details multiple opportunities for improvement. Our future work aims to focus on the development of open-source system benchmarks and LLM models. This will create more avenues for consistent data collection methods and allow for more direct access to logs and monitoring metrics so that all enterprises can address parts of their infrastructure with these methods.

References

1. Kayenta (2024). https://github.com/spinnaker/kayenta. Accessed 13 Apr 2024
2. Spinnaker (2024). https://spinnaker.io/. Accessed 13 April 2024
3. Ancestry PR Team: Company facts (2024). https://www.ancestry.com/corporate/about-ancestry/company-facts
4. Schmidt, C.T., Ganesha Venkatesha, S., Heymann, J.: Empirical insights into the perceived benefits of agile software engineering practices: a case study from sap. In: ICSE Companion 2014: Companion Proceedings of the 36th International Conference on Software Engineering, pp. 84–92 (2014)
5. CNCF SURVEY: CNCF Annual Survey 2022: 2022 The year cloud native became the new normal (2022). https://www.cncf.io/reports/cncf-annual-survey-2022/
6. Bjarnason, E., Wnuk, K., Regnell, B.: A case study on benefits and side-effects of agile practices in large-scale requirements engineering. In: AREW 2011: Proceedings of the 1st Workshop on Agile Requirements Engineering (2011)
7. Ghofrani, J., Lübke, D.: Challenges of microservices architecture: a survey on the state of the practice. ZEUS **2018**, 1–8 (2018)
8. Le, V.H., Zhang, H.: Log parsing: How far can chatgpt go? In: 2023 38th IEEE/ACM International Conference on Automated Software Engineering (ASE), pp. 1699–1704. IEEE (2023)
9. Lewis, J., Fowler, M.: Microservices: a definition of this new architectural term. MartinFowler. com **25**(14–26), 12 (2014)
10. Li, Y., et al.: Exploring the effectiveness of LLMs in automated logging generation: an empirical study. arXiv:2307.05950 (2024)
11. Nickolov, P.: Opsani public github repo. https://github.com/opsani
12. Paganini, F., Goldsztajn, D., Ferragut, A.: An optimization approach to load balancing, scheduling and right sizing of cloud computing systems with data locality. In: 2019 IEEE 58th Conference on Decision and Control, pp. 1114–1119 (2019)
13. Schoeberl, M., Vitek, J.: Garbage collection for safety critical java. In: JTRES 2007: Proceedings of the 5th International Workshop on Java Technologies for Real-time and Embedded Systems, pp. 84–92 (2007)
14. Shen, D., et al.: Stochastic modeling of dynamic right-sizing for energy-efficiency in cloud data centers. Future Gener. Comput. Syst. **48**, 82–95 (2015)

15. Tarvo, A., Sweeney, P.F., Mitchell, N., Rajan, V., Arnold, M., Baldini, I.: CanaryAdvisor: a statistical-based tool for canary testing (demo). In: Proceedings of the 2015 International Symposium on Software Testing and Analysis, pp. 418–422. ISSTA 2015, Association for Computing Machinery, New York, NY, USA (2015)
16. Velayutham, S., Shanmugam, G.: Artificial intelligence assisted canary testing of cloud native ran in a mobile telecom system. Master's thesis, Uppsala University (2021)

The Nature of Questions that Arise During Software Architecture Design

Neil B. Harrison[1]([✉])[iD] and Ademar Aguiar[2][iD]

[1] Utah Valley University, Orem, UT 84058, USA
neil.harrison@uvu.edu
[2] INESC TEC, Faculdade de Engenharia da Universidade do Porto, Porto, Portugal
ademar.aguiar@fe.up.pt

Abstract. During the process of software architectural design, numerous questions arise which must be answered. These questions may be about requirements on the proposed system (the problem space) or about how the system should be designed and developed (the solution space). As questions arise they may be answered immediately, deferred until later, or provisionally answered with an assumption about the answer. The objective of this work was to explore the nature of questions that arise during architecture. We explored the types of questions, how they are organized, how they are tracked, and how and when they are answered. We started by surveying highly experienced architects about their practices with respect to architectural questions. We also performed a controlled experiment with master students about organizing architectural questions that clarified and substantiated the survey data. We learned that architectural questions include slightly more questions about the problem space than the solution space, as well as a minority of questions related to the managing of the project. We found that architects often use ad hoc methods to organize and track them, although they typically organize them along more than one dimension. We learned also that, about a third of the time, architects make assumptions about the answers to architectural questions in order to make progress on the architecture. This suggests that some projects may have risks of incorrect design or later costly rework due to inadequate tracking or incorrectly answered architectural questions.

Keywords: Software Architecture · Requirements · Quality Attributes

1 Introduction

Architecture has been characterized as, "The set of principal design decisions governing a system" [9]. Clearly, decisions are a central piece of architecture. A decision is the result of answering one or more questions. Thus, in one sense, the process of architectural design is one of asking and answering questions. Therefore the questions that are answered during architectural design are an important aspect to consider if we want to improve the overall process. The questions that are asked during the conception phase of a project determine the shape of the product.

By definition, early in the architectural design of a product, many things are unknown. During the course of architectural design, and even throughout more detailed design, the unknowns become known, or in other words, questions which arise early are answered. Thus, the process of developing a system includes asking and answering these questions in a systematic manner. In this work we concern ourselves with those questions which are asked during architectural design, how they arise, and how they are characterized.

1.1   Breadth of Architectural Questions

Architectural questions cover a large variety of areas. Clements et al. state, "Software architecture is composed of elements, connections or relations among them, and, usually, some other aspect or aspects, such as ... configuration; constraints or semantics; analyses or properties; or rationale, requirements, or stakeholders' needs" [6]. During architecture, questions may arise about each of these items. Note that some of these are about the problem space (such as requirements and stakeholders' needs), and others are in the solution space (such as elements and connections) [16].

In addition to questions in the problem and solution spaces, questions may arise which are concerned with the management of the project itself. For example, there may be questions about the number of people available and how teams are organized. Project development methodologies may be of interest to the architects; some Agile methodologies, for example, advocate the commencement of development early, even before the architecture is complete [11]. Adoption of an agile methodology can influence which parts of the system are designed first [13].

1.2   Intermingling of Architectural Questions

During the course of architectural design, problem space questions and solution space questions are intermingled. Harrison et al. observed that architects would work in a pattern of design, which would cause questions about quality attribute requirements to arise [7]. Another study showed that questions that could not be immediately answered were noted in a list of open issues [8].

Requirements analysis is a well-established practice, and has been shown to reduce unnecessary costs, confusion and complexity in the later phases of software development [15]. Its focus is firmly in the problem space [3]. However, Castro et al. [4] note that "most agree that requirements that are free of architectural influence tend to be very difficult to elicit and analyse". This illustrates the problem of focusing entirely on the problem space. Likewise, a software architecture or design approach that focuses solely on the solution space has mirroring problems. Such a design approach is based on the premise that the requirements have been completely defined.

Yet, because questions about the problem space and the solution space are intermingled, techniques that focus only on one side or the other do not match what actually happens [7]. This can lead to missed opportunities to tie requirements and design together, and to effectively design the system. We give hypothetical examples:

- Architects note that the system under design is similar to an existing system which has high availability requirements, and that the new system may have similar requirements. The existing system uses a hot spare, which raises questions about whether a similar solution is practical in this case. These are technological questions (solution space) as well as budgetary questions (problem space).
- Architects designing the data architecture of a new system are concerned about the security of the data. They realize that the specification is deficient in security information, so they recommend a quality attribute analysis on security. This shows how architectural design questions can lead to a focused quality attribute analysis.
- Architects realize that in order to make progress on the design, they need more clarity about certain proposed features. However, because of schedule pressure, they cannot wait for answers, and must make educated guesses. This presents a dilemma for architects. Making an educated guess creates a risk for costly rework later on, but not making a decision causes the project to stall. Part of our study includes learning how often this situation occurs and how architects handle it.

As work in both the problem space and the solution space are concerned with making decisions based on questions, and that many questions arise during architectural design, we wish to understand the nature of such questions. We have explored the types of questions and their relationship to each other. We have explored when questions must be answered, and have examined how architects handle questions. These insights into architectural questions should enable architects to manage and answer questions better, reducing risk of mismatches between the architecture and requirements.

2   Research Questions and Methods

We have based our research work about architectural questions on the following research questions:

RQ1. What is the nature of questions that software architects ask during architectural design?
RQ2. When must the questions that arise during architectural design be answered?
RQ3. How do architects deal with questions that must be answered immediately, but they do not have the answers available?

RQ4. How are architectural questions, especially those that cannot be answered immediately, organized?

To answer the research questions above we followed a qualitative research approach using two complementary empirical methods: (i) a survey with experienced software architects in the industry and (ii) an experiment with software architecture students. We found that they complement each other, and together they paint a more complete picture of the nature of architectural questions.

3   Survey

With the goal of reaching a significant number of experienced software architects in the industry, we developed a questionnaire to learn about how questions were used. We distributed the questionnaire by email through a list of relevant authors' contacts and professional networks.

The questionnaire aimed to gather quantitative and qualitative data that we can mine for hints about the nature of architectural questions. The questionnaire starts by characterizing the personal and organizational context of the respondents, their experience in terms of software development and software architecture, and then asks for several aspects about architectural questions, namely: how they come up, what kinds of questions, and how they are managed.

3.1   Demographics

The questionnaire had twenty-seven responses. All respondents had more than three years of software development experience, and all but three had more than ten years experience. Most also had significant experience as software architects. Only three had less than three years experience as software architects. Twenty of the twenty-seven had at least five years experience. After elimination of erroneous responses to the numerical questions, twenty-one remained. Free-form question responses were unaffected.

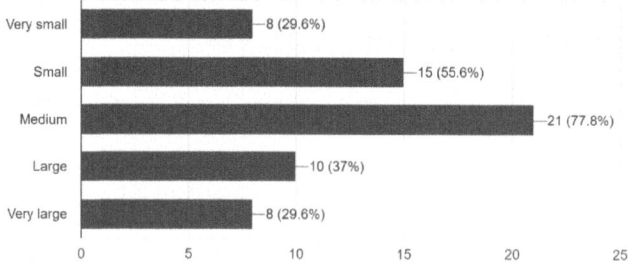

Fig. 1. Size of projects you designed (Check all that apply).

As architects, they have worked on various sized projects, as shown in Fig. 1. The data collected was not skewed toward either large or small projects.

3.2   Quantitative Data

In the following questions, the survey participants were asked to estimate the percentages in each category. For example, "What percentage of the questions that arise came from incomplete or unclear specifications?" Due to rounding, the totals do not always add to 100.

Figure 2 shows that the origin of the majority of the questions is the problem space, but the solution space (design) is still significant. The category of question, Fig. 3, is consistent with the origin, suggesting that the origin matches the category.

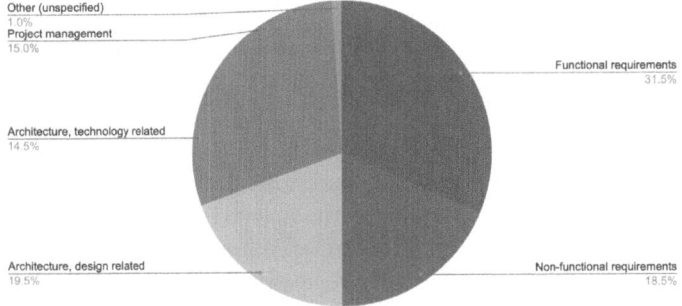

Fig. 2. Where do architecture questions come from?

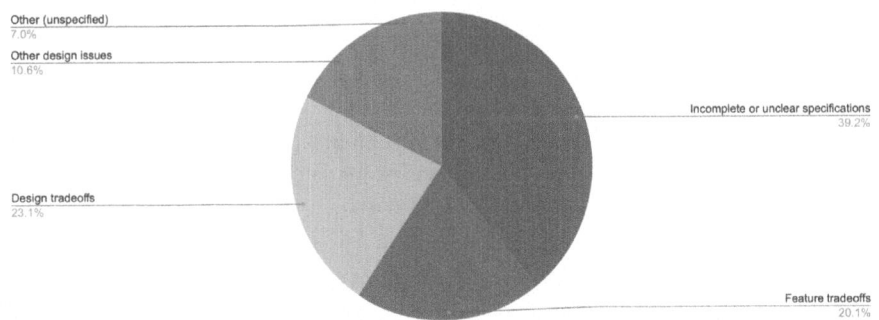

Fig. 3. Categories of architectural questions

The next questions were what architects do with questions that cannot be answered immediately, and how they track them. Table 1 shows that architects make an assumption about a question about a third of the time when they cannot answer it. Figure 4 shows that except for the three respondents who did not organize the questions, they organize them along multiple dimensions.

Table 1. When you cannot answer a question, what do you do?

Action	Minimum %	Maximum %	Average %
Find the answer immediately	10	60	30
Defer the question until later	10	80	32
Make an assumption about the answer	10	80	32

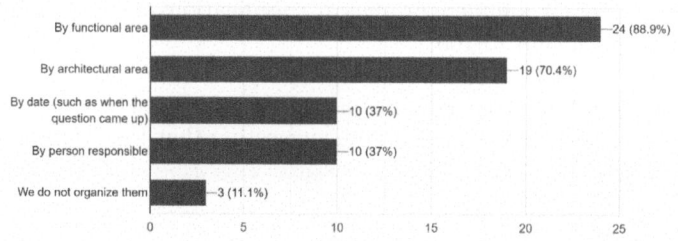

Fig. 4. How do you organize the questions (Check all that apply).

3.3 Qualitative Data

The next questions were free response questions. Both researchers analyzed the responses independently, and organized the answers into general categories. We compared our analyses and found them to be consistent with each other. In some cases, a response contained more than one answer, such as the following response to the question about how to get the answer: "First search online, second reach out to [the] SME." Since this was actually two responses, the total number of comments for each question exceeds the number of respondents.

If You Seek an Answer Immediately, How Do You Get the Answer? Most of the 44 total comments involved working with others. These included "Work with others" (unspecified) (19), "Work with an expert" (8), Brainstorming (2), and working with people in the problem space such as customers or stakeholders (6). The remaining 9 responses were solo work, including searching the web, research, and analyzing documentation.

How Do You Track Questions that You Answer Immediately? About half the respondents (12) used some software such as Jira or Trello to formally track the questions. Thirteen used informal means such as to do lists, notes, or email to track them. Two responses were not to track. The informal methods of tracking may be less visible to the team. They may also be more vulnerable to loss of information.

How Do You Track Questions that You Defer? The responses were virtually identical to the previous question, indicating that they use the same method to track both open and closed questions.

Under What Conditions Do You Defer Answering Architectural Questions? A common condition given was insufficient information available, either generally or due to unavailability of subject matter experts (SMEs) (9 occurrences). Eight people stated if the answer is expected to have low impact. Six said when the design can be deferred or the design is isolated from the rest of the system. (This might also be considered cases of low impact, which would give a total of 14 responses for low impact.) Five answers mentioned insufficient time right now or that other things have higher priority. One person said that when they want to investigate multiple design options and need time later to do it. Two said they never defer questions.

When are Deferred Questions Answered? The free-response answers showed significant different dimensions for the timing of answering the questions. Some were based on time, namely ASAP, 24 h, or a week (6 occurrences). Others were based on completion or starting gates, namely by the time the architecture is complete or before development on a particular part can begin (7 occurrences). Priority was the most common time factor (8 occurrences). In most of these cases, the questions would reach the top of the question stack. In two of the eight, it was that it became necessary to answer the questions (so the questions themselves became high priority.) Another category was that resources became available to answer the question: either architects or SMEs became available (7 occurrences). One survey respondent wrote "where there is sufficient interest" and explained that in Agile development, some questions might be put on the backlog and not answered until there is sufficient interest in that feature. "Risk" (no further explanation) was mentioned once.

Under What Conditions Do You Assume an Answer Yourself? If answers to architectural questions are unavailable, sometimes the architects must decide on an answer in order to proceed. The answers included the following: "I know the material sufficiently to make an assumption" (13 occurrences), "Impact or risk associated with the decision is expected to be low" (6), and "When the question must be answered now" (5). In most cases, the respondents gave a combination of two of the criteria. All combinations of the three criteria appeared. This indicates a good deal of care and caution associated with assuming answers. In addition, four respondents stated that they reviewed these decisions with others, and made corrections as necessary.

What Types of Questions Do You Assume the Answers To? Nearly all the answers were related to design and implementation. There were a total of 19 responses in this category, and included use of patterns, design, configuration, technology choices, details, and decisions not impacting the architecture. Three respondents mentioned quality attributes, and one simply said "requirements". If these numbers were high, it would be somewhat troubling. One person mentioned decisions that impact the schedule. These responses were consistent with expectations.

How Do You Track Questions that You Assume the Answers To? Several responses included more than one approach. The responses are consistent with the earlier questions about tracking. Ten responses included informal tracking methods and ten responses mentioned formal tracking tools such as Jira. In addition, twelve responses stated that the questions (and responses) are captured in architecture or design documents.

4   Experiment

The controlled experiment served to provide a different perspective on the nature of architectural questions. Whereas the survey revealed architects' recollections of experiences in general, the experiment was focused entirely on architectural questions, and provided the questions directly.

The experiment followed guidelines of an experiment as given in [17] and was run in the context of a masters' course on Software Systems Architecture at University of Porto [1], with fifty students self-organized in eleven teams, and having all being given the same tasks. Because the study was exploratory rather than evaluation of a treatment, no control group was needed.

The experiment focused on asking participants to come up with architectural questions and put them in categories. They were asked to estimate the timing of each question. They were not told which categories to use, nor were they told how to determine the timing of the questions. The system to be designed was an automatic traffic signal control system.

The eleven teams came up with 334 questions in all, an average of around 30 questions per team. The average number of categories per group was 7.4.

Table 2. Most common categories.

Category (as written)	Num. questions (total)	General category
Design	41	Architectural design
Technology	37	Architecture, technology
Safety	27	Requirements, quality attribute
Project Management	27	Project management related
Security	24	Requirements, quality attribute
Functional requirements	24	Requirements, functional

Table 2 shows the most common categories, by number of questions. All other categories had ten or fewer questions. We note the prominence of two quality attributes, safety and security. This is a characteristic of the system under design, where these two quality attributes are particularly critical.

Table 3. Summary of Categories and Questions.

General category	Num. categories	Question Count	Percent
Requirements, quality attribute	37	97	29.0%
Requirements, functional	10	80	24.0%
Architectural design	14	79	23.7%
Architecture, technology	5	48	14.4%
Project management related	14	30	9.0%
Summary			
Requirements	47	177	53.0%
Architecture	19	127	38.1%
Project Management	14	30	9.0%

Table 3 shows a summary of all the questions and their categories. We see a large number of categories of quality attributes. From Table 2 we saw that this area is dominated by two quality attributes; all the rest appeared very few times and had few questions. It does indicate that the system under design is very rich in quality attribute issues.

Other than the strong presence of safety and security, the results are quite similar to those from the survey: problem space questions are around half and solution space questions are somewhat less. Twelve to fifteen percent of the questions are about technology (solution space). Project management issues are a small but noticeable fraction; ten to fifteen percent. The participants were asked to estimate when each question must be answered. They showed this by arranging questions visually with distances from the center, which represented "now". The results shown in Table 4 gives a rough scale, but more importantly can show the questions relative to each other: when must a question be answered relative to other questions. We suggest that this is common in industry projects, namely that the time when a question must be answered is predominately a function of when other questions are answered. There are a large number of questions that are to be answered in the medium term or even later, indicating the need to track questions.

Table 4. When questions must be answered.

When?	Now	Soon	Medium term	Later
	5	67	153	90

We observed several instances where a question depended on the answers to another question. For example, three questions in one entry were, "What types of sensors will be needed?", "How many sensors will there be (one per signal, one per direction, one per lane)?", and "What is the expected range and accuracy of the sensors?" We did not look for this specifically, but it could be a subject of future research.

5   Interpretation

In this section we discuss how the data gives insight into the research questions.

RQ1. What is the Nature of Questions that Software Architects ask During Architectural Design?

For the most general categories of questions, problem space (requirements), solution space (design), and related to the management of the project, we found that a slight majority of the questions were related to requirements, somewhat less related to design, and about twelve percent related to management of the project. Requirements questions appeared more than architecture, but both are substantial. Management questions are a small, but meaningful slice.

Table 5. Summary of Categories and Questions.

Category	Survey	Experiment	Comments
Requirements, quality attribute	19%	29.0%	largest difference
Requirements, functional	32%	24.0%	
Architectural design	20%	23.7%	
Architecture, technology	15%	14.4%	very consistent
Project management related	15%	9.0%	
Summary			
Requirements	51%	53.0%	
Architecture	35%	38.1%	
Project Management	15%	9.0%	

Table 5 shows strong consistency between the data from the survey and from the experiment, in spite of the differences between them. The largest difference is that the experiment produced a higher percentage of questions about quality attributes than the survey. This could be due to the nature of the domain used in the experiment, which is rich in quality attributes and lower in features (functional requirements).

On the solution side, questions were mainly about the structure and design, with a small but meaningful set of questions related to technology. The survey and experiment were reasonably consistent with each other. We expect that there is some variation depending on the project, but not nearly as much as with quality attributes.

There is some indication from the experiment that the questions do arise in a mixed order, namely mixed between problem space and solution space. Since the participants were instructed not to worry about sorting the questions in any way, the order in which the questions are written may reflect the order in which they occurred to the teams. This aspect of the experiment was not controlled, thus the conclusion is not strong, but is consistent with Harrison et al. [7]. It requires further validation.

We noted that many different quality attributes were represented (37). This indicates the richness of the domain. On the other hand, the relatively few design categories (14) is likely more indicative of fact that the experiment was situated very early in the architecture cycle. As he architecture begins to mature, questions are likely to relate to the architectural design, and more design categories may emerge as architectural partitioning continues. Research into question types over time could shed additional light on this topic.

RQ2. When must the Questions that Arise During Architectural Design be Answered?

In the experiment, a very small number of questions needed to be answered immediately; less than one per team. In a real project, questions that must be answered now are typically answered now, and therefore are likely to not even be tracked. This was verified by comments from the survey, where some respondents noted that they didn't track questions that were immediately answered.

The experiment showed a large number of questions that are to be answered in the medium term or even later. This may be partly a result of the way the experiment was structured, where eliciting questions was a distinct activity. We suggest further validation. Nevertheless, even if the numbers are lower, it is still a significant number of questions that are answered later, and thus must be tracked. The diversity of the question types, discussed earlier, suggests a tracking system capable of tracking along multiple dimensions; at least by type and by date.

We recognize that the experiment focused on when questions must be answered, while the survey asked about whether questions were answered immediately or deferred. These are different, as questions can be answered before they must be. This was supported by the fact that the survey showed that many deferred questions are answered when time becomes available or when they pop to the top of the list. We suggest that the methods of handling deferred questions (from the survey), while expedient, may lead to crisis-driven approaches to answering questions. Additional support for managing open questions may be beneficial.

RQ3. How do Architects Deal with Questions that must be Answered Immediately, but they do not have the Answers Available?

The data show that architects while architects on the whole defer an answer, or assume an answer commonly, there is a very wide variation among the architects.

In both categories, the range was from 10 % to 80 % of the time, from 10 % of the time (for each) up to 80 % of the time. The variations may be driven by not only the nature of the domain, but by other factors, including maturity of the domain and architect's experience in it. This is supported by the reasons given why they assume the answers. One might extrapolate that where architects are highly experienced in the domain, they are able to make many decisions themselves, leading to a more efficient architectural design process.

We note that responses show that the architects do not take the decision to assume an answer lightly, and they often seek additional validation of these answers. We also see evidence of the care that they take in their answers that they generally seek answers to questions by consulting with others. However, it is notable that few responses specifically mentioned working with people in the problem space. While the unspecified "work with others" and "working with an expert" undoubtedly contain some communication with individuals in the problem space, it was not specifically mentioned. Yet over half of the questions that arise clearly fall in the problem space. This may indicate that architects tend to turn inward first when questions arise, and not consult requirements engineers and others in the problem space as much as they ought.

RQ4. How are Architectural Questions, Especially those that cannot be Answered Immediately, Organized?

We observed that there are multiple dimensions along which the questions are organized, and that architects track them along more than one dimension. This indicates the need for a way of tracking them that can manage multiple dimensions simultaneously. The survey also indicated, however, that about half the tracking is informal, such as "to do" lists and even email. Such approaches are not likely to robustly manage complex tracking needs. Thus there may be a risk of questions getting lost or not addressed in a timely manner. Support of practices or tools specialized for organizing and tracking architectural questions may be useful. It should also include support for dependencies among questions.

One reason for the large use of informal tracking methods is likely ease of use: architects may not wish to spend time using a formal tracking tool. Any support for tracking questions should be as easy to use as informal methods architects now use. Furthermore, if architects are already using tools, an additional software architecture tool for tracking questions may not be used instead of project management tools they are already using.

We note the criteria for answering deferred questions is quite diverse, and perhaps somewhat ad hoc. Support for understanding when questions are to be answered may also be beneficial.

6   Related Work

Along the Software Development Life Cycle, from the initial phase of requirements elicitation to maintenance phase, software developers face many different types of problems for which they need to find solutions. Questions are a frequently used form to state those problems, namely to: elicit requirements, derive high-level architectures, select technologies, guide low-level design, support business decisions, and product management, to mention a few. There are several authors proposing the use of questions to support different development activities.

Muller et al. propose a systematic approach to software architecture design based on architectural styles and patterns annotated with questions to efficiently identify potential architecture solutions [12]. Questions are used as annotations in the solutions and help developers identify solution candidates. The approach uses questions but does not attempt to understand their origins.

Anish et al. describe how architects use probing questions to get information about functional requirements and architecturally significant functional requirements [2]. They state that to their knowledge, this is the first time the concept of Probing Questions has been introduced, although questions may be used in classical requirements analysis. [15]. Unlike our study, probing questions are limited to requirements elicitation.

Questions are used in Architectre Tradoff Analysis (ATAM) [10]. Deeper insight into the nature of questions that emerge during architecture could bolster the effectiveness of such analysis methods.

Chen et al. present a framework for characterizing architecturally significant requirements [5]. Some (but not all) questions that arise during architecture relate directly to such requirements, so these studies complement each other.

Salehin performed a case study to explore the impact of missing requirements on architecture and found significant effort spent on recreating requirements [14]. Our work augments this study: the survey shows how architects handled questions that cannot be answered (which includes missing requirements); this work estimated the cost of some unanswered questions.

Van Landuyt et al. explored the potential impact of architectural assumptions on modular structure of an architecture [16]. They also distinguished between problem-space and solution-space assumptions. Our work complements this work in that we explored the nature of questions and how they lead to assumptions.

7   Limitations and Threats to Validity

The experiment participants were not experienced architects. However, they were not asked to design an architecture, they were asked to come up with questions. They did have experience in software design, so they could ask questions about software. And they know about traffic signals, so they could ask questions about traffic signals. Thus, their lack of software architecture experience may have affected their sets of questions in a limited way, we do not believe that it materially affected the questions in general categories.

Experiment results with respect to types and number of different quality attributes may not generalize to all domains. We do not claim that the profile of quality attribute questions matches other domains.

Survey participants self-selected, as is the case with most surveys. On the other hand, the survey respondents had strong experience in both software architecture and software development in general. Furthermore, they represented experience in a large variety of project sizes. These factors suggest that the survey results may be representative of a software architects. We recognize that the domains of architects' experience may not reflect all industries.

Analysis of the free-response questions in the survey required some interpretation, which may lead to questions about subjectivity of interpretation. However, the survey comments were generally very clear (e.g. "use Trello") and were straightforward to organize. In addition, both authors analyzed the survey comments independently, and came up with consistent organization.

As described, the survey and the experiment are different in what they asked about timing of architectural questions, so the data do not line up between the survey and experiment. We do not try to do so. Nevertheless, the data around timing of questions is somewhat limited.

We did not explore different levels of importance of architectural questions. The importance may affect how they are organized and when they are answered. This limits the results somewhat, and should be explored further.

Data on when architectural questions must be answered is preliminary and as yet imprecise. We present it as preliminary and recommend further study.

8 Recommendations and Future Research

We recommend tight coupling of architects and stakeholders, particularly during analysis and architectural design. The data showed the importance of doing so. But we note that the data did not conclusively show a problem in this area, so this is a recommendation to continue, rather than start.

Because informal and ad hoc methods of tracking questions are at risk of losing information, practitioners should consider using more formal ways of organizing and tracking them, and/or specialized tracking tools. Ease of use must be a priority, and they must fit within an organization's development processes. Consider developing an approach or tool to support eliciting, organizing, and tracking architectural questions. It would have to support multiple dimensions of organizing questions. It would have to support timing and scheduling of questions.

Because architectural questions are mixed between problem space and solution space, early-phase activities that focus on the problem space and the solution space should integrate closely together. For example, methods of requirements and quality attribute analysis might support the capture and recording of solution-space architectural questions that come up during those activities.

We recommend further research into the timing of architectural questions, including relating types of questions and their timing. We also recommend studying dependencies of questions on each other.

Longitudinal studies of questions throughout the software development cycle may provide more insight into the nature of questions. For example, one might study how questions emerge during the period of architectural design. In addition, priorities and importance of questions may change over time.

We recommend research into the importance of architectural questions. Differing importance of questions may impact when they are answered, how they are answered, and how they are tracked. This understanding will help complete the picture of architectural questions.

9   Conclusions

Key contributions of this work include the following. Questions the come up in architecture are roughly evenly split between problem space and solution space. It is well understood that issues are both spaces, but until now the proportion has not been understood. Questions that arise have different timelines for being answered. The general concept is understood, this research adds limited additional understanding about the nature of timing of questions. While it is known that architectural issues are likely tracked, we learned that architects often organize and track questions along multiple dimensions, using standard tools not specialized for architectural questions. Some research has been done about the impact of assumptions made during architecture; this research shows how architects handle questions, including how often they make assumptions. We invite further research to explore how to support architects handling questions.

Data Availibility Statement. Data from the survey and the controlled experiment are available in the ECSA Zenodo community (https://zenodo.org/communities/ecsa) (https://doi.org/10.5281/zenodo.12521472).

Disclosure of Interests. The authors have no competing interests to declare that are relevant to the content of this article.

References

1. Software Systems Architecture course, Master in Informatics and Computing Engineering, Faculty of Engineering, University of XXXX. https://sigarra.up.pt/feup/en/UCURR_GERAL.FICHA_UC_VIEW?pv_ocorrencia_id=518812
2. Anish, P.R.: Towards an approach to stimulate architectural thinking during requirements engineering phase. In: 2016 IEEE 24th International Requirements Engineering Conference (RE), pp. 421–426 (2016). https://doi.org/10.1109/RE.2016.30
3. Bastani, B.: A requirements analysis framework for open systems requirements engineering. SIGSOFT Softw. Eng. Notes **32**(2), 1–19 (2007). https://doi.org/10.1145/1234741.1234753
4. Castro, J., Kramer, J.: From software requirements to architectures. In: Proceedings of the 23rd International Conference on Software Engineering, ICSE 2001, pp. 764–765. IEEE Computer Society, USA (2001)
5. Chen, L., Ali Babar, M., Nuseibeh, B.: Characterizing architecturally significant requirements. IEEE Softw. **30**(2), 38–45 (2013). https://doi.org/10.1109/MS.2012.174

6. Clements, P., et al.: Documenting Software architectures. Pearson Education (2011)
7. Harrison, N.B., Avgeriou, P., Zdun, U.: Focus group report: capturing architectural knowledge with architectural patterns. In: EuroPLoP, p. 691 (2006)
8. Harrison, N.B., Rudolph, G., Tang, J., Thackeray, L., Wagstaff, D.: Capturing notable architectural decisions. In: 2023 IEEE 20th International Conference on Software Architecture Companion (ICSA-C), pp. 179–182 (2023). https://doi.org/10.1109/ICSA-C57050.2023.00047
9. Jansen, A., Bosch, J.: Software architecture as a set of architectural design decisions. In: Proceedings of the 5th Working IEEE/IFIP Conference on Software Architecture, WICSA 2005, pp. 109–120. IEEE Computer Society, USA (2005). https://doi.org/10.1109/WICSA.2005.61
10. Kazman, R., Klein, M.: Performing architecture tradeoff analysis. In: Proceedings of the Third International Workshop on Software Architecture, ISAW 1998, pp. 85–88. Association for Computing Machinery, New York, NY, USA (1998). https://doi.org/10.1145/288408.288430
11. Miyachi, C.: Agile software architecture. SIGSOFT Softw. Eng. Notes **36**(2), 1–3 (2011). https://doi.org/10.1145/1943371.1943388
12. Müller, M., Kersten, B., Goedicke, M.: A question-based method for deriving software architectures. In: Crnkovic, I., Gruhn, V., Book, M. (eds.) ECSA 2011. LNCS, vol. 6903, pp. 35–42. Springer, Heidelberg (2011). https://doi.org/10.1007/978-3-642-23798-0_4
13. Nord, R.L., Brown, N., Ozkaya, I.: Architecting with just enough information. In: Proceedings of the 6th International Workshop on SHAring and Reusing Architectural Knowledge, SHARK 2011, pp. 9–12. Association for Computing Machinery, New York, NY, USA (2011). https://doi.org/10.1145/1988676.1988680
14. Salehin, R.: Missing requirements information and its impact on software architectures: a case study. In: Masters Thesis, University of Western Ontario (2013). https://api.semanticscholar.org/CorpusID:28003460
15. Sommerville, I., Ransom, J.: An empirical study of industrial requirements engineering process assessment and improvement. ACM Trans. Softw. Eng. Methodol. **14**(1), 85–117 (2005). https://doi.org/10.1145/1044834.1044837
16. Van Landuyt, D., Truyen, E., Joosen, W.: On the modularity impact of architectural assumptions. In: Proceedings of the 2012 Workshop on Next Generation Modularity Approaches for Requirements and Architecture, NEMARA 2012, pp. 13–16. Association for Computing Machinery, New York, NY, USA (2012). https://doi.org/10.1145/2162004.2162008
17. Wohlin, C., Höst, M., Henningsson, K.: Empirical research methods in web and software engineering. In: Mendes, E., Mosley, N. (eds.) Web Engineering, pp. 409–430. Springer, Berlin, Heidelberg (2006). https://doi.org/10.1007/3-540-28218-1_13

Automated Architecture Recovery for Embedded Software Systems: An Industrial Case Study

Domenico Amalfitano[1], Marco De Luca[1(✉)], Domenico Francesco De Angelis[1,2], and Anna Rita Fasolino[1]

[1] University of Naples Federico II, Naples, Italy
{domenico.amalfitano,marco.deluca2,domenicofrancesco.deangelis,
fasolino}@unina.it
[2] Micron Technology, Inc., Boise, USA
ddeangelis@micron.com

Abstract. The software architecture documentation of embedded systems is often overlooked in industry, due to time pressure, project budget constraints, and lack of culture. However, adequately documenting the architecture from different points of view is mandatory to reach the expected maintainability, testability, and safety requirements. This paper presents a software architecture recovery (SAR) process for automating the documentation process of embedded system software architectures. The approach uses static code analysis to extract detailed information about the systems and reconstruct architectural models. It has been implemented in a tool that automatically generates different UML models, including package diagrams, component diagrams, component and connector diagrams, and state machine diagrams. To evaluate the effectiveness of our approach, we conducted a survey with industrial experts within Micron, that allowed us to assess the accuracy and usefulness of the generated documentation.

Keywords: Software Architecture Recovery · Static Analysis · Reverse Engineering · Embedded Software Architectures

1 Introduction

Software documentation, as a critical aspect of software development processes, faces numerous challenges, particularly in industrial settings where code rapidly evolves through frequent development iterations. The changes in the code throughout multiple development cycles create a complex set of demands for software documentation. In such a situation, keeping the documentation accurate, corresponding to the contemporary state of the code, and not getting lost during development processes is challenging. Indeed, as software projects evolve or change ownership, their documentation can become outdated or fragmented. Industries engage in redocumentation efforts to ensure that the documentation remains accurate, comprehensive, consistent, and reflective of the current state of the project. Moreover,

M. Galster et al. (Eds.): ECSA 2024, LNCS 14889, pp. 53–68, 2024.
https://doi.org/10.1007/978-3-031-70797-1_4

Continuous Integration/Continuous Deployment (CI/CD) practices have revolutionized software development lifecycles by automating build, testing, and deployment processes. Integrating documentation tools into the CI/CD pipelines is essential for establishing a robust mechanism to ensure continuous harmony between the codebase and its corresponding documentation. This proactive approach minimizes the risk of inconsistencies and fosters agility and efficiency in the development lifecycle. While various reverse engineering techniques exist for recovering software documentation from traditional software [11], few are tailored to embedded software systems [23]. These systems, designed to work on dedicated hardware platforms, have to satisfy strict real-time and other quality requirements, are often implemented in low-level coding languages, and pose unique comprehension challenges. Therefore, there's a pressing need for techniques to extract structural and behavioral models from embedded systems, enhancing comprehension and simplifying maintenance.

This paper introduces a tool-supported Software Architecture Recovery (SAR) process aimed at automatically generating Software Architecture Documentation (SAD) from the code of embedded software. The SAR process utilizes a reverse engineering approach, leveraging static analysis of the C code of embedded software systems to recover models that can be integrated into the architectural views of the documentation template proposed in [1]. The resulting SAD includes various UML models, such as package diagrams, component diagrams, component and connector (C&C) diagrams, and state chart diagrams. Finally, we conducted a questionnaire-based industrial survey involving 14 participants from Micron developing embedded software systems. The survey had a dual purpose: to evaluate the accuracy and utility of the UML diagrams automatically produced by the tool and to gather feedback on the limitations of the SAR process for further refinement. The contribution of this paper is threefold:

1. proposing a tool-supported SAR process that automatically produces UML models from static analysis of C code for embedded software systems;
2. conduction of an industrial survey to validate the effectiveness of the SAR process within real-world contexts;
3. evaluation of the accuracy of the UML diagrams generated by the tool and its usefulness through a survey involving developers of the industry in which the study was conducted.

The paper is structured as follows, Sect. 2 describes the studies related to this work, Sect. 3 presents the proposed SAR process, Sect. 4 discusses implementation details, Sect. 5 presents the industrial survey we conducted, and finally, conclusions and future research directions are summarized in Sect. 6.

2   Related Studies

Reverse engineering techniques apply to software and hardware systems, to extract and identify system structure and design for better comprehension. This process enhances program understanding in software systems by extracting design information about components and their relationships, presenting them at a higher abstraction level. Reverse engineering techniques are particularly valuable for maintaining and comprehending poorly documented or legacy systems, which are often affected by Architectural Technical Debt due to resource limitations and inadequate documentation [10,26]. As outlined by Nelson *et al.* [13] reverse engineering can have different goals, i.e., *Design Rediscovery*, *Reengineering*, and *Redocumentation*. *Design Rediscovery* aims to develop a system model at a higher level of abstraction. *Reengineering* combines reverse and forward engineering techniques to understand which functionalities need to be refactored, deleted, or added. *Restructuring* aims to refactor the whole system to improve maintainability, readability, or other quality attributes by keeping the system's functionalities intact. Lastly, *Redocumentation* aims at generating new documentation of updating existing ones depicting the system behavior. Moreover, reverse engineering processes rely on the execution of three steps [20]: *Extraction*, to gather data from the source code and existing documentation to retrieve design and construction artifacts; *Abstraction*: to synthesize and abstract the extracted information in a format less dependent on its implementation; *Presentation*: to convert the abstracted information into a format that is user-friendly and easy to comprehend.

2.1   Software Architecture Recovery (SAR)

SAR processes are an extension of reverse engineering and focus on abstraction and presentation, highlighting the architectural structure [22]. SAR addresses issues like architectural discrepancies during software evolution [7,17,25]. SAR techniques are critical for software maintenance, especially in understanding and modifying legacy systems. They assist in conformance checking, reconstructing descriptions, and analysis for meeting new requirements [5]. Different SAR techniques have been developed, ranging from static and dynamic analysis to machine learning approaches. Static analysis does not require the executing of the code and extracts architectural structures and dependencies [21]. Dynamic analysis, involving code execution, aims to create high-level system models. Revealer [14], using lexical and syntactic code analysis, abstracts high-level architectural views using regular expressions and XML-based patterns. Sora *et al.* [19] introduced a static analysis-based approach using PageRank algorithm to identify important classes. NEGAR, a machine learning algorithm by Chem et al. [4], uses graph representations of software dependencies for clustering related files, aiding maintenance. These approaches vary in accuracy and effectiveness but collectively contribute to a deeper understanding and maintenance of software systems.

2.2   Reverse Engineering of State Chart Diagrams

Recovering state machine behavior involves two main strategies: dynamic analysis, which logs features and execution flow through code instrumentation and static analysis, based on source code, allowing for customization. Static analysis as the one proposed by Walkinshaw *et al.* [12] utilizes symbolic execution and program conditioning on source code. It identifies state transitions and annotates them with corresponding code segments, observing that transition points often correlate with specific syntax elements. This model is especially effective for small, structured systems like object-oriented ones. Tonella *et al.* [24] consider each method call on an object as a transition, an assumption not applicable to embedded firmware with implicit state machines. Similarly, Kung *et al.* [9] and Sen and Mall [16] employ symbolic execution for automatic state machine extraction from C++ and Java sources, focusing on small methods and state variables influencing branching in object-oriented systems. Bae *et al.* [2] suggest generating state charts from contract-based specifications, using method pre and postconditions instead of path conditions, and strengthening transition conditions when necessary. Researchers like van den Brand *et al.* [3], Knor *et al.* [8], and Somé *et al.* [18] have worked on extracting state machines from procedural code with specific patterns, like nested-choice structures. However, these techniques are limited and not widely applicable to more diverse code categories.

3 The Proposed Reverse Engineering Process

In this section, we present the proposed Software Architecture Recovery (SAR) process, which is based on static analysis and used to recover both the structural and behavioral views of embedded software. According to the five-axis taxonomy for SAR characterization presented by Ducasse *et al.* [6], our SAR focuses on documentation and conformance. Our SAR was used to analyze an embedded firmware developed for managing NAND memory systems. This firmware provides different functionalities for memory access to external hosts and it's designed to be configurable for running on different hardware platforms. Moreover, the firmware follows a layered architectural style, where the lower layers are strictly related to the hardware architecture. To ensure this portability requirement, the firmware is developed according to company-specific coding practices that make extensive usage of macros and pre-processor directives. As a consequence, the resulting code is difficult to analyze and comprehend. The SAR outputs include various UML diagrams, including Package, Component, Component and Connector (C&C), and State Chart models. The recovered models can fill the views of the documentation template proposed in [1], designed for documenting ISO26262-compliant automotive software systems. This template is suitable for documenting embedded software systems as well. The SAR process, as illustrated in Fig. 1, consists of four phases. It has been implemented in a tool that exploits the features of both Enterprise Architect[1] (EA) for UML modeling and reverse engineering, and Pycparser[2], a Python library for analyzing C code and generating Abstract Syntax Trees (AST). In the following, we provide an overview of these phases and their inner activities.

---

[1] Enterprise Architect: https://sparxsystems.com/.
[2] Pycparser: https://github.com/eliben/pycparser.

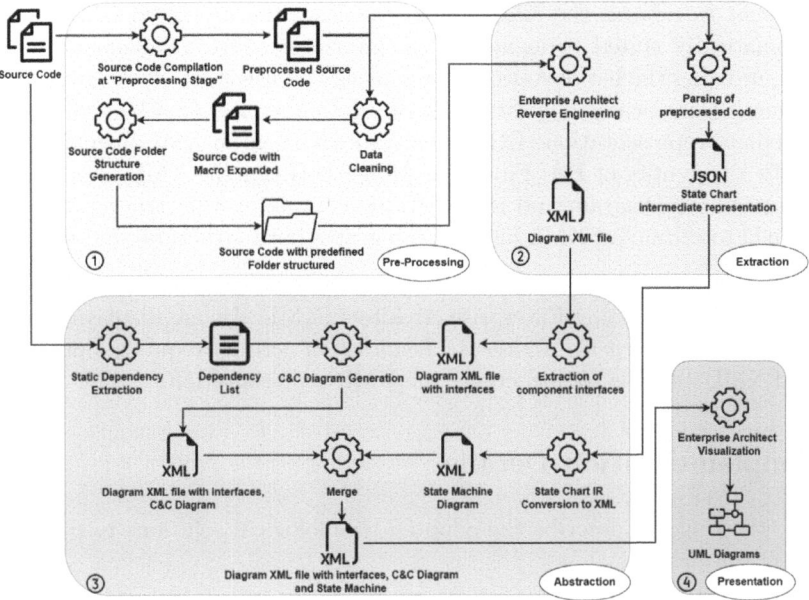

Fig. 1. The proposed SAR process

1. Pre-processing: This phase includes several activities: first, during the _Source Code Compilation at "Preprocessing Stage"_, the compilation is stopped early to expand macros, resolve #include directives and headers, and remove comments, producing the _preprocessed source code_. This step is essential for processing with Pycparser and Enterprise Architect. Next, in _Data Cleaning_, the source code is cleaned up of all the code generated by the expansion of header files and directives, resulting in a clean version free of macros or keywords. Finally, in _Source Code Folder Structure Generation_, the source code is reorganized into a new configuration of folders and subfolders that reflects the architectural design of the software.

2. Extraction: This extraction process involves two main activities: first, the _Enterprise Architect Reverse Engineering_ feature is used to generate structural models of the software system, that include package and component diagrams. Second, the _Parsing of preprocessed code_ involves adapting and parsing the _preprocessed code_ to produce an abstract-syntax tree (AST). This AST is subsequently analyzed to create an intermediate representation (IR) of a state chart, which includes detailed information about states and transitions, including guards, signals, and triggers.

3. Abstraction: The models produced by EA are enhanced by extracting component interfaces and generating C&C diagrams. This process includes several key activities: First, _Static Dependency Extraction_ involves analyzing the source code to identify static dependencies between components. Next, in _Extraction of_

Component Interfaces, the XML file representing the extracted UML diagram is automatically edited to include UML interfaces. *C&C Diagram Generation* follows, utilizing the list of static dependencies and the XML file to generate the diagrams automatically. The *State Chart IR Conversion to XML*, converts the intermediate representation (IR) of the state chart in an XML format compatible with the syntax of EA. Finally, a *Merge* between the XML file containing the State Chart Diagram and the XML file containing the structural model is performed to obtain an XML file that integrates both structural and behavioral models.

4. Presentation: Using Enterprise Architect, XML files are displayed as UML diagrams. The *Enterprise Architect Visualization* activity requires inputting the merged XML into the tool to create the UML representation.

4   Implementation Details

In the following, we describe the adopted technological solutions to recover the architectural models.

Pre-processing is a crucial activity in our recovery process, designed to address company-specific coding and programming practices, especially the extensive use of macros that complicate analysis and understanding. This phase standardizes the code's syntax to align with the common syntax standards of the programming language. Our starting point is a collection of source code files organized in a single folder, including both .c files and associated header files (.h). To align the source code with the requirements of EA and Pycparser, it undergoes a pre-processing phase via the C compiler. This is ensured by executing the compiler command `gcc -E` which interrupts the compilation of the code in the preprocessing phase, resulting in what we define as *preprocessed source code*. This phase expands macros, resolves `#include` directives and headers, and removes comments from the source code. The pre-processed code has a dual function: it is routed to Pycparser for the identification of the finite-state machine and is simultaneously submitted to the *Data Cleaning* step. After this step, a variant of the source code identical to the original is obtained, but with expanded macros and keywords and resolved directives.

Package Diagram Generation. In Micron, the names of the .c files and their corresponding .h files reflect the layered structure of the embedded software system. We implemented a Python script that analyzes the names of all pairs of .c and .h files and creates a package structure, in terms of folders and sub-folders, mirroring the architecture of the software system. The newly-organized source code is then input into EA to leverage its static code analysis capabilities and generate UML diagrams. Consequently, the tool produces a package diagram of the entire project, reflecting the folder and sub-folder structure provided as input, as shown in Figs. 2 and 3.

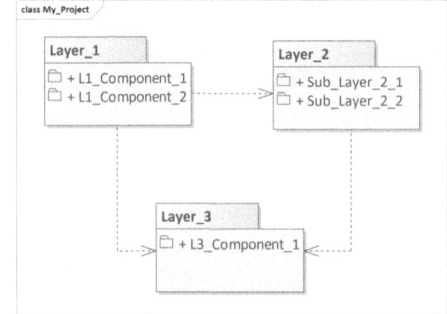

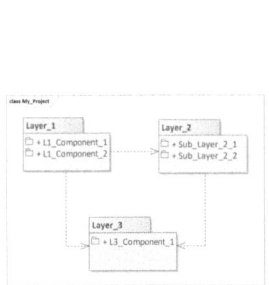

Fig. 2. Package Diagram browser view   **Fig. 3.** Reconstructed Package Diagram

Component Diagram Generation. For each .c and .h file pair, EA automatically generates a UML class diagram, like the one shown in Fig. 4a. These diagrams include attributes, operations, structs, and enumerations. Attributes and operations are classified as public if declared in the .h file and implemented in the .c file; otherwise, they are private if only defined in the .c file. EA allows these UML diagrams to be converted into an XML format for easy modification. These XML files are structured into tables such as **t_package**, **t_object**, and **t_diagram**, each one holding different UML elements like packages, classes, and diagrams respectively, as shown in listing 1.1). Moreover, each table row represents a UML element with attributes including `Name`, `Object_Type`, and `ea_guid`, the latter serving as a unique identifier for the element. To enhance the diagrams generated by EA and align them with the reference metamodel, we developed a Python script to manipulate the XML files representing UML diagrams. Firstly, the script updates the `Object_Type` attribute of each *Class* object to convert it into a *Component* object. Subsequently, an *Interface* object is created and added to the `t_object` table. To achieve this, we follow the XML file structure, setting the `Object_Type` attribute to *Interface* and adding the *ea_guid* of the package where we intend to place the interface within the `<Extension>` tag. This ensures that the Interface and its corresponding Component reside in the same package. Next, the tool generates a Realization `Object_Type` to establish the relationship between the Interface and Component. This is achieved by specifying the respective *ea_guid* of the interface and component within the `<Extension>` tag of the realization object being created. Finally, the script analyzes the `t_operation` table, appropriately modifying the *ea_guid* within the `<Extension>` tag to bind the interface with the operation it should contain. In Fig. 4b, the diagram reconstructed by our tool is depicted, where we can observe the interface extracted from the component.

Listing 1.1. Sample of the XML file generate by EA

```
< Package name="My_Project" guid="{350E9CE0-4958-42c0-8F83-1E449CB016AC}">
  < Table name="t_package">
          < Row> [...] < /Row>
  < /Table>
  < Table name="t_object">
    < Row>
      < Column name="Object_id" value="4"/>
      < Column name="Object_Type" value="Component"/>
      < Column name="Diagram_id" value="2"/>
      < Column name="Name" value="My_Component"/>
      < Column name="Author" value="marco"/>
      < Column name="Package_id" value="2"/>
      < Column name="GenType" value="C"/>
      [...]
      < Column name="Scope" value="Public"/>
      < Column name="ea_Guid" value="{A7FCE46C-3C14-48c6-8BFB-07FAB08F22D7}"/>
      [...]
        < Extension Package_ID="{350E9CE0-4958-42c0-8F83-1E449CB016AC}"/>
    < /Row>
    < Row> [...] < /Row>
  < /Table>
  < Table name="t_diagram"> [...] < /Table>
  < Table name="t_operation"> [...] < /Table>
  < Table name="t_connector"> [...] < /Table>
< /Package>
```

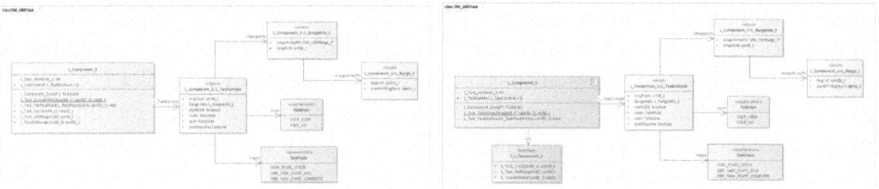

(a) Class Diagram reconstructed by Enter-   (b) Component Diagram reconstructed by
prise Architect   the tool

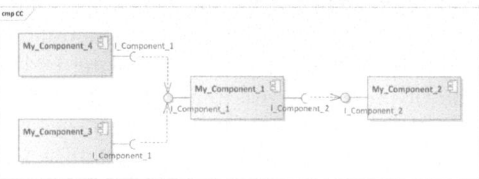

(c) C&C Diagram reconstructed by the tool

Fig. 4. Diagrams Examples

Component and Connector (C&C) Diagram Generation. Using gcc to compile code it is possible to generate a `MapFile`, which details the memory layout of code including functions and variables. This file also includes *cross-references* among symbols to identify function calls, enabling the extraction of static dependencies list between different functions within the system. We have developed an additional Python script to automatically reconstruct C&C diagrams by taking as input the list of static dependencies extracted from the *Map-File* and the XML files of the UML diagrams. For each component present in the XML file, the script adds *Provided* and *Required* interface objects, specifying their `Object_Type` and inserting the `ea_guid` of the corresponding component in the <Extension> tag. Subsequently, the script inserts a new diagram object into the *t_diagram* table of the XML file. Leveraging the list of static dependencies, the script can: i) incorporate into the diagram the components with which the component in question interacts, ii) reconstruct the dependency relationships between the various components, which are appropriately created and inserted into the `t_connector` table. At the end of the process, the C&C diagram for each component is reconstructed, and the new XML file, representing the UML representation, is provided as output, depicted in Fig. 4c.

State Chart Diagram Generation. Our SAR process also reconstructs State Chart diagrams from the source code, to represent the system's states and transitions for a better understanding of its behavior. To achieve this, we rely on the task table, a table containing the function pointer of the APIs scheduled by the embedded operative system as a task. This table is used to identify tasks in the pre-compiled source code, ensuring proper parsing. The tasks are units of execution and are implemented as state machines. After processing the code into a parsing tree, we generate an intermediate representation (IR) in JSON format, which is then converted into XML format, to be given as input to EA. The recovery of state chart diagrams relies on several assumptions. It presupposes that each task consistently produces a return value indicating its status and employs a `switch-case` construct to define a state machine. The tool also requires well-formed `switch-case` statements composed solely of `case-statements` to define behavior. The state variables are identified as the condition expression in the `switch-case` statement. Signals and triggers are discerned through the use of APIs provided by the embedded operating system. Furthermore, tasks may include multiple exit points, and the presence of spaghetti code allows for the use of jump instructions such as `goto`. All these cases are handled appropriately by our tool. The proposed reverse engineering approach relies on the execution of the following 4 steps:

Prepare state-machine before starting: the *preprocessed source code* is elaborated to simplify the abstraction of state-machines. This adaptation is necessary because the C language allows various patterns for describing state machines, primarily using the `switch-case` structure. In some cases, `switch-case` statements are embedded within selective or iterative flows, which are not well-formed for straightforward conversion into state chart diagrams. Therefore, these `switch-case` statements require refactoring to align the C constructs with the state chart diagram representation.

State variable definition: according to the criteria outlined by Said [15], our approach is based on a specific pattern of `switch-case` implementation. In this context, the state variable is defined as the condition expression of the `switch` statement.

Guard & transition definition: drawing from Walkinshaw's strategy [12], our method extracts guards from source code analyzing two key variables: the state variable, discerned within `switch-case` statements, and the return variable, typically located at the begging of variable declarations. These transitions between states are further classified into three categories: event-driven, `goto`, and rescheduling transitions. Event-driven transitions correspond to specific events, while `goto` transitions occur independent of triggers, and rescheduling transitions follow guard condition verification, subsequently rescheduling tasks in subsequent cycles. Furthermore, to enhance readability, we prioritize transitions, inspired by Said et al. [15], over a sequential read approach, and conduct boolean reduction on conditions. Additionally, recurring sub-expressions or API calls in the code are identified and simplified with the use of symbolic names.

Generate XML diagram EA compatible: the step for generating the XML output file involves several steps. Initially, Entry and Exit point UML state chart objects are created to mark the beginning and end of the state machine. An additional Exit point may also be generated to handle error conditions if necessary. Subsequently, node states are defined for each state in the state machine, transitions between states are established, and guards, triggers, and signals are applied as needed.

In Fig. 5, we present an example of a task-based switch-case statement. Figure 6 shows the corresponding reconstructed state chart diagram.

```
TaskReturn_t task_example(void* args) {
    TaskReturnState_em ret;
    switch(taskContext.state){
    STATE_CMD_GET: case STATE_CMD_GET:{
        CmdPtr_t* CmdHandler = HAL_GetCommand();
        if (CmdHandler != NULL){
            taskContext.cmd = CmdHandler;
            goto STATE_CHECK_PRIORITY;
        }
        return TASK_READY;
    }
    STATE_CHECK_PRIORITY: case STATE_CHECK_PRIORITY: {
        CmdHeader_t* cmd = taskContext.cmd->header;
        if (cmd->flag & ATTR_MASK) == PRIORITY_MASK){
            if (cmd->hit){
                OS_WaitEvent(EVENTID_HIGH_PRIORITY_DONE);
                return TASK_SUSPEND;
            }
            else
                cmd->flag = Analyze_Command(CmdHandler);
        }
    }
    STATE_EXECUTE_CMD: case STATE_EXECUTE_CMD: {
        uint8_t cmdIdx = taskContext.cmd->cmdIdx;
        CmdPayload_t* cmdEntry = taskContext.cmd->payload;
        if (CmdAbortStatus(cmdIdx) == true){
            if(OS_ResourceLocked() == true)
                OS_RealeseResource();
            goto STATE_CMD_GET;
        }
    }
    }
}
```

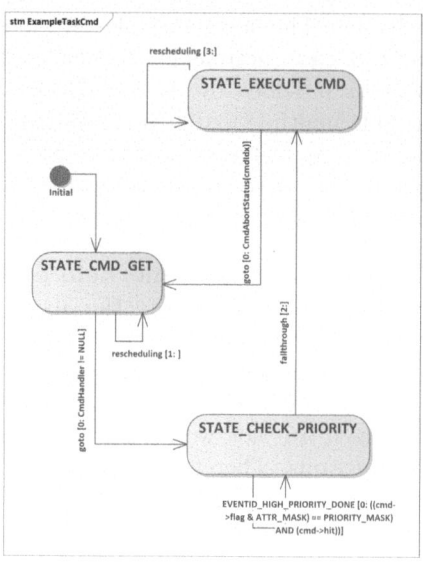

Fig. 5. Task Code Example

Fig. 6. Reconstructed Task State Chart Diagram

5 Experimental Evaluation

We aimed to assess the effectiveness of the proposed documentation process in terms of extracting accurate information from the source code and representing it in UML diagrams, as well as its usability in aiding system comprehension. To achieve this, we conducted a survey involving software engineers from the embedded systems domain. The survey aimed to answer the following Research Questions (RQs):

1. *RQ1 - To what extent the recovered software architecture documentation is considered accurate by the practitioners?*
2. *RQ2 - If important information is missing, what kind of information is it?*
3. *RQ3 - To what extent have the practitioners found the proposed software architecture recovery (SAR) tool useful?*

Interviewees Selection. We recruited 14 embedded software engineers currently employed in Micron where the case study was conducted. Each participant has been with the company for at least three years and is experienced in working on software documentation and development tasks. Additionally, all the engineers understand the principles of software architecture documentation and are familiar with using UML notation.

Questionnaire Design. The 14 participants were surveyed through a questionnaire we appositely designed to answer the three RQs. The questionnaire was organized into three sections, to evaluate the accuracy and comprehensibility of diagrams generated by the tool, along with evaluating the tool's usability. An additional fourth section was introduced to collect data about the strengths and the limitations of the tool. It consisted of 30 questions, mainly closed-ended, with respondents given the opportunity to provide additional feedback through open-ended questions. The closed-ended questions were measured using a 5-point Likert scale. The questionnaire was implemented in Microsoft Forms. The data analysis relied on both quantitative and qualitative methods to answer the research questions.

Survey Execution. We conducted a preliminary training phase with all the participants, by executing two focus group sessions, each one lasting two hours. The first focus group session introduced the participants to the tool and included small reverse engineering tasks to familiarize them with its use. Three days later, the second focus group session was devoted to exploring more in detail the tool's features and addressed any doubts that arose during the reverse engineering tasks performed in the first session. Following this training, participants were tasked with a software documentation recovery exercise to be executed with the tool. Each task involved two components of the system they had previously manually designed and implemented. All components were of medium complexity according to the company standards, and each respondent analyzed two different components. We assigned them one hour to complete the task. After that, the respondents filled out the Microsoft Forms questionnaire.

Answer to RQ1 - to What Extent the Recovered Software Architecture Documentation is Considered Accurate by the Practitioners? The box plot in Fig. 7 shows the accuracy distribution for the diagrams reconstructed by the tool, based on the responses from practitioners. Higher values indicate a good perception of accuracy. From the chart, we can observe that the distribution of the State Chart Diagram is skewed on the higher part of the graph, indicating that it is considered to be reconstructed with a high level of accuracy.

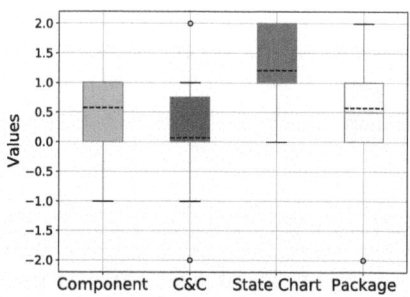

Fig. 7. Box plot showing the accuracy distribution of diagrams reconstructed automatically by the tool

Both the Component Diagram and Package Diagram have a positive median, suggesting that the accuracy of these two diagrams is generally regarded as good. The C&C diagrams have a distribution concentrated on lower values, and the presence of outliers with low values indicates a negative perception of their accuracy. From the analysis of open-ended answers, we observed that some respondents mentioned that the tool sometimes struggled to accurately identify interfaces, enumeration variables, or functions, which they had to manually add to the diagram. They believe this issue may be attributed to the wide variety of programming styles present in the code.

Our analysis revealed a generally positive perception of accuracy, particularly with State Chart diagrams being deemed the most accurate. Component and Package diagrams were also considered sufficiently accurate, while improvements were identified as necessary for the C&C diagrams.

Answer to RQ2 - If Important Information is Missing, What Kind of Information is it? To identify the types of missing information, two authors manually analyzed the responses to the open-ended questions, extracted keywords from the responses, and finally classified the missing information into the four categories shown in Fig. 8. Another author validated the proposed classification. The histogram rendered in Fig. 9 shows the distribution of the categories of missing information. In Fig. 9, the "Relationship" category was identified as the most commonly missing element, reported by 32.4% of respondents. This category highlights the tool's difficulty in capturing interactions within the diagrams, such as the realization relationships between components and interfaces, dependency relationships between provided and required interfaces, and transitions between states. The "Module Structure" category was the next most frequently mentioned, with 26.5% of responses indicating that the tool does not adequately capture structural details of the recovered Components such as methods or attributes. Both the "Model Element" category and "No Missing" were noted by 20.6% of respondents each. These results indicate that about one-fifth of users find shortcomings in how the tool reconstructs system components, packages, and interfaces, while another-fifth see no missing information.

Category	Keyword
Relationship	Realization relationship between component and Interface
	Dependency relationship between component and enumeration
	Dependency relationship between provided and required interfaces
	Transitions between State
	Dependency between package
Component Properties	Method
	Enumeration
	Required Interfaces
	Provided Interfaces
	Attribute Type
Model Element	Interface
	Component
	Package
No Missing	No missing information

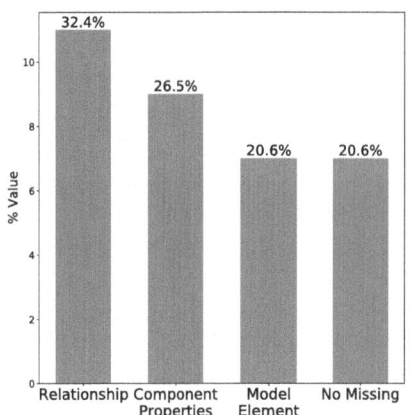

Fig. 8. Identified categories of missing information and their relative keyword

Fig. 9. Distribution of the missing information categories

Answer to RQ3 - to What Extent Have the Practitioners Found the Proposed Software Architecture Recovery (SAR) Tool Useful? The pie charts in Figs. 10 and 11 display the results of responses to specific questions regarding the tool's usefulness in aiding the comprehension of the embedded system architecture. As shown in Fig. 10, the majority of the interviewed participants (71.4%) agreed that the tool is useful in supporting the comprehension of the system. Similarly, as shown in Fig. 11, a significant portion of the respondents (61.5%) expressed a preference for the models automatically generated by the tool to understand the system architecture. To answer *RQ3*, we found strong evidence indicating a positive response regarding the effectiveness of the tool in facilitating the understanding of architectural elements and relationships. These aspects are often complex and challenging to discern from code alone. However, some respondents remained neutral, suggesting that certain aspects of the reconstructed documentation may not fully meet the needs or expectations of all practitioners. These respondents prefer consulting the source code over UML diagrams for understanding the system's structure. Others noted that while auto-generated documentation aids in comprehension, it lacks in capturing the system's "semantics". These limitations suggest potential areas for future development, such as implementing customized rules to better suit various working contexts.

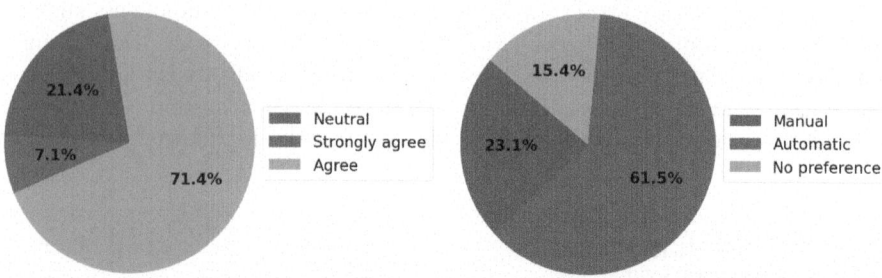

Fig. 10. Pie chart on the closed-ended question: *"Based on my experience in using the tool, I find it useful in supporting the comprehension of the system."*

Fig. 11. Pie chart on the closed-ended question: *"Which reconstructed models, whether generated automatically or manually, do you find most useful for understanding the system?"*

Threats to Validity. Our study faces several potential threats to validity. The primary external threat arises from the fact that the SAR process proposed is tailored specifically to the company programming practices in which the case study was conducted, resulting in a static analysis of the process customized for the C code developed by Micron. We plan to mitigate this threat by extending the case study to other industries developing embedded software systems. Another threat to external validity is the heavy reliance on predefined reverse engineering rules within the EA tool. We intend to mitigate this threat by implementing tool-independent reverse engineering rules. Finally, the narrow scope of our company survey, involving only a small number of participants from a single organization, may also impact the generalizability of the study. To mitigate this threat, we plan to interview more personnel from Micron and expand the survey to include other companies in the embedded systems software development sector.

6   Conclusion and Future Works

This paper presented a tool-supported Software Architecture Recovery (SAR) process for the automated generation of Software Architecture Documentation (SAD) from C code in embedded systems. The evaluation was conducted via a questionnaire with 14 industry professionals, demonstrating the tool's utility in enhancing documentation practices and providing significant insights into its performance and areas for improvement. In conclusion, our tool demonstrated to be accurate in reconstructing documentation and supporting system comprehension, addressing critical challenges within the embedded systems industry.

We've identified several areas for future work based on the insights from this initial study. To enhance the applicability of our tool, we'll extend the SAR process to other companies in the embedded software industry, including those using different programming languages, ensuring adaptability to diverse development environments. We also plan to develop independent reverse engineering rules to reduce reliance on EA-specific ones, making the SAR process more versatile across various software development contexts. Additionally, we'll expand our industrial survey to include more participants from the surveyed company and to encompass additional companies, gathering broader feedback to refine the tool and better cater to different organizational needs.

Acknowledgements. This work is supported by the PhD Program of Micron in collaboration with the University of Naples Federico II on "Functional Safety in Managed NAND Embedded Systems."

Data Availibility Statement. Data and tools are not publicly available as they are Micron's intellectual property and are subject to confidentiality agreements.

References

1. Amalfitano, D., De Luca, M., Rita Fasolino, A.: Documenting software architecture design in compliance with the ISO 26262: a practical experience in industry. In: 2023 IEEE 20th International Conference on Software Architecture Companion (ICSA-C) (2023)
2. Bae, J.H., Chae, H.S.: Systematic approach for constructing an understandable state machine from a contract-based specification: controlled experiments. Softw. Syst. Model. **15**(3), 847–879 (2016)
3. van den Brand, M., Serebrenik, A., van Zeeland, D.: Extraction of state machines of legacy C code with Cpp2XMI. In: Proceedings of 7th Belgian-netherlands Software Evolution Workshop, pp. 28–30 (2008)
4. Chen, J., et al.: Negar: network embedding guided architecture recovery for software systems. In: 2022 29th Asia-Pacific Software Engineering Conference (APSEC), pp. 367–376 (2022)
5. Chen, J.C., Huang, S.J.: An empirical analysis of the impact of software development problem factors on software maintainability. J. Syst. Softw. **82**(6), 981–992 (2009)
6. Ducasse, S., Pollet, D.: Software architecture reconstruction: a process-oriented taxonomy. IEEE Trans. Software Eng. **35**(4), 573–591 (2009)
7. Garlan, D., Allen, R., Ockerbloom, J.: Architectural mismatch: why reuse is still so hard. IEEE Softw. **26**(4), 66–69 (2009)
8. Knor, R., Trausmuth, G., Weidl, J.: Reengineering C/C++ source code by transforming state machines. In: van der Linden, F. (ed.) ARES 1998. LNCS, vol. 1429, pp. 97–105. Springer, Heidelberg (1998). https://doi.org/10.1007/3-540-68383-6_15
9. Kung, D., Suchak, N., Gao, J., Hsia, P., Toyoshima, Y., Chen, C.: On object state testing. In: Proceedings Eighteenth Annual International Computer Software and Applications Conference (COMPSAC 94), pp. 222–227 (1994)

10. Luna-Herrera, Y.A., Pérez-Arriaga, J.C., Ocharán-Hernández, J.O., Sanchéz-García, Á.J.: Comprehension of computer programs through reverse engineering approaches and techniques: a systematic mapping study. In: Mejia, J., Muñoz, M., Rocha, Á., Hernández-Nava, V. (eds.) New Perspectives in Software Engineering. CIMPS 2022. Lecture Notes in Networks and Systems, vol. 576, pp. 126–140. Springer, Cham (2022). https://doi.org/10.1007/978-3-031-20322-0_9

11. Maqbool, O., Babri, H.: Hierarchical clustering for software architecture recovery. IEEE Trans. Software Eng. **33**(11), 759–780 (2007)

12. Walkinshaw, N., Bogdanov, K., Ali, S., Holcombe, M.: Automated discovery of state transitions and their functions in source code. Software Test. Verification Reliab. **18**(2), 99–121 (2008)

13. Nelson, M.L.: A survey of reverse engineering and program comprehension. arXiv arXiv:cs/0503068 (2005)

14. Pinzger, M., Fischer, M., Gall, H., Jazayeri, M.: Revealer: a lexical pattern matcher for architecture recovery. In: Ninth Working Conference on Reverse Engineering, 2002. Proceedings, pp. 170–178 (2002)

15. Said, W., Quante, J., Koschke, R.: Mining understandable state machine models from embedded code. Empir. Softw. Eng. **25**(6), 4759–4804 (2020)

16. Sen, T., Mall, R.: Extracting finite state representation of java programs. Softw. Syst. Model. **15**, 497–511 (2016)

17. Silva, L., Balasubramaniam, D.: Controlling software architecture erosion: a survey. J. Syst. Softw. **85**, 132–151 (2012)

18. Somé, S.S., Lethbridge, T.C.: Enhancing program comprehension with recovered state models. In: Proceedings of the 10th International Workshop on Program Comprehension, IWPC 2002. IEEE Computer Society (2002)

19. Şora, I.: Helping program comprehension of large software systems by identifying their most important classes. In: Maciaszek, L.A., Filipe, J. (eds.) ENASE 2015. CCIS, vol. 599, pp. 122–140. Springer, Cham (2016). https://doi.org/10.1007/978-3-319-30243-0_7

20. Stringfellow, C., Amory, C., Potnuri, D., Andrews, A., Georg, M.: Comparison of software architecture reverse engineering methods. Inf. Softw. Technol. **48**(7), 484–497 (2006)

21. Systä, T.: Static and Dynamic Reverse Engineering Techniques for Java Software Systems (2000)

22. Telea, A., Maccari, A., Riva, C.: An open visualization toolkit for reverse architecting. In: Proceedings 10th International Workshop on Program Comprehension (2002)

23. Thomas, S.L., Van den Herrewegen, J., Vasilakis, G., Chen, Z., Ordean, M., Garcia, F.D.: Cutting through the complexity of reverse engineering embedded devices. IACR Trans. Cryptographic Hardware Embed. Syst. **2021**(3), 360–389 (2021)

24. Tonella, P.: Reverse engineering of object oriented code. In: Proceedings of the 27th International Conference on Software Engineering, ICSE 2005, pp. 724–725. Association for Computing Machinery, New York, NY, USA (2005)

25. Uzun, B., Tekinerdogan, B.: Architecture conformance analysis using model-based testing: a case study approach. Softw. Pract. Experience **49**(3), 423–448 (2019)

26. Verdecchia, R., Kruchten, P., Lago, P.: Architectural technical debt: a grounded theory. In: Jansen, A., Malavolta, I., Muccini, H., Ozkaya, I., Zimmermann, O. (eds.) ECSA 2020. LNCS, vol. 12292, pp. 202–219. Springer, Cham (2020). https://doi.org/10.1007/978-3-030-58923-3_14

An Analysis of MLOps Architectures: A Systematic Mapping Study

Faezeh Amou Najafabadi[(✉)] [ID], Justus Bogner[ID], Ilias Gerostathopoulos[ID], and Patricia Lago[ID]

Vrije Universiteit Amsterdam, Amsterdam, The Netherlands
{f.amou.najafabadi,j.bogner,i.g.gerostathopoulos,p.lago}@vu.nl

Abstract. *Context.* Despite the increasing adoption of Machine Learning Operations (MLOps), teams still encounter challenges in effectively applying this paradigm to their specific projects. While there is a large variety of available tools usable for MLOps, there is simultaneously a lack of consolidated architecture knowledge that can inform the architecture design. *Objective.* Our primary objective is to provide a comprehensive overview of (i) how MLOps architectures are defined across the literature and (ii) which tools are mentioned to support the implementation of each architecture component. *Method.* We apply the Systematic Mapping Study method and select 43 primary studies via automatic, manual, and snowballing-based search and selection procedures. Subsequently, we use card sorting to synthesize the results. *Results.* We contribute (i) a categorization of 35 MLOps architecture components, (ii) a description of several MLOps architecture variants, and (iii) a systematic map between the identified components and the existing MLOps tools. *Conclusion.* This study provides an overview of the state of the art in MLOps from an architectural perspective. Researchers and practitioners can use our findings to inform the architecture design of their MLOps systems.

Keywords: Machine Learning Operations · MLOps · Architecture · Components · Tools · Systematic Mapping Study

1 Introduction

The use of Machine Learning (ML) continues to grow in industry, and developing high-quality ML models is important to sustain it. However, when creating ML-based systems, another major concern of ML engineers and operations teams is the effective deployment and maintenance of ML models in production [11]. To address this, the Machine Learning Operations (MLOps) paradigm has formed in industry [7]. Similar to DevOps [3], MLOps comprises a set of best practices and related methods, technologies, and tools that aim to bridge the gap between the development of ML models and their deployment, maintenance, and evolution. Despite the increasing adoption of MLOps [14], it is still challenging for practitioners to effectively apply the paradigm to their projects [9]. First, there

ⓒ The Author(s), under exclusive license to Springer Nature Switzerland AG 2024
M. Galster et al. (Eds.): ECSA 2024, LNCS 14889, pp. 69–85, 2024.
https://doi.org/10.1007/978-3-031-70797-1_5

is a large variety of available tools usable for MLOps, which makes it hard for practitioners to analyze and compare all the options at their disposal [6]. Second, while reusable design decisions for certain parts of ML-enabled systems are starting to emerge [22], there is still a lack of consolidated MLOps architecture knowledge that could guide architectural decisions. Third, MLOps evolves at a fast pace and simultaneously in different domains, which makes it difficult to discern the generalizable concepts and technologies from the domain-specific ones.

In this paper, we therefore aim to provide an overview of the state of the art in MLOps from an architectural perspective. We extract and analyze the components that comprise typical architectures of MLOps systems and identify several variants of such architectures based on the existing variability points, e.g., the optional presence of an online training pipeline. We analyze the dependencies between components and synthesize them in the form of a UML component diagram. To align terminology in a rather scattered domain, we provide several known aliases for each component we identify. We also extract and analyze the tools that are mentioned in the MLOps literature and map them to the architecture components. This provides insight into the implementation options for each component and sheds light on untapped R&D opportunities in the form of less tool-supported components.

We accomplish this by performing a Systematic Mapping Study (SMS) of the scientific literature of MLOps, which grounds our analysis in the scientific state of the art. As such, we complement other attempts that analyzed gray literature for MLOps best practices and architecture design decisions [6,22]. As our long-term goal, we aim to provide a comprehensive reference architecture [17] for MLOps that covers both the *structural* perspective (the focus of this paper), and the *process* and *stakeholder* perspectives.

In summary, the contributions of this study are (i) a synthesis and categorization of 35 MLOps architecture components, (ii) a description of several MLOps architecture variants, and (iii) a systematic map between the identified components and the existing MLOps tools.

The target audience of our study is (i) researchers in software architecture for ML, who can build on our results to derive and consolidate architecture knowledge in the form of MLOps best practices and patterns, and (ii) ML practitioners, who can use our findings to inform the architecture design of their MLOps systems.

2 Related Work

MLOps definitions, practices, and guidelines have been the subject of numerous secondary studies including scoping reviews, systematic literature reviews, and multivocal literature reviews. Many secondary studies on this topic aim at **clarifying the definition of MLOps**. Mboweni et al. [16] state that there is still no official standard definition for MLOps. Based on their systematic review to disambiguate the definition of MLOps in the literature, they claim that they did not find evidence of a common understanding among scholars and experts on how MLOps should be implemented and institutionalized across the industry to create a common vision. Lima et al. [13] systematically reviewed 30 papers aiming at deriving practices, standards, roles, maturity models, challenges, and tools for MLOps. Based on the addressed challenges and assessment of models, they draw the conclusion that "research on MLOps is still in its initial stages."

Some literature reviews provide methodologies for **effectively approaching MLOps projects**. Testi et al. [21] provide a taxonomy of the current approaches toward and propose a methodology for addressing MLOps projects. Kolltveit and Li [10] specifically focus on the operationalization of ML models with regard to tools and infrastructure that are deployed in different stages of MLOps workflows. Recupito et al. [19] take a different perspective and provide an overview of the most common tools and their characteristics that support the creation of MLOps pipelines, without a clear mapping to components.

Several papers on the **architecture of ML-based systems** are also closely related to our study. For example, Warnett and Zdun conducted two studies in which they used practitioner gray literature to synthesize architectural design decisions (ADDs) for ML workflows [23] and ML deployment [22], with several of their sources being blog posts about MLOps. As a result, several of their ADDs are related to architecture components that we synthesize in our study. However, they do not combine this knowledge into a holistic architecture and also do not cover several parts of MLOps, such as inference and monitoring. In a controlled experiment, Warnett and Zdun [24] also compared the understandability of informal textual and graphical MLOps architecture representations with semiformal MLOps architecture diagrams. They conclude that the understandability of MLOps architecture descriptions is significantly larger with supplementary semiformal architecture diagrams. Lastly, Kumara et al. [12] strive towards a reference architecture of MLOps by eliciting requirements and components from the gray literature. In their preprint, they provide a layered architecture that focuses on requirements that the MLOps environment needs to provide.

Our own study complements the above studies by using scientific literature to synthesize MLOps architecture components, their relationships, and supporting tools to implement them. The results of this synthesis address gaps identified by previous studies, namely "no common understanding of MLOps definition" and "no clear mapping between the tools and the related components". Moreover, unlike existing works, we also synthesize and discuss several architectural variants. Researchers and practitioners can use the proposed architectures to identify a suitable variant for their requirements.

3   Methodology

In this section, we describe the research goals and process of our study. We design and follow a rigorous protocol, by following established guidelines for systematic secondary studies [8,18].

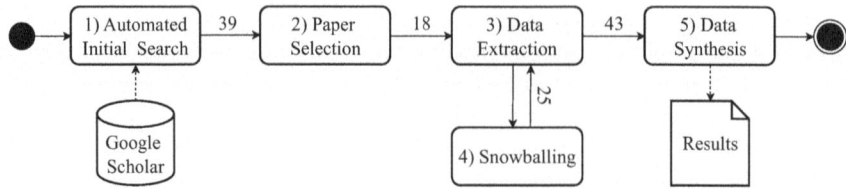

Fig. 1. Overview of the research process

3.1 Goal and Research Questions

The main goal of this SMS is to identify, classify, and analyze the architectures of existing MLOps systems described in scientific literature. In particular, we focus on their structural view, namely the individual components, their dependencies, and their responsibilities. The target audience of this study is (i) practitioners, e.g., ML engineers and software architects, who need to obtain an overview of the common MLOps landscape to make informed decisions, and (ii) researchers aiming to improve the state of the art in architectures of MLOps systems. To approach this goal, we have phrased our overall Research Question (RQ) as **"How are MLOps architectures described in the scientific literature?"**, and broke it down into the following sub-RQs:

RQ1: Which are the different components and their dependencies? This RQ helps us categorize the various components within an MLOps architecture and comprehend how different researchers and practitioners interconnect these components.

RQ2: Which tools are used to support or implement the identified components? This RQ helps us identify tools that can be used for implementing an architecture component, as well as identify the most and the least tool-supported components.

3.2 Research Process

We visualize our research process in Fig. 1 and describe the different steps in this section. Essentially, we first obtain an initial set of papers via an automated title-based search on Google Scholar, then filter this to obtain a starting set of primary studies. After extracting data from the first set of primary studies, we augment this set via bi-directional one-step snowballing [25] and extract data from the new set of primary studies. Using Google Scholar as a meta-engine allows us to avoid bias towards specific publishers [2]. Finally, the extracted data is used in a rigorous synthesis process.

Step 1: Automated Initial Search. By extracting relevant search terms from MLOps literature known to us and adding synonyms, we iteratively construct and test several search strings. As a result, we finally arrive at the following query to search in titles of the available literature:

allintitle: (MLOps OR "machine learning operations") AND (model OR models OR pipeline OR pipelines OR architecture OR architectures OR architecting OR workflow OR workflows OR process OR processes)

The advantages of this final title-focused variant are its manageable number of results and the low number of false positives. Additionally, its potential limitation regarding the extensiveness of its results is compensated by snowballing. The automated query was executed via Google Scholar in September 2023 and yielded 39 potentially relevant papers.

Step 2: Paper Selection. The following inclusion (I) and exclusion (E) criteria are used during the selection of primary studies.

I1 The paper describes the architecture of an MLOps system or part thereof.
I2 The described MLOps architecture is an original contribution of the paper, not simply cited related work.
E1 The paper is not written in English.
E2 The paper is a shorter or earlier version of a paper that is already included.
E3 The paper is a secondary or tertiary study.
E4 The full text of the paper is not available.
E5 The paper is not peer-reviewed and published.

To be included in the list of primary studies, a paper has to fulfill all inclusion and no exclusion criteria. To arrive at the above selection criteria, we conduct a selection pilot in which two reviewers independently review five papers and discuss their selection strategy in a consensus meeting. After fine-tuning the selection process, it is applied to all potentially relevant papers. Each paper is independently evaluated by two researchers for inclusion, with a consensus being necessary for final inclusion. Applying the selection criteria results in 18 papers that form our starting set used in the first round of data extraction and in snowballing.

Step 3: Data Extraction. In this step, we systematically analyze the primary studies and extract data related to the RQs. To refine our data extraction strategy, we first conduct a data extraction pilot on five randomly chosen papers. Data from each paper is extracted by two authors independently and discussed in a consensus meeting leading to the final data extraction framework depicted in Table 1. For RQ1, we extract the items "Architecture or process figures" and "Architecture components", while we extract "Tools" and "Tool-component mapping" for RQ2.Experimentation in software engineeringThe "Application domain" and the "Author affiliations" are extracted as generic information to use for further analysis. From this point on, each of the remaining papers is assigned to two authors for data extraction. The extracted data is validated through bilateral discussions and consensus between the authors.

Table 1. Data Extraction Framework

RQ	Data Item	Notes
Generic	Application domain	The respective domain the architecture is proposed for
Generic	Author affiliations	Origin of the architecture: academia, industry, or collaboration
RQ1	Architecture or process figures	The list of relevant figures and their types (values can be architecture, process, or combined)
RQ1	Architecture components	The list of components and their relationship
RQ2	Tools	The list of tools that are used or suggested in the paper
RQ2	Tool-component mapping	The list of tools mapped to the components

Step 4: Snowballing. We apply backward and forward snowballing to enrich the results obtained via automated search, as suggested by Wohlin et al. [25]. During this step, all papers that either cite or are cited by a paper from the starting set are examined for inclusion in the final set of primary studies. We apply both the same selection criteria and data extraction (steps 2 and 3) as for the initial set of papers and the same process: each paper is examined independently by two researchers and consensus needs to be reached for including it, then the data from each paper extracted by two researchers and the results are discussed. After conducting a first round of bidirectional snowballing, we add 25 more papers to our set of primary studies. We limit the snowballing to a single round; this decision stemmed from the observation that subsequent data extraction yielded minimal additional components and tools compared to the initial seed collection.

Step 5: Data Synthesis. In this phase, we harmonize and classify the extracted data per parameter (architecture components, tools). To achieve this synthesis, we use *card-sorting*, a lightweight, collaborative, qualitative analysis technique [27].

In particular, we use *hybrid* card-sorting [5], a combination of open card-sorting (where categories emerge from the data) and closed card-sorting (categories are defined beforehand based on existing taxonomies). We define the initial set of categories based on our background knowledge of MLOps and software architecture and iteratively refine and enrich this set. We conduct the card-sorting in three phases: in the preparation phase, we print all the extracted components on cards; in the execution phase, we sort the cards into meaningful categories and groups and name them; in the analysis phase, we identify the relationships between the identified components and categories. We also disambiguate and group together the extracted tools and derive their mapping to the synthesized architecture components. The final output is a general architecture of the MLOps workflows from a structural perspective (represented in a UML component diagram), along with a map between tools and components. Lastly, we synthesize the information about the mentioned domains for which the MLOps models are proposed.

4   Results

The results of our study are based on extracted and synthesized data from 43 papers published between 2020 and 2024. This timeframe aligns with the emergence and maturation of the MLOps discipline, which is placed in early 2019 [14]. We observe a peak in the publications on this topic in 2022 (19/43 papers).

Generally, we observe that most papers (20/43) propose domain-agnostic architectures. At the same time, six papers focus on the domain of edge computing, three on manufacturing, and each of the remaining 14 papers focus on different domains including healthcare, psychomotor learning, etc.[1] Our data also shows that 23/43 papers are authored by academic researchers, 10/43 by industrial practitioners, and 10/43 as a collaboration between the two communities. This is a testament to the strong interest of *both* academia and industry in MLOps.

In the remainder of this section, we provide the results of the study regarding the first and second RQs.

4.1   MLOps Architecture Components and Their Dependencies (RQ1)

In total, we synthesized 35 unique architecture components by systematically going through all the architecture figures and descriptions contained in our 43 primary studies. These components are domain-agnostic, i.e., they are not tied to a specific application domain. We also identified domain-specific components such as *IoT Sensors* (IoT domain) or *User Feedback Collector* (psychomotor learning domain), but excluded them for the sake of general applicability. We group the 35 identified architecture components into 6 categories:

- *Data Curation* entails components responsible for gathering and processing data for the MLOps system.
- *Storage and Versioning* comprises components responsible for storing, versioning, and managing the data and models in the system.
- *ML Training* includes components responsible for training and evaluating the ML models, both in the experimentation and production phases.
- *CI/CD* refers to the category of components responsible for continuously building and deploying ML pipelines, models, and components.
- *Inference* entails the components responsible for providing predictions, making subsequent decisions, and monitoring the system.
- *Infrastructure and Supporting Services* comprises infrastructure components that provide system support, e.g., Container Manager, Orchestrator, etc.

Table 2 displays the 35 components by category, their responsibilities, important aliases that we identified for each component name, and the number of occurrences of each.

---

[1] To observe the complete list of domains, please refer to the replication package [1].

Table 2. MLOps architecture components, their aliases, responsibilities, and number of occurrences in the primary studies

	Component Name	Aliases	Responsibilities	#
Data Curation	Data Source	external data sources	Produces and exposes data from a real-world environment, e.g., domain events, IoT sensors, human inputs, etc.	4
	Data Collector	data acquisition, data loading	Collects raw data like events from various data sources.	10
	Data Preprocessor	data processing, data cleaning, data validation, data curation pipeline	Validates, cleans, and prepares collected data for storing as ML training datasets.	9
Storage and Versioning	Dataset Catalogue	–	Stores metadata of datasets in an organized inventory, allows users to browse datasets.	1
	Dataset Repository	data store, data repository, data versioning	Stores and versions the datasets used for ML workflows.	10
	Raw Data Store	data store	Stores the raw data that are collected from sources.	2
	Feature Store	–	Computes and stores reusable features, serves the computed features with low latency.	8
	Code Repository	source code management, source code repository	Stores and versions the training, deployment, and application source code.	9
	Model Repository	model registry, model store	Stores and versions the trained ML models along with basic metadata, e.g., their versions, etc.	21
	Artifact Repository	image repository, container registry	Stores a packaged or containerized ML component that incorporates an ML model for inference.	3
	ML Metadata Repository	experiment tracking DB, ML metadata store	Stores metadata related to model training for experiment tracking purposes, e.g., model performance metrics, etc.	15
	Feedback Database	feedback store	Stores stakeholder feedback and experiences, e.g., from domain experts or engineers, which are manually considered during iterative model development.	2
ML Training	Data Labelling Component	data annotation, ground truth annotation	Adds the ground truth labels for supervised learning models to dataset instances.	3
	Feature Engineering Pipeline	feature selection	Selects and transforms the features of the used dataset for model training.	4
	ML Experiment Pipeline (Offline)	ML pipeline (offline), manual ML pipeline, data science experiments	Develops and trains ML models at design time (more experimental and manual).	14
	ML Training Pipeline (Online)	ML pipeline (online), continuous training pipeline, incremental online learning, MLOps pipeline	Continuously trains ML models at runtime in a production environment (completely automated).	12
	Model Evaluator	–	Evaluates the prediction performance of the models during training.	4
CI/CD	ML Pipeline Builder	build and test pipeline, CI tool	Builds, tests, and packages, e.g., in containers, the code of the ML pipeline.	2
	ML Pipeline Deployer	pipeline deployment, ML training pipeline deployment, CD tool	Deploys the built and packaged code of the ML pipeline to staging or production environments.	3
	ML Model Deployer	model deployment, deployment pipeline	Deploys the trained model packaged with the dependencies, e.g., required libraries, preprocessing code, etc. to the production environment.	4
	ML Component Builder	build automation pipeline, continuous integration (CI)	Builds and tests ML components, i.e., deployment-ready containerized ML models wrapped in an API.	3
	ML Component Deployer	ML deployment, continuous delivery, continuous deployment, CI/CD pipeline	Deploys the ML components to staging or production environments.	7

(*continued*)

Table 2. (*continued*)

	Component Name	Aliases	Responsibilities	#
Inference	Inference Service	model inference, production ML service, model server	Serves the trained models to provide predictions on new data (ML component).	10
	Inference Engine	pool inference, local inference engine, model serving component	Includes an ML runtime into which trained models can be continuously deployed to serve predictions.	9
	Runtime Model Monitor	performance monitor, monitoring component, model runtime monitor	Continuously observes the model-serving performance and infrastructure in real-time.	12
	Trigger	retraining triggering webhook, retraining trigger	Triggers retraining of the ML models based on predefined events and intervals or a predefined performance threshold observed via Runtime Model Monitor.	3
	Model Comparison Runner	model comparison runner, model metrics evaluator	Compares the newly trained model to the old model and deploys the better performing one.	2
	Decision Processor	decision processing	Derives decisions based on the predictions of the model. The decisions are then acted upon by the actors inside or outside the system.	2
Infrastructure and Supporting Services	Resource Manager	resource leasing, model engine	Provides foundational hardware and software computational resources. The provided computational resources can be distributed or non-distributed and scalable or non-scalable.	5
	Communication Middleware	message queue, event streaming bus	Distributes the received requests and model predictions to resources.	4
	Container Manager	container service	Manages, e.g., starts and stops, the containerized ML components that are built with the ML Component Builder.	2
	Orchestrator	adaptive scheduler, workflow orchestration, job tracking module	Provides system-wide orchestration, and decides the execution schedule of multiple models balancing throughput and latency.	5
	Log Master	logging, info collector, object store, predictions store	Records and saves information regarding all the actions in the system, e.g., running the services, training, user requests, predictions, etc.	3
	API	API gateway	Provides interaction between the components within the platform and also between the platform and external entities.	5
	MLOps User Interaction Manager	Ops dashboard, front-end	Provides interaction between the MLOps team and the MLOps platform.	2

We combined the synthesized components and their dependencies into a holistic UML component diagram depicted in Fig. 2. In this diagram, we introduce five types of architecture components: **baseline components** are mandatory components. These components form the baseline of the architecture (available in all variants of MLOps architectures) and need to be complemented with the components of either the **inference service** or the **inference engine** group to form a complete MLOps architecture. **Optional components** represent non-essential components that can be situationally useful, e.g., the *Model Comparison Runner*. As a special type of optional components, the ones of the **online training** group can be added to an MLOps architecture, but always together as a group. Note that the *Infrastructure and Supporting Services* components are not included in the diagram, since they provide support to the whole system.

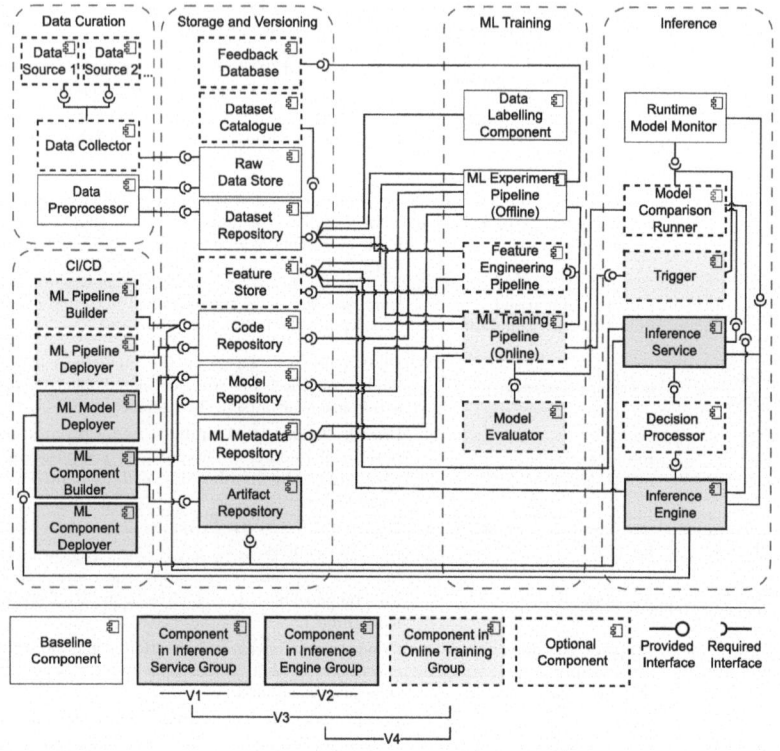

Fig. 2. UML component diagram of MLOps architecture variants

The baseline components include the *Data Preprocessor* which reads its input from the *Raw Data Store* and stores its results to the *Dataset Repository*. The latter provides data to both the *Data Labelling Component* and the *ML Experiment Pipeline*. At the same time, the *ML Experiment Pipeline* uses the datasets from the *Dataset Repository* and the ML algorithms from the *Code Repository* to train ML models, and then stores the trained ML models to the *Model Repository*. After deployment, the *Runtime Model Monitor* provides real-time model performance data.

Overall, the blueprint for assembling a complete MLOps system involves the above-mentioned baseline components, as well as incorporating either the inference service or the inference engine group, and potentially the online training group and/or other optional components. In the following, we describe four characteristic architecture variants (V1 to V4) depicted in Fig. 2.

V1: This architecture variant describes an architecture containing an *Inference Service* to serve ML models in a production environment. The *Inference Service* represents an *ML component* [15], i.e., a containerized, deployment-ready ML model wrapped into an API that is usable for predictions. Thus, in this variant, these ML components are built, tested, and packaged by the *ML Component Builder*, deployed through the *ML Component Deployer*, and stored in the *Artifact Repository*.

V2: Contrary to the *Inference Service* of V1, this architecture variant involves an *Inference Engine* to serve the trained ML models. The *Inference Engine* contains a runtime for ML models that allows the deployment of new models through the *ML Model Deployer*. Hence, this variant efficiently updates only the ML model instead of always replacing the complete ML component. Alternatively, the *Inference Engine* can also check the *Model Repository* periodically or in an event-based fashion to fetch a new model version if certain criteria are met.

V3: The third variant combines the *components in the online training group* with V1. The presence of *Trigger* entails the presence of *ML Training Pipeline (Online)*, *ML Pipeline Builder*, *ML Pipeline Deployer*, and the optional presence of *Model Evaluator*. The *Trigger*, using data provided by the *Runtime Model Monitor*, submits a periodic or event-based retraining request to the *ML Training Pipeline*. This automatically retrains and deploys a new model in production.

V4: The fourth variant, similar to the third, combines the *components in the online training group* with V2, which results in the automatic retraining and deploying of ML models in the *Inference Engine*.

Vx: In addition to these four described variants, the selection of any combination of the optional components can result in several additional variants. Table 2 can be consulted for a detailed description of all the other components. As an example, we describe the addition of the *Model Comparison Runner* here. Adding this component to either V3 or V4 allows more informed model update decisions. It compares the performance of a newly trained model in the production environment to the currently deployed model and keeps the one that performs better. In the absence of this component, the newly trained model is always deployed, even if its performance would be inferior to the current ones.

Considering the eight optional components and four described major variants, the selection of any combination between them results in a large number of different architecture variants ($4 \times (2^8 - 1) = 1020$). Selecting the most suitable variant may depend on factors like specific design decisions, resource availability, scalability considerations, required update frequency, or technological expertise.

4.2 Tools Used to Support or Implement the Components (RQ2)

Among the 43 reviewed papers, we identified 76 tools in total. Figure 3 depicts a heatmap of the tools and platforms that are mentioned at least 3 times among the papers, mapped to the components that they support. In the figure, it is evident that *Jenkins* is the most frequently mentioned tool. This tool is used to support the components within the CI/CD category. *AWS SageMaker* is also one of the most popular tools. This tool is observed to support the highest number of components in the architecture of MLOps workflows (10 distinct components among 5 categories of our architecture). *MLflow* follows by supporting six different components mentioned in several papers.

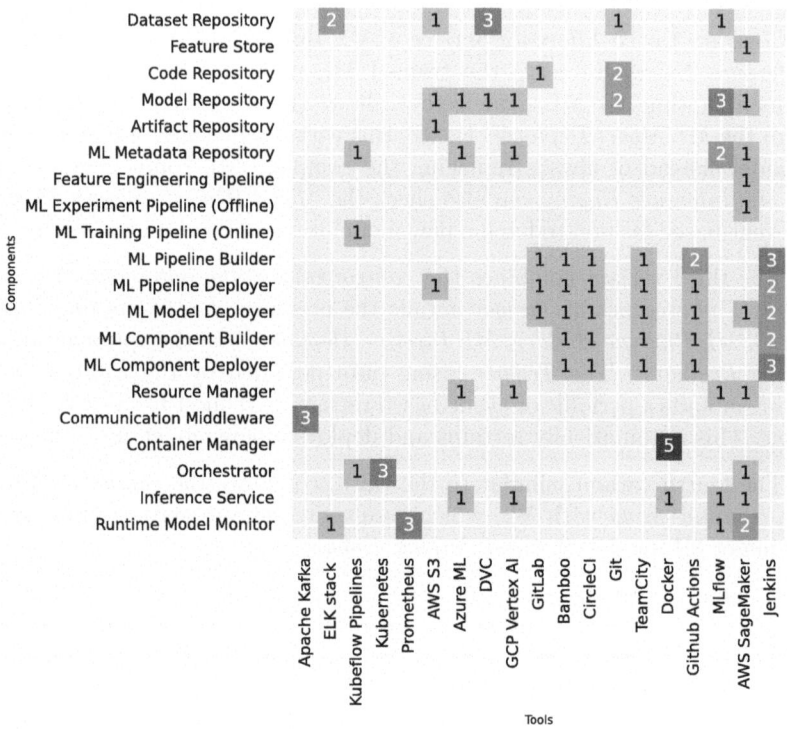

Fig. 3. The most popular tools and platforms, mapped to the components

Within the six categories of architecture components, the greatest variety of tools is mentioned for *Storage and Versioning*, followed by *CI/CD*. The tools supporting the entire *CI/CD* category encompass Jenkins, GitHub Actions, TeamCity, CircleCI, and Bamboo. Regarding the components, however, the most diverse variety of tools is mentioned for *Model Repository*, *ML Model Deployer*, and *ML Pipeline Deployer*. There are also several components for which no tool is mentioned. These include *Dataset Catalogue, Raw Data Store, Feedback Database, Data Collector, Data Source, Model Evaluator, Log Master, MLOps User Interaction Manager, Trigger*, and *Decision Service*. The complete mapping between the tools and the components can be accessed through the replication package [1].

Lastly, some papers generally mention tools and services that support the end-to-end MLOps workflows, without mapping them to any specific component.

These tools are AWS SageMaker, MLflow, Kubeflow, Weights and Biases, Clear ML, MLReef, Iguazio, Polyaxon, Vertex AI, Azure ML, and Snorkel.

5   Discussion

In this section, we discuss the implications of the derived results of this SMS. The UML component diagram and component-tool mapping provided in this study serve as valuable references for practitioners and researchers who aim to design or enhance MLOps systems. Nonetheless, the findings derived from this study suggest several noteworthy points, which we will discuss in the following.

Complexity and Mixed views in Architecture Diagrams. A majority of the analyzed figures and descriptions of MLOps workflows were a combination of architecture, process, and stakeholder roles, thereby combining several different architectural views and concerns in a single diagram. Even though this approach is a common practice and makes the provided information more compact, the increased complexity reduces the clarity of the provided architectures.

Non-standard Notations to Represent Architectures. Almost all of the analyzed figures used informal box-and-line diagrams as a notation. Among the 43 reviewed papers, only one paper (S23) uses a standard notation, the Fundamental Modelling Concepts (FMC) [4]. Combined with the complexity of the mixed views and following the results of Warnett and Zdun [24], the understandability of these MLOps architecture representations is substantially impacted.

Level of Abstraction in Architectures. Among the extracted figures and descriptions of the papers, the level of provided details and abstraction varied over a large spectrum. During the study, we extracted and synthesized data from figures only representing as few as six high-level architecture components (S10, S43) to complex "combined figures" with over 40 concrete entities including components, actions, and stakeholder roles (S7). The different levels of abstraction and multitude of dependencies between components also allow many fine-grained possibilities for variations in component dependencies and interface directions. For simplicity, we modeled only the most common dependencies in the diagram.

Tools in Place of Architectural Entities. In addition to the mixed architectural views and different levels of abstractions, the extracted figures in some papers also included a mixture of architectural entities and the employed tools as a stand-alone entity (see, e.g., S31). This complexity in the views makes the architectural understanding and comparison difficult.

Inconsistently Named Components and Activities. The terminology for the same individual components or activities could vary substantially among the papers. We identified two types of inconsistent naming: (a) some papers use common component names that imply different responsibilities, e.g., using "Feature Store" for a component that provides the training data for the ML training pipeline instead of "Data(set) Store" or using "Artifact Store" instead of "ML

Model Store", and (b) other papers use unique names to represent common ML-based software components, e.g., "knowledge base manager" for a component that stores and versions the trained ML models (*Model Repository* in our synthesis). These inconsistencies highlight why communication and collaboration in MLOps projects is often difficult.

Domain-specific Architecture Components. In almost 50% of the papers, the provided architectures are domain-specific and therefore contain components that are only situationally applicable or focused on a specific application scenario. For example, the presence of "additive manufacturing" components in S36, or "Blockchain"-related components in S40. In the analysis, we generalized these components to a relevant category or discarded the components that represented a very specialized entity.

End-to-End Tools only Mentioned for a Certain Set of Components. A tool-related observation is that most of the end-to-end MLOps tools like MLflow, Kubeflow, or AWS SageMaker are recommended only for a subset of components, rather than for the entire workflow. For example, in the reviewed papers, MLflow is mentioned only for components in the *Storage and Versioning, Inference,* and *Infrastructure and Supporting Services* among our categories. Interestingly, this tool is not mentioned for any component in the *ML Training* category.

6   Threats to Validity

Following the categorization by Wohlin et al. [26] and the checklist by Ampatzoglou [2], we outline the threats that may have affected the validity of our research and outline the actions that we take to mitigate the threats.

Threats to *internal validity* undermine the conclusion about a possible causal relationship between the study and the outcome [26]. To mitigate this threat, we assign two researchers for the study selection and data extraction phases, who perform the selection and extraction independently, and then three researchers discuss the results in consensus meetings. Another possible threat to internal validity in this study stems from the selection of papers from various domains. Varying author expertise across fields, particularly outside ML-based systems, may influence the accuracy of the architecture and process descriptions in papers, which influences our data extraction. To mitigate this threat, we have mapped every component and activity name and description to the authors' intentions, e.g., in some papers authors refer to a dataset repository as a feature store.

External validity is concerned with the generalization extent of the findings of the study [26]. A possible threat to the external validity of our study is the selection of the papers among the peer-reviewed resources, which may limit our initial set of papers. This is a research design decision to ensure the reliability of the selected papers. We mitigate this threat by applying our search query on Google Scholar as a meta-engine, which results in papers from different venues, as well as applying one round of bidirectional snowballing. However, since we focus on peer-reviewed scientific papers, we might miss works from the many practitioners that certainly focus on this very popular topic. This may introduce a threat to the generalizability of the results.

Reliability is concerned with the extent to which the data and the analysis are dependent on the specific researchers [2,20]. A possible threat in our study concerning reliability can be the authors' bias in synthesizing data due to the nature of the extracted data and the different levels of abstraction of the architecture in the papers. To mitigate this threat, three authors participate in analyzing and synthesizing the data.

7 Conclusions

Based on an SMS with 43 scientific papers, we synthesized architectural MLOps components, their dependencies, and tools to implement them. Furthermore, we combined these components into a holistic MLOps architecture and discussed several architectural variants that emerge from the literature. Our results contribute to understanding the architecture aspects of MLOps systems and potentially support communication in this complex and still maturing domain.

Regarding future work, a structural perspective is not the only architectural view that is important in MLOps. Therefore, we plan to synthesize a process view of MLOps with a similar research design and to map activities to MLOps roles. Moreover, synthesizing architectural decisions, best practices, and antipatterns in this domain may also support practitioners, which we plan to provide via our long-term goal, a reference architecture for MLOps. In the end, practitioners could be best supported if we could clearly link architectural MLOps variants to functional and quality requirements, so that practitioners can easily choose the variant that best suits their needs. To allow such endeavors and to increase transparency, we share our research artifacts on Zenodo [1].

Acknowledgments. This research is supported by ExtremeXP, a project co-funded by the European Union Horizon Programme under Grant Agreement No. 101093164.

Data Availability. The data and artifacts of this study are available as a replication package [1].

References

1. Amou Najafabadi, F., Bogner, J., Gerostathopoulos, I., Lago, P.: An Analysis of MLOps architectures: a systematic mapping study. [Data set], Zenodo (2024). https://doi.org/10.5281/zenodo.11067770
2. Ampatzoglou, A., Bibi, S., Avgeriou, P., Verbeek, M., Chatzigeorgiou, A.: Identifying, categorizing and mitigating threats to validity in software engineering secondary studies. Inf. Softw. Technol. **106**, 201–230 (2019)
3. Bass, L., Weber, I., Zhu, L.: DevOps: A Software Architect's Perspective. 1st edn. Addison-Wesley Professional, Boston (2015)
4. The fundamental modeling concepts. http://www.fmc-modeling.org

5. Hudson, W.: Card sorting. The Encyclopedia of Human-Computer Interaction (2014). https://www.interaction-design.org/literature/book/the-encyclopedia-of-human-computer-interaction-2nd-ed/card-sorting, Accessed 31 Mar 2024
6. Idowu, S., Strüber, D., Berger, T.: Asset management in machine learning: a survey. In: Proceedings of the 43rd International Conference on Software Engineering: Software Engineering in Practice, pp. 51–60. ICSE-SEIP 2021, IEEE Press (2021)
7. John, M.M., Olsson, H.H., Bosch, J.: Towards MLOps: a framework and maturity model. In: 2021 47th Euromicro Conference on Software Engineering and Advanced Applications (SEAA), pp. 1–8. IEEE (2021)
8. Keele, S., Kitchenham, B., et al.: Guidelines for performing systematic literature reviews in software engineering (2007)
9. Kolar Narayanappa, A., Amrit, C.: An analysis of the barriers preventing the implementation of MLOps. In: International Working Conference on Transfer and Diffusion of IT, pp. 101–114. Springer, Cham (2023)
10. Kolltveit, A.B., Li, J.: Operationalizing machine learning models: a systematic literature review. In: Proceedings of the 1st Workshop on Software Engineering for Responsible AI, pp. 1–8 (2022)
11. Kreuzberger, D., Kühl, N., Hirschl, S.: Machine Learning Operations (MLOps): overview, definition, and architecture. IEEE Access **11**, 31866–31879 (2023)
12. Kumara, I., Arts, R., Di Nucci, D., Van Den Heuvel, W.J., Tamburri, D.A.: Requirements and reference architecture for MLOPs: Insights from industry. Authorea Preprints (2023)
13. Lima, A., Monteiro, L., Furtado, A.P.: Mlops: practices, maturity models, roles, tools, and challenges-a systematic literature review. ICEIS **1**, 308–320 (2022)
14. Mark Treveil, T.D.T.: Introducing MLOps. O'Reilly Media, Inc, Sebastopol (2021)
15. Martínez-Fernández, S., et al.: Software engineering for AI-based systems: a survey. ACM Trans. Softw. Eng. Methodol. **31**(2), 1–59 (2022)
16. Mboweni, T., Masombuka, T., Dongmo, C.: A systematic review of machine learning devops. In: 2022 International Conference on Electrical, Computer and Energy Technologies (ICECET), pp. 1–6. IEEE (2022)
17. Nakagawa, E.Y., Antonino, P.O.: Reference Architectures for Critical Domains: Industrial Uses and Impacts. Springer, Cham (2023). https://doi.org/10.1007/978-3-031-16957-1
18. Petersen, K., Vakkalanka, S., Kuzniarz, L.: Guidelines for conducting systematic mapping studies in software engineering: an update. Inf. Softw. Technol. **64**, 1–18 (2015)
19. Recupito, G., et al.: A multivocal literature review of MLOPs tools and features. In: 2022 48th Euromicro Conference on Software Engineering and Advanced Applications (SEAA), pp. 84–91. IEEE (2022)
20. Runeson, P., Höst, M.: Guidelines for conducting and reporting case study research in software engineering. Empir. Softw. Eng. **14**, 131–164 (2009)
21. Testi, M., et al.: MLOPs: a taxonomy and a methodology. IEEE Access **10**, 63606–63618 (2022)
22. Warnett, S.J., Zdun, U.: Architectural design decisions for machine learning deployment. In: 2022 IEEE 19th International Conference on Software Architecture (ICSA), pp. 90–100 (2022)
23. Warnett, S.J., Zdun, U.: Architectural design decisions for the machine learning workflow. Computer **55**(3), 40–51 (2022), publisher: IEEE
24. Warnett, S.J., Zdun, U.: On the understandability of MLOps system architectures. IEEE Trans. Softw. Eng. **50**, 1–25 (2024)

25. Wohlin, C.: Guidelines for snowballing in systematic literature studies and a replication in software engineering. In: Proceedings of the 18th International Conference on Evaluation and Assessment in Software Engineering, EASE 2014, ACM (2014)
26. Wohlin, C., Runeson, P., Höst, M., Ohlsson, M.C., Regnell, B., Wesslén, A.: Experimentation in Software Engineering. Springer, Berlin (2012)
27. Zimmermann, T.: Card-sorting: From text to themes. In: Perspectives on Data Science for Software Engineering, pp. 137–141. Elsevier (2016)

Attention-Based Method for Design Pattern Detection

Rania Mzid[1,2]([✉]) [iD], Ilyes Rezgui[1], and Tewfik Ziadi[3] [iD]

[1] ISI, University Tunis-El Manar, 2 Rue Abourraihan Al Bayrouni, Ariana, Tunisia
`rania.mzid@isi.utm.tn, ilyes.rezgui@etudiant-isi.utm.tn`
[2] CES Lab ENIS, University of Sfax, Sfax, Tunisia
[3] Sorbonne Université CNRS, LIP6, 75005 Paris, France
`tewfik.ziadi@lip6.fr`

Abstract. Design patterns are standard solutions to recurrent software engineering problems. The use of design patterns helps developers improve software quality. However, when integrating design patterns into their systems, software developers usually do not document their use. To this end, the use of an automatic approach for their detection may accelerate program comprehension, assist developers in software refactoring, and reduce efforts during the maintenance task. In this paper, we propose an attention-based approach for design pattern detection. Specifically, we utilize an automatic feature extraction step with a transformer-based model incorporating the attention mechanism. Based on an unsupervised approach, this step learns from source code to identify code attributes and then produces embedding vectors. These vectors capture syntactic and semantic information related to design pattern implementations and serve as input to train a classifier for the design pattern detection task. The attention mechanism is used to produce important representative features of design pattern implementations and improve the accuracy of the classification model. The evaluation shows that our classifier detects GoF design patterns with an accuracy score of 86%, precision of 87%, recall of 86%, and F1-score of 86%. The comparison of our findings with state-of-the-art methods shows an improvement in (i) precision of 25%, (ii) recall of 6%, and (iii) F1-score of 8%.

Keywords: Design pattern detection · Feature extraction · classification · Transformer architecture

1 Introduction

Design patterns are typically defined as a solution to a recurring problem applicable in a specific context [18]. In software engineering, the application of design patterns significantly simplifies the work of software developers and enhances the quality of software systems across various domains. Patterns may aid in the reuse of existing knowledge regarding similar development difficulties. They propose previously used and optimized solutions to specific problems in a variety

M. Galster et al. (Eds.): ECSA 2024, LNCS 14889, pp. 86–101, 2024.
https://doi.org/10.1007/978-3-031-70797-1_6

of contexts. The lack of pattern-related data after the application of a design pattern in source code is a common problem. Indeed, software developers apply design patterns without frequently documenting their use. However, knowing their presence in a software system improves the developer's comprehension and provides essential information for the developers to facilitate software refactoring and maintenance tasks. One way to reduce refactoring and maintenance costs is to detect design patterns. However, manually identifying and locating these patterns can be complex and time-consuming. Developers must not only recognize where a pattern is applicable but also ensure that it is applied correctly to reap its full benefits.

Different approaches have been proposed in the literature for design pattern detection. Some of those approaches rely on intermediate representations [5,17] and aim to convert the source code of the system implementing the design pattern into an intermediate representation, like a graph or a matrix followed by a comparison to find common structural elements. These methods show low performance, and frequently fail to distinguish between design patterns with similar structures. Metric-based approaches [6,9] for design pattern detection uses quantitative measures to identify and assess the presence of design patterns in source code. However, the possibility of implementing a design pattern in various ways makes the use of software metrics challenging to capture all instances accurately. Indeed, software metrics are generally based on structural analysis, aiming to quantify certain structural characteristics and relationships between the code elements. To address the limitations of the previous two approaches, feature-based methods for design pattern detection have recently emerged. Detecting design patterns in code refers to the process of checking whether a source code implements the features that define that pattern. Code features correspond the specific characteristics of code that define a particular pattern. These methods are usually supported by machine learning techniques and mainly consist of two stages: examining the code statically to extract syntactic and semantic code features (known as feature extraction), then constructing a machine learning model for classification. The feature extraction step is challenging since the extracted attributes have a strong impact on the classification model's performance. In addition, this step is judged to be hard and time-consuming [11,15] due to the various representations of design patterns (i.e., variants). Another challenge facing feature-based approaches is embedding source code into representative vectors in the dimensional space and that is due to the complex structure of source code itself. Embedding helps capture semantic representation, which consequently contributes to enhancing the accuracy of the classification model. To perform this step, existing approaches perform a transformation of source code to a textual representation that represents code features [11,16], then employ word embedding techniques such as word2vec [3] to produce the embedding vectors.

Motivated by the limitations of existing works, we propose in this paper a fully automated approach for GoF design pattern [8] detection named DPD_{Att}. The proposed approach, based on the transformer architecture and enriched with the attention mechanism [21], intends to automatically extract code features using an unsupervised approach. Transformer is a neural network architecture first introduced by Ashish et al. [21] in 2017. It adheres to an encoder-decoder structure, with the encoder tasked with transforming the input into an intermediate representation that the decoder can subsequently employ to generate output. Existing architectures built on top of this transformer often adopt one or more of its building blocks, as some rely on an encoder-only network, whereas others employ a decoder-only or encoder-decoder architecture. In this paper, the feature extraction module is built on the transformer's encoder. The attention mechanism within a transformer-based auto-encoder enables the capture of contextual information within an input sequence. It produces an output vector that encodes the relationships between words or sub-words known as tokens in the input. This method ensures that tokens sharing semantic similarities are represented by vectors with smaller distances in the vector space, making it useful for learning representations of sequential data such as code. Furthermore, to eliminate treating a source code as plain text for embedding, we utilize a code embedding model in this paper. The resulting contextualized embeddings retrieved by our feature extraction module serve as rich representations that include syntactic and semantic information, making them helpful for design pattern detection. The main contributions of this paper can be summarized as follows:

- We provide an attention-based approach for design pattern detection called DPD$_{Att}$. The proposed approach uses the transformer's encoder to automatically learn features with syntactics and semantics from encoded token vectors extracted from source code, and then uses the attention mechanism to generate key features (i.e., embedding vectors) for training a more precise design pattern detection model. The approach is fully automated, and a tool called the DPD$_{Att}$ detector is created.
- We build a large corpus DPD$_{Att}$ corpus covering 13 GoF design patterns and consisting of 1645 labeled Java files.
- We demonstrate the effectiveness of our approach by comparing it with state-of-the-art methods in terms of precision, recall, and F1-score.

The rest of this paper is organized as follows: In Sect. 2, we provide background information on the concept of features in source code. Section 3 describes the proposed approach. In Sect. 4, we present implementation details to create our DPD$_{Att}$ detector, we discuss evaluation results, and we compare the proposed approach with the state-of-the-art methods. Section 5 overviews related work. Conclusions and future work are discussed in Sect. 6.

2   Code Features

Code features are code attributes or characteristics that are extracted or computed from source code for a specific purpose. The code features used are determined by the task at hand. Code features might be syntactic or semantic. Syntactic features are the structural elements and arrangement of the code, whereas semantic features are the meaning and behavior enclosed within those structures. In object-oriented programming, syntactic features may include class and method declarations, variable names, inheritance relationships, and the overall organization of code blocks. The inclusion of access modifiers like "public" or "private," or the use of particular phrases like "extends" for inheritance are some examples of syntactic features. On the other side, semantic characteristics examine the functionality and logic contained in the code. Important semantic features include, for instance, the creation of objects, method calls, and interactions between various classes and objects. For feature extraction, various strategies may be used. Static source code analysis, which involves examining the code without executing it, may be used to extract static features. This includes techniques such as parsing the code to construct an abstract syntax tree (AST) or using data flow analysis to understand how data moves through the program. Furthermore, advanced static analysis techniques such as symbolic execution [19], formal methods [27], or machine learning [16] can be used to derive more in-depth semantic features, revealing insights into the code's intended behavior. In this paper, we are interested in machine learning techniques to extract syntactic and semantic features from source code. In particular, we aim to automatically extract code features using an unsupervised approach.

3 Proposed Approach

The proposed approach as depicted in Fig. 1 includes three main phases: the data collection and preparation, the feature extraction, and the classification. Each of these phases is explained in detail in the following subsections.

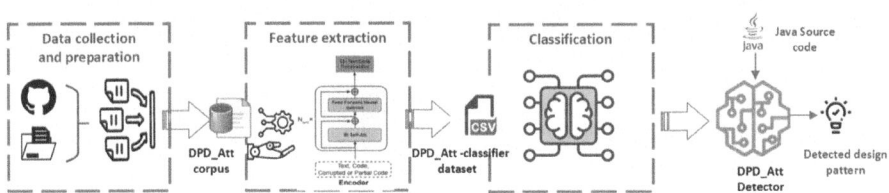

Fig. 1. Proposed approach in a nutshell

3.1 Data Collection and Preparation

The effective training of any machine learning model requires a substantial volume of high-quality data. In our specific research context and to identify GoF design patterns within Java source code, we opted for class-level detection as it is the lowest possible level of granularity that may capture design patterns. We started by conducting an extensive search for an open-source dataset. Our search led us to Nazar et al.'s [16] corpus, DPD_F, one of the few open-source available design pattern detection datasets. This dataset was considered a starting point for our work and a benchmark corpus. The DPD_F comprises 1300 files sourced from 216 projects, covering 12 design patterns. These files were carefully selected from the GitHub Java Corpus [1] and expertly labeled through crowd knowledge. We further processed this dataset, starting with eliminating redundant files and keeping only the ones with valid GitHub URLs and a ".java" file extension, resulting in a reduced dataset size of 943 labeled examples. This data curation step is essential since having duplicates in the dataset can lead to inflating performance metrics. Following this, we continued to work with the refined DPD_F, expanding it to encompass 1645 Java files and to include 13 design patterns, thereby covering a wider spectrum of design patterns and implementations. These additional files were also collected from GitHub and labeled manually by experts in the field. Subsequently, we named this refined corpus DPD$_{Att}$, which played a fundamental role in our feature extraction and the training of our multi-class classification models. As depicted in Fig. 2, the DPD_F dataset encompasses 12 design patterns in addition to the "unknown" label. Whereas, our dataset extends its label set to include the "Strategy" design pattern [8], bringing the total to 14, including the "unknown" label.

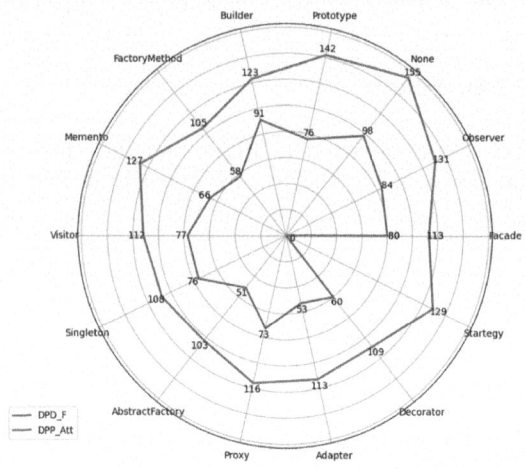

Fig. 2. Comparison between the DPD_F and DPD$_{Att}$ corpora by the number of collected instances per design pattern

3.2 Feature Extraction

In this paper, we aim to utilize a machine learning classifier for the detection task. To do so, the classifier must be trained to distinguish between the features that represent each design pattern. As a result, we need to supply it with data that implements these GoF pattern features so that it learns to map source code to the design pattern it implements. However, supervised learning classifiers cannot process code in its raw format, making an intermediary step of feature extraction required. The feature extraction step as depicted in Fig. 1, aims to automatically extract code features from the collected Java files (DPD_{Att} corpus) and represent them as fixed-size vectors (DPP_{Att}-classifier dataset). The feature extraction starts with the tokenization step. This tokenization refers to the process of mapping a given input into a vector of integers based on a specific vocabulary, where each integer corresponds to a word or sub-word from the input code. An example of a Java class called "Database" to which the Singleton design pattern is applied is shown in Fig. 3. This code enables clients to create a connection pool for accessing the database, ensuring the reuse of the same connection across all clients. A possible tokenization process result is shown in Fig. 4. In this figure, the token "1" represents a special character initiating the start of the sentence, and the token "2" represents a special token for the end of the input. Then, the feature extraction module automatically extracts code features from the tokenized DPD_{Att} corpus given as input. The attention mechanism is used to capture contextual information and perform semantic analysis. Our feature extraction module utilizes this mechanism to learn how much a token from the input sentence relates to all the other tokens from the same sentence. That way, tokens that contribute most to the context of others will be looked at with more "attention" during the learning. This contextualization is very important in the context of pattern detection since it allows distinguishing between code elements that are relevant to the implementation of the design pattern and other tokens from the source code that are irrelevant. As a final step, these learned dependencies are then mapped to an output vector, which is our embeddings. It is worth noting that the size of this vector, called embedding size, has an impact on the detection task, which will be discussed in more detail in the evaluation section. Figure 5 depicts the embedding vector for the Java class "DataBase" with an embedding size equal to 256. Each element in the embedding vector represents a specific feature characterizing the design pattern Singleton in the source code.

```
class Database {
    private static Database dbObject;
    private Database() {
    }
    public static Database getInstance() {
        // create object if it's not already created
        if (dbObject == null) {
            dbObject = new Database();
        }
        // returns the singleton object
        return dbObject;
    }
    public void getConnection() {
        System.out.println("You are now connected to the database.");
    }
}
```

Fig. 3. An example of a Java class "DataBase" to which the Singleton pattern is applied

```
tensor([1,      203,  1106, 5130,  288,  203,   282, 3238,  760,  5130, 1319, 921,   31,   203,   282, 3238,
        5130, 1435,  288, 4202,  203,  282,  289,  203,  282, 1071,  760, 5130, 3694, 1435,  288,  203, 1377,
         368,  752,  733,  309,  518, 1807,  486, 1818, 2522,  203, 1377,  309,   12, 1966,  921,  422,  446,
          13,  288,  203,  540, 1319,  921,  273,  394, 5130, 5621,  203, 1377,  289,  203, 4202,  368, 1135,
         326, 6396,  733,  203, 4202,  327, 1319,  921,   31,  203,  282,  289,  203,  282, 1071,  918, 6742,
        1435,  288,  203, 4202, 2332,   18,  659,   18, 8222, 2932, 6225,  854, 2037, 5840,  358,  326, 2063,
        1199, 1769,  203,  282,  289,  203,   97,  203,    2]))
```

Fig. 4. Tokens of the Java class "DataBase"

```
tensor([[-0.0599, -0.0019, -0.1362, -0.0059, -0.1154,  0.0384,  0.0832, -0.0376, -0.0993,  0.0214,  0.0576,  0.0866,  0.0443,  0.0262,  0.0683,
         0.0198,  0.0836, -0.1733, -0.0918,  0.0761,  0.0331,  0.0234, -0.0019,  0.0564,  0.0160,  0.0435, -0.0038,  0.0191, -0.0854, -0.0252,
        -0.0348, -0.0619, -0.0560, -0.0234, -0.0905,  0.0341,  0.0479, -0.0481,  0.0293,  0.0505, -0.0272,  0.0239,  0.1233,  0.0406,  0.0708,
         0.0073, -0.0586,  0.0429,  0.0112, -0.1061,  0.0861,  0.0571,  0.0533, -0.0291, -0.0571,  0.0004,  0.0285,  0.0192,  0.0083,  0.0727,
        -0.0278, -0.0619, -0.0561, -0.0448, -0.0364,  0.0140,  0.0142, -0.0319, -0.0248, -0.0987, -0.0095,  0.0792,  0.0146, -0.0824, -0.1464,
         0.0493, -0.0365, -0.1038,  0.0748, -0.1273,  0.0096,  0.0127,  0.0376, -0.0701, -0.0534, -0.0139, -0.0733, -0.0380,  0.1140, -0.0623,
         0.0147,  0.0357,  0.0839,  0.0499, -0.0235,  0.0419,  0.0270, -0.0733,  0.0156, -0.0276, -0.0074, -0.0528, -0.0212, -0.0434,  0.0233,
         0.0309, -0.0446,  0.0289, -0.0390, -0.0726, -0.0382,  0.0447,  0.0352,  0.0756, -0.0075, -0.0513, -0.1216, -0.0088,  0.0157, -0.0249,
         0.0972,  0.0932,  0.0408,  0.1057,  0.0070, -0.0331, -0.0617,  0.0050,  0.0713, -0.0198,  0.0048, -0.0483, -0.0229,  0.0209, -0.0004,
        -0.0154, -0.0235, -0.1061,  0.1601,  0.0173, -0.0198,  0.0494,  0.0565,  0.0537, -0.0054, -0.0800,  0.0901, -0.0204, -0.0797, -0.1150,
         0.0884, -0.0690, -0.1142, -0.0339,  0.0111, -0.0533, -0.0654,  0.0312,  0.0156,  0.1225,  0.0139,  0.0977,  0.0066, -0.0013, -0.0131,
         0.0762, -0.0556,  0.0423,  0.0634,  0.1082, -0.0373,  0.0078, -0.0478,  0.0431,  0.0101, -0.1972, -0.0150, -0.0603,  0.0469, -0.0080,
         0.0867,  0.0667, -0.0568, -0.1139,  0.0926,  0.0226, -0.0058,  0.0367, -0.1321, -0.0385,  0.0581,  0.0676,  0.0019, -0.0905, -0.0394,
        -0.0797, -0.0053,  0.0525,  0.1161,  0.0251, -0.0484, -0.0360,  0.0391,  0.0485, -0.0285, -0.0750,  0.0178,  0.0649,  0.0151, -0.0130,
        -0.0383,  0.1017,  0.0646,  0.0963, -0.0029,  0.0035,  0.0368,  0.0024,  0.0705, -0.0175, -0.1066, -0.0705,  0.0345, -0.0398,  0.0171,
         0.0783,  0.0085,  0.0318,  0.0706,  0.0262,  0.1060,  0.0507, -0.0347, -0.0416,  0.1388, -0.1385, -0.0072, -0.0366,  0.0398, -0.0862,
         0.0602,  0.0041, -0.0139, -0.0036,  0.0715, -0.1764,  0.0325,  0.1186,  0.0024,  0.0252, -0.0263, -0.0007,  0.0133,  0.0036,  0.0775,
         0.0557], grad_fn=<SelectBackward0>)
```

Fig. 5. Embedding vector for the Java class "DataBase"

3.3 Classification

Different paradigms exist when it comes to training machine learning models. Supervised learning is one of them. Supervised learning involves training machine learning models on labeled data. During its training, the model takes inputs in addition to what it is supposed to predict given these inputs. That way, the supervised learning algorithm can learn to map like-wise data points to corresponding target values. Supervised learning can solve two types of problems by providing the correct data. These problems can be regression or classification problems. When the output we want to predict takes continuous values, we are dealing with regression, while when it takes a finite set of discrete values, it is a classification problem. In our work, we aim to detect design patterns given the extracted features from the source code. Since the target we want to predict takes a value from a discrete set of values, being the 13 design patterns our collected dataset DPD_{Att} covers, we can then conclude that the design pattern detection task is a classification problem. For the classification step, we first constructed the DPD_{Att}-classifier dataset as shown in Fig. 1 based on the embedding vector produced in the feature extraction step. This dataset will be used to train our classifier. Subsequently, we conducted experiments with various classifiers and assessed their performance. Notably, we trained and evaluated the quality of a Support Vector Machine (SVM), a Multi-Layer Perceptron (MLP), and a Multinomial Logistic Regression (MLR) multi-class classifiers [12]. Once the choice of the classification model is made, dividing the data for training and testing is a crucial step. Testing data should be around 20% to 30% of the total size of the data. The training data is the one passed to the classification model with its corresponding label so that it learns the design pattern features from it. The testing data is then used to evaluate how well the model has learned from its training process. Each Java source code is passed without its corresponding label, as the trained classifier is required to predict that. Then the evaluation is done based on the number of correctly predicted patterns and the wrongly detected ones.

Cross-validation: One way to assure the quality of the classification and assess the model's ability to generalize is to use the cross-validation technique [13]. As we said, a sample of the data is used for testing. This sample could be situated in any part of the corpus. For example, if the testing data is 25% of the total size of the dataset, these 25% could be situated in the first Fold of the data, the second, the third, or the fourth. Cross-validation takes that into consideration by evaluating each of these scenarios individually and then taking the mean value as the final evaluation score.

4 Experimental Results

This section first presents the implementation details. Then, an evaluation and a comparison with related work are given. At the end of this section, we discuss the computational complexity of the proposed approach.

4.1 Implementation

For the feature extraction step, our design pattern detector ($DPD_{Att-Detector}$) uses the CodeT5+'s [23] encoder, an open-source transformer-based code Large Language Model (LLM). To produce embedding vectors, a code embedding model called codet5p-110 m-embedding model [23] made available on the Hugging Face Transformers library in May 2023 is used[1]. For the detection task, we used the Scikit-learn library in Python to train the SVM, MLP, and MLR classifiers.

4.2 Evaluation

In this section, we first introduce the considered evaluation criteria in this paper. Then, we describe the evaluation results and compare our findings with related work. The computational complexity is given at the end of this section.

Evaluation Criteria. For the statistical evaluation of our approach, we report four different metrics: Precision, Recall, F1-Score, and Accuracy.

---

[1] https://huggingface.co/Salesforce/codet5p-110m-embedding.

- **Accuracy** measures the proportion of correctly classified instances (both true positives and true negatives) out of all instances in the dataset and is calculated as:

$$\textbf{Accuracy} = \frac{\text{Correct Predictions}}{\text{Total Predictions}} \tag{1}$$

Where Correct Predictions is the sum of true positives and true negatives, the denominator is the total number of instances in the dataset.

- **Precision** inform you on the model's ability to classify positive instances correctly while minimizing false positives. We calculate precision as:

$$\textbf{Precision} = \frac{\text{True Positives}}{\text{True Positives} + \text{False Positives}} \tag{2}$$

The term True Positive represents the instances correctly identified. Whereas the term False Positives indicates the number of instances that were classified as correct but were falsely predicted.

- **Recall** indicates the fraction of instances belonging to a certain class and being identified as such by the classifier. We calculate the recall metric as:

$$\textbf{Recall} = \frac{\text{True Positives}}{\text{True Positives} + \text{False Negatives}} \tag{3}$$

False Negatives informs about the instances classified as negative when they are actually positive.

- **F1-score** quantifies the harmonic mean between precision and recall as it balances between them. We determine F1 as:

$$\textbf{F1} - \textbf{score} = \frac{2 \cdot \text{Precision} \cdot \text{Recall}}{\text{Precision} + \text{Recall}} \tag{4}$$

Evaluation Results. By considering the evaluation metrics described in the previous section, we conducted experiments with our proposed approach. We started by analyzing the impact of the feature extraction module on the classification process, especially the embedding size. To explore this, we adjusted the length of the extracted features (i.e., embedding size) and examined its effect on the detection task. Our findings indicated that maintaining a size of 300 produced the most favorable outcomes. Specifically, it led to superior accuracy, precision, recall, and F1 scores compared to other sizes. Figure 7 depicts the ability of our tool to detect design patterns given different numbers of features (100, 250, 300, and 350) generated by the feature extraction step. As depicted in Fig. 7, increasing the embedding improved results. This enhancement can be attributed to the increased capacity provided to the encoder, allowing it to better capture features from the Java source code and minimize information loss. We can also see that over-increasing the size of the vector to 350 resulted in decreased performance, which can be explained by overfitting. Overfitting occurs when the model becomes too complex for the available data and starts memorizing the data instead of learning from it. In this case, when we increased the size of our learned vector to 350 features, we provided the model with more parameters to learn, which made it more capable of fitting the training data. That can be explained by its inability to generalize to new, unseen data. As a result, we configured our model to generate vectors of size equal to 300. These vectors were represented as torch tensors, which we then used to build $DPD_{Att-Classifier}$ dataset. The $DPD_{Att-Classifier}$ dataset is the csv file that is going to be passed to the classifier for training and testing and is composed of 303 columns. The first column serves as the name of the Java project. The second is the name of the class, while the third represents the design pattern detected (label). The last 300 columns are the features extracted by CodeT5+'s encoder, capturing the necessary information about each Java source code from each project.

We also compared the different classifiers for detecting GoF design patterns in terms of accuracy, precision, recall, and F1-score. The different classifiers utilized 80% of the $DPD_{Att-Classifier}$ dataset for training and 20% for testing. To ensure that the models do not overfit the training data. We opted for the K-fold cross-validation technique with K = 5 by utilizing the K-fold module in the sklearn library. Figure 6 shows the mean values for the defined metrics. As depicted in Fig. 6, the SVM classifier results in slightly better performance than the other classifiers. This is the reason why we opted for the SVM classifier for our detection tool ($DPD_{Att-Detector}$). Indeed, using our $DPD_{Att-Classifier}$ dataset and the SVM classifier, we obtained an accuracy score of 0.86, an average precision score of 0.87, an average recall score of 0.86, and a harmonic mean of 0.86.

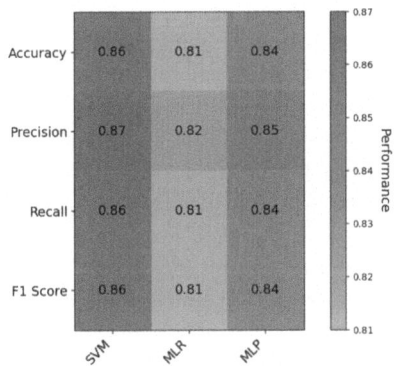

Fig. 6. Performance comparison between the used classifiers

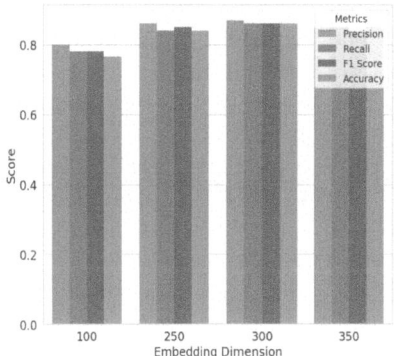

Fig. 7. Average experimental results with respect to the embedding size

Table 1 gives the evaluation of our detection tool per label using the SVM classifier and DPD_{Att}-classifier dataset. As shown in Table 1, our approach detects 13 design patterns with promising scores. The strategy and proxy patterns were predicted with precision higher than 95%. Five other design patterns were detected with a precision higher than 90%.

Table 1. Evaluation per each design pattern on DPD_{Att}

Class	Precision	Recall	F1-Score
Singleton	0.88	0.79	0.84
Observer	0.94	0.91	0.92
Memento	0.87	0.90	0.89
Proxy	0.97	0.91	0.94
Prototype	0.94	0.97	0.95
Builder	0.76	0.73	0.75
Abstract Factory	0.94	0.94	0.94
Factory Method	0.90	0.95	0.92
Facade	0.94	0.82	0.87
Adapter	0.80	0.97	0.88
Decorator	0.81	0.83	0.82
Visitor	0.78	0.95	0.86
Unknown	0.64	0.59	0.61
Strategy	1.00	0.87	0.93

Comparison with Related Work. The objective of this section is to compare our findings with related work that deal with design pattern detection. To this end, we consider a machine learning-based approach, called DPD_F, that uses word embedding (i.e., word2vec) to represent their features for the classifier [16], in addition to two metric-based approaches, MARPLE-DPD [25], and FeatureMap [20]. To perform this comparison, we specifically refer to Nazar et al. [16]'s corpus (i.e., DPD_F corpus). Indeed, in their paper, the authors in [16] used the DPD_F corpus for comparing DPD_F, MARPLE-DPD, and FeatureMap approaches. As shown in Fig. 8, testing the DPD_{Att} approach on the same DPD_F corpus yields better results in terms of mean precision, mean recall, and mean F1-score. These results indicate superior performance compared to state-of-the-art approaches [16,20,25], demonstrating the effectiveness of our approach in addressing the problem of GoF design pattern detection in source code. The comparisons of our approach with the state-of-the-art per design pattern in terms of F1-score is presented in Table 2. It is worth noting that this comparison concerns only nine design patterns, which are Singleton, Observer, Abstract Factory, Factory Method, Adapter, Builder, Decorator, Visitor, and the "unknown" target since MARPLE-DPD [25], and FeatureMap [20] deal only with these patterns. As we can see from the table, we achieve more promising results than FeatureMap [20] and MARPLE-DPD [25], in addition to DPD_F [16]. Indeed, our approach, DPD_{Att}, outperforms them in detecting eight out of nine GoF design patterns in Java. This is due to the capacity and performance of the feature extraction method for automatically extracting contextual information that encompasses the GoF design pattern features.

Table 2. Comparison of the DPD_{Att} in term of F1-score per design pattern with the state-of-the-art

Design Pattern	FeatureMap [20]	MARPLE-DPD [25]	DPD_F [16]	DPD_{Att}
Singleton	0.66	0.72	0.74	0.83
Observer	0.49	0.51	0.85	0.89
Builder	0.61	0.55	0.83	0.58
Abstract Factory	0.52	0.76	0.93	1.0
Factory Method	0.55	0.81	0.78	0.98
Adapter	0.33	0.82	0.69	0.88
Decorator	0.23	0.59	0.78	0.83
Visitor	0.65	0.63	0.94	0.94
Unknown	0.72	0.54	0.73	0.92

Computational Complexity. In this section, we evaluate the computational complexity of the feature extraction module. LLMs are generally evaluated by the number of parameters they have [2]. The count of parameters is intricately linked to kernel sizes, input and output channels, and serves as an indicator of computational resource utilization, particularly memory, during both model training and detection processes. In our feature extraction process, we employ the encoder of the CodeT5+ model, which has 110 million frozen parameters. In addition, we evaluate the operational performance of our feature extractor using Floating Points of Operations (FLOPs). FLOPs represent the total number of multiplication and addition operations within the model, providing a measure of its computational complexity. The testing was conducted on a system equipped with an Intel Core i7 processor running at 2.30 GHz, 16 GB of RAM, and a NVIDIA Ge-Force GTX 1650 with Max-Q Design for video memory. Figure 9 shows the variation of FLOPs with respect to DPD$_{Att}$ corpus. The maximum number of floating operations by the encoder to extract features and construct the embedding vector is 998.92G, and it reflects on the Java class with the highest number of tokens from the DPD$_{Att}$ dataset. Whereas the minimum number of FLOPs is equal to 1.002G, and it corresponds to the Java code with the least number of tokens in DPD$_{Att}$.

5 Related Work

This section overviews attention-based methods proposed in the literature to perform software engineering task. Then, it discusses existing design pattern detection approaches and concludes with a discussion.

5.1 Attention-Based Methods in Software Engineering

For software defect prediction, different attention-based approaches have been proposed in the literature. In [7], the authors employed attention mechanism to

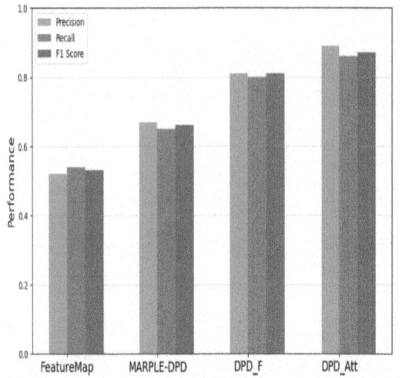
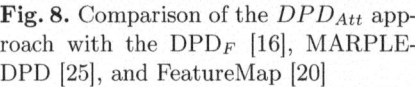
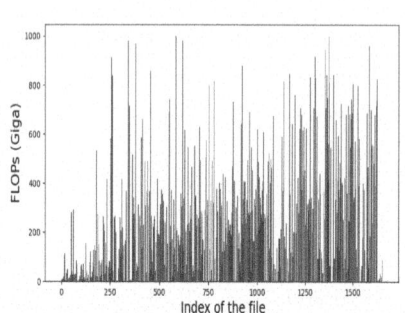

Fig. 8. Comparison of the DPD_{Att} app-roach with the DPD_F [16], MARPLE-DPD [25], and FeatureMap [20]

Fig. 9. FLOPs variations on the DPD_{Att} dataset

capture syntactic and semantic features of programs and use them to improve defect prediction. The proposed approach parses the abstract syntax trees (ASTs) of programs and extracts them as vectors. Then, it encodes vectors and employs the attention mechanism to further generate significant features for accurate defect prediction. In [14], the authors proposed an attention-based approach for statement-level software defect prediction. In this work, the authors define a set of 32 statement-level metrics. Then each statement is labeled to make a three-dimensional vector for automatic learning. The attention mechanism is used to generate important features and improve accuracy. In [22], the authors proposed an attention-based method for self-admitted technical debt detection. Indeed, Self-Admitted Technical Debt (SATD) is a kind of technical debt that seeks to capture technical debts that are intentionally introduced by developers during the software development process. To extract the sequential properties of self-admitted technical debts from code comments, the proposed solution uses a positional encoder and a Bi-directional Long Short-Term Memory (Bi-LSTM) network [26]. Then, it leverages a variety of attention techniques to emphasize the importance of the automatically generated features that contributed to the detection of SATDs.

5.2 Machine Learning-Based Pattern Detection

The use of machine learning techniques for pattern detection was widely addressed in the literature. In [24], the authors proposed a machine learning-based approach for the detection of security patterns in code. To be independent of the programming language, the proposed solution split the code in two vectorial representations: control flow and data flow. The control flow section is treated with word embedding to preserve the semantics of the code, and by clustering of the resulting embedding vectors to reduce the dependency on the lexical structure of the program. The authors in [10] presents a machine learning-based approach for the detection of architectural patterns, namely MVP (Model-View-Presenter) and MVVM (Model-View-ViewModel). In the paper, the authors performed analysis using nine popular machine learning models and established a set of source code metrics that can be used to detect MVVM and MVP architectural patterns using machine learning. Various studies have been interested in GoF design pattern detection. In [4], the authors proposed a probabilistic approach for GoF design pattern detection. In this work, the authors used neural networks and regression analysis to measure the possibility of the presence of the design pattern in the source code. The authors in [4] applied a correlation feature selection method to match the system design to the design pattern. Indeed, the authors describe in the paper a graph matching algorithm that uses a correlation-based feature selection technique to identify design pattern instances in system design. Different studies aim to train machine learning classifiers to detect GoF design pattern participants [11,16]. These approaches used code features to construct a text representation of Java classes. Indeed, they extract semantic and structural features from call graphs (SCG), which are in turn embedded in a text file as a natural language in a syntactic and semantic representation (SSLR) document. Then, a text embedding algorithm on the SSLR representation is applied, such as word2vec [11,16], to produce a vector representation for the Java source code. This vector served to train a supervised machine learning classifier.

5.3 Discussion

In this paper, we aim to propose a fully automated approach for GoF design pattern detection. Compared to related work [11,16], which include a pre-processing step for source code embedding, the proposed solution in this paper is based on a code embedding model, which eliminates this step. Indeed, as far as we know, all the existing approaches that deal with feature-based pattern detection used neural processing language techniques for the embedding step, such as word embedding [16], which requires feature engineering efforts to increase the accuracy of the design pattern detector. In fact, code features relevant to design pattern implementations, in some cases manually performed [11,15], must be defined. This stage can be hard and time-consuming since for a single design pattern that can have multiple disjunctive variations, a set of features that describe each single variant must be defined and a method to extract them must be implemented. For instance, Nacef et al. [15] define 33 features for the singleton design pattern only.

6   Conclusion

In this paper, we have presented a method for automatic detection of design patterns. The proposed approach includes a transformer-based feature extraction step. Features refer to code attributes that represent design pattern implementations and will help in detecting design patterns during the classification task. To this end, we build the DPD_{Att} corpus, which consists of 1645 labeled Java files covering 13 GoF design patterns. Then, following an unsupervised approach and considering the DPD_{Att} corpus as input, the feature extraction module learns from the code to automatically extract syntactic and semantic features. The features are then encoded as embedding vectors to train a classifier for the detection task. To evaluate the efficacy of the proposed approach, we apply four commonly used statistical measures, namely accuracy, precision, recall, and F1-Score. We also build a heat-map to showcase the efficiency of our chosen classifier. Empirical evaluation shows that the DPD_{Att} approach shows promising results with an accuracy score of 86%, precision of 87%, recall of 86%, and F1-score of 86%. The comparison with the related work shows that the proposed approach outperforms three existing methods.

While our results are promising, as future work, we aim to extend the DPD_{Att} corpus to consider more variants and extend to more design patterns. We also aim to use feature localization approaches to locate design pattern implementations among participants in a source code.

Data Availability. The data that supports the findings is available through Zenodo (https://doi.org/10.5281/zenodo.11584286).

References

1. Allamanis, M., Sutton, C.: Mining source code repositories at massive scale using language modeling. In: The 10th Working Conference on Mining Software Repositories, pp. 207–216. IEEE (2013)
2. Bae, H., Deeb, A., Fleury, A., Zhu, K.: Complexitynet: Increasing llm inference efficiency by learning task complexity. arXiv preprint arXiv:2312.11511 (2023)
3. Church, K.W.: Word2vec. Nat. Lang. Eng. **23**(1), 155–162 (2017)
4. Dewangan, S., Rao, R.S.: Design pattern detection by using correlation feature selection technique. In: 2022 IEEE 11th International Conference on Communication Systems and Network Technologies (CSNT), pp. 641–645. IEEE (2022)
5. Dong, J., Zhao, Y., Sun, Y.: A matrix-based approach to recovering design patterns. IEEE Trans. Syst. Man Cybern.-Part A: Syst. Hum. **39**(6), 1271–1282 (2009)
6. Dwivedi, A.K., Tirkey, A., Rath, S.K.: Applying software metrics for the mining of design pattern. In: 2016 IEEE Uttar Pradesh Section International Conference on Electrical, Computer and Electronics Engineering (UPCON), pp. 426–431. IEEE (2016)
7. Fan, G., Diao, X., Yu, H., Yang, K., Chen, L., et al.: Software defect prediction via attention-based recurrent neural network. Scientific Programming **2019**, 6230953 (2019)
8. Gamma, E., Helm, R., Johnson, R., Vlissides, J.: Design patterns: elements of reusable object-oriented software. Pearson Deutschland GmbH (1995)
9. Issaoui, I., Bouassida, N., Ben-Abdallah, H.: Using metric-based filtering to improve design pattern detection approaches. Innov. Syst. Softw. Eng. **11**, 39–53 (2015)
10. Komolov, S., Dlamini, G., Megha, S., Mazzara, M.: Towards predicting architectural design patterns: a machine learning approach. Computers **11**(10), 151 (2022)

11. Kouli, M., Rasoolzadegan, A.: A feature-based method for detecting design patterns in source code. Symmetry **14**(7), 1491 (2022)
12. Li, L., Wu, Y., Ye, M.: Experimental comparisons of multi-class classifiers. Informatica **39**(1), 71–75 (2015)
13. Misra, P., Yadav, A.S.: Improving the classification accuracy using recursive feature elimination with cross-validation. Int. J. Emerg. Technol. **11**(3), 659–665 (2020)
14. Munir, H.S., Ren, S., Mustafa, M., Siddique, C.N., Qayyum, S.: Attention based GRU-LSTM for software defect prediction. PLoS ONE **16**(3), e0247444 (2021)
15. Nacef, A., Bahroun, S., Khalfallah, A., Ahmed, S.B.: Features and supervised machine learning based method for singleton design pattern variants detection. Proceedings of the 18th International Conference on Evaluation of Novel Approaches to Software Engineering, ENASE 2023, Prague, Czech Republic, 24–25 April 2023, pp. 226–237 (2023)
16. Nazar, N., Aleti, A., Zheng, Y.: Feature-based software design pattern detection. J. Syst. Softw. **185**, 111179 (2022)
17. Rasool, G., Philippow, I., Mäder, P.: Design pattern recovery based on annotations. Adv. Eng. Softw. **41**(4), 519–526 (2010)
18. Richards, M., Ford, N.: Fundamentals of software architecture: an engineering approach. O'Reilly Media (2020)
19. Ruaro, N., et al.: Syml: guiding symbolic execution toward vulnerable states through pattern learning. In: Proceedings of the 24th International Symposium on Research in Attacks, Intrusions and Defenses, pp. 456–468 (2021)
20. Thaller, H., Linsbauer, L., Egyed, A.: Feature maps: a comprehensible software representation for design pattern detection. In: 2019 IEEE 26th International Conference on Software Analysis, Evolution and Reengineering (SANER), pp. 207–217. IEEE (2019)
21. Vaswani, A., et al.: Attention is all you need. In: Advances in neural Information Processing Systems, vol. 30 (2017)
22. Wang, X., Liu, J., Li, L., Chen, X., Liu, X., Wu, H.: Detecting and explaining self-admitted technical debts with attention-based neural networks. In: Proceedings of the 35th IEEE/ACM International Conference on Automated Software Engineering, pp. 871–882 (2020)
23. Wang, Y., Le, H., Gotmare, A.D., Bui, N.D., Li, J., Hoi, S.C.: Codet5+: Open code large language models for code understanding and generation. arXiv preprint arXiv:2305.07922 (2023)
24. Zaharia, S., Rebedea, T., Trausan-Matu, S.: Machine learning-based security pattern recognition techniques for code developers. Appl. Sci. **12**(23), 12463 (2022)
25. Zanoni, M., Fontana, F.A., Stella, F.: On applying machine learning techniques for design pattern detection. J. Syst. Softw. **103**, 102–117 (2015)
26. Zhou, P., et al.: Attention-based bidirectional long short-term memory networks for relation classification. In: Proceedings of the 54th Annual Meeting of the Association for Computational Linguistics (volume 2: Short papers), pp. 207–212 (2016)
27. Zhu, H., Bayley, I., Shan, L., Amphlett, R.: Tool support for design pattern recognition at model level. In: 2009 33rd Annual IEEE International Computer Software and Applications Conference, vol. 1, pp. 228–233. IEEE (2009)

Architecture Evaluation

Cause-Effect Chain-Based Diagnosis of Automotive Onboard Energy Systems

Stefan Kugele[1][(✉)] [iD], Lorenz Schreyer[2], and Martin Lamprecht[2]

[1] Research Institute AImotion Bavaria, Technische Hochschule Ingolstadt,
Ingolstadt, Germany
Stefan.Kugele@thi.de

[2] BMW Group, EE-440, Munich, Germany
Martin.LB.Lamprecht@bmw.de

Abstract. *Context:* Vehicle diagnostics are critical tools for identifying, locating, and resolving automobile faults. However, the increasing connectivity within vehicles poses challenges to seamless diagnostic processes. *Aim:* This paper aims to improve the rectification of faults following a diagnostic trouble code entry in a vehicle's electrical power system. *Method:* The approach involves designing a graph based on the cause-effect chain from the 'flexible Energy and Power Management' (fEPM) detailed model, defining areas for each signal to identify potential causes for diagnostic trouble code entries using simulated signal traces. The method, coupled with graph reduction techniques, was evaluated through an interview study with engineers who provided feedback on its practical applicability and efficacy in real-world scenarios. *Results:* The application of this method results in a clear fault image, graphically representing the origin of the diagnostic trouble code entry. This reduced graph can be interpreted comprehensively for each component and for each diagnostic trouble code entry, with the possibility of automating the interpretation process. The interview study confirmed the applicability and efficiency of the approach. *Conclusion:* This research introduces a method to identify the root causes of faults in automotive electrical power systems, thereby improving the efficiency and accuracy of vehicle diagnostics.

Keywords: Automotive · Architecture · Diagnosis · Cause-effect chains

1 Introduction

1.1 Context

In the contemporary automotive landscape, characterised by electrification, automation, and connectivity, conventional modes of transportation have evolved into complex electronic systems [14]. The prevailing trend towards automation necessitates the deployment of high-performance inference hardware, particularly for perception methods and, more broadly, for intelligent AI-driven functionalities. This requires integrating increasingly high-performance and energy-intensive hardware components into automotive systems and, thus, triggers a

M. Galster et al. (Eds.): ECSA 2024, LNCS 14889, pp. 105–120, 2024.
https://doi.org/10.1007/978-3-031-70797-1_7

significant demand on the vehicle's onboard energy network, which is required to ensure safe and customer-friendly operation at all times [9]. Particularly, the inter- and intra-vehicle networking, along with the integration of automated driving assistance systems, pushes the limits of electronic control units (ECUs) that process or transmit data in real-time, as well as storage systems such as fault memory, due to the vast amount of data involved. In the realm of fault diagnosis, traditional approaches often focus on isolated fault identification and storage using pre-stored expert knowledge, which could be relevant to the cause of the problem [8].

The motivation for this work stems from the recognition that conventional *diagnostic methods* contribute only limited to *fault identification and resolution* in addressing the increasing complexity of modern vehicles implementing thousands of functions, primarily realised through software [2]. This intricate interplay across ECU boundaries induces highly complex signal dependencies involving several tens of thousands of signals in modern electric/electronic (E/E) architectures. Tracing back from check control messages (CCMs) to original signals in such a distributed software architecture presents a significant challenge and is the subject of this work. In addition to diagnostics within a network of ECUs, the diagnosis of a single ECU presents no less of a challenge and is methodologically significant. This is attributed to the current trend where functions are distributed across multiple control units, as well as the integration of a myriad of disparate software components—developed by suppliers or different departments—within a complex internal ECU *software architecture*.

To address this issue, this work focuses on the promising concept of *cause-effect chains* in automotive *architectures*. Traversing these chains provides a holistic view of CCM fault symptoms, leading to the interpretation of problematic signals. A significant advantage is that cause-effect chains can deduce issues without human intervention, greatly enhancing the diagnostic process.

1.2   Objective

This work aims to develop a method for *cause-effect chain-oriented* vehicle diagnosis in the onboard energy system architecture, identifying the *root causes* of faults in CCMs within the energy network. The software architecture accommodates modern vehicle diagnostics complexity, enhancing vehicle diagnosis, maintenance, repair, and customer service efficiency. This innovative approach could transform the automotive industry's diagnostic systems, making modern vehicle maintenance and repair more cost-effective and efficient.

1.3   Research Question

The central and guiding research question of this work is:

How effective are cause-effect chain-oriented approaches in diagnosing the onboard energy network at improving fault detection and resolution efficiency in modern vehicles?

To answer this question, various aspects will be examined, including the acquisition and storage of diagnostic knowledge and the implementation of an efficient backwards-chaining algorithm that traverses the signal graph, i. e., a graph containing all signals of the software components in a chained format, deducing possible causes of faults. Finally, the study will explore and discuss the implications of potentially multiple root causes on the triggering fault source.

Contributions. This work has the following scientific and practical contributions:

(i) Development of a cause-effect chain-based diagnostic approach;
(ii) Fully automated fault evaluation approach;
(iii) Real-world evaluation of BMW's fEPM function.

Outline. The remainder of this work is structured as follows: In Sect. 2, we briefly discuss related work, followed by a summary of background information in Sect. 3. Section 4 describes the main approach of this work. Two use cases conducted at the industry partner are summarised in Sect. 5 and the results of an interview study are presented in Sect. 6. Finally, the paper ends with a discussion and conclusion in Sects. 7 and 8.

2 Related Work

Cause-effect chains fundamentally link different artefacts together. This linkage allows for the derivation of dependencies, the more detailed analysis of temporal program or system properties, or the conduct of system archaeology by examining repositories. A very important field of application, as we also use it in this paper, is the analysis of cause-effect relationships to deduce the causes of system failures. Another significant application area of cause-effect chains is the timing analysis, especially of safety-critical real-time systems (e. g., Günzel et al. [5]).

The principle of cause-effect chains is particularly used in the analysis of technical and, especially, software-based systems. Earlier works by Zeller [15] focused on the analysis of software and, in particular, on the automatic generation of cause-effect chains that led to a fault state. These works are also known as *Delta Debugging*. Besides the use of cause-effect chains, AI-based approaches for diagnosis are described in the literature. In their paper, Nguyen et al. [12] introduce a temporally recurrent neural graph network model tailored for fault diagnosis. Their novel approach offers superior generalisation capabilities compared to prior methods. Abdelgayed et al. [1] propose a fusion of wavelet functions and machine learning techniques for fault classification in microgrids, utilising decision trees, k-nearest neighbours, support vector machines, and naïve Bayes for comparison. However, unlike these works, our method excels in capturing intricate system relationships. Nevertheless, our approach categorises signals into discrete areas, while machine learning techniques enable evaluation with continuous values, presenting a notable distinction. The work by Min et al. [11] introduces a framework for autonomous vehicles featuring sensor self-diagnosis, emphasising hardware fault detection. However, it's important to note that faults can stem from software issues or human error, which this approach may not address. A machine learning-based anomaly detection is utilised for fault identification.

3   Background

In the context of automotive engineering, the vehicle's electrical power network plays a pivotal role in its operational capability and performance. The capability to initiate the engine, a fundamental attribute of the low-voltage electrical power network, ensures that the network can reliably provide sufficient energy to power the starter motor for engine ignition. This process, termed a cold start, occurs when a vehicle that has been parked for an extended duration is started. During a cold start, all components not essential for engine ignition are deactivated. Additionally, a balanced charging equilibrium is crucial to ensure the battery is charged efficiently and provides adequate energy for both vehicle operation and the start-up process. Voltage stability is another vital characteristic, ensuring the electrical power network consistently delivers a constant voltage to guarantee the reliable functioning of electrical components. Voltage drops in the network can occur during the start-up process or due to the positioning of the low-voltage battery, especially if the cables exhibit excessive resistance. This stability is particularly critical for ECUs operating within a narrow voltage range. Lastly, the aspect of storage load plays a significant role in the distribution and utilisation of electrical energy within the network. This is particularly relevant in hybrid and electric vehicles where the battery is the primary energy storage. The cycling of a battery is directly linked to its ageing process. Efficient storage load management is imperative for the energy supply of the vehicle, even after years of operation. These characteristics are intricately interrelated and collectively contribute to ensuring the reliability and efficiency of the vehicle's electrical power network.

In this work, we apply the diagnosis approach, detailed in Sect. 4, on the central function called *fEPM* (flexible Energy and Power Management), which realises the above-mentioned properties and consists of dozens of software components orchestrated in a *layered software architecture* (cf. Sect. 3.4).

3.1   Data, Signals, Buses, and Encoding

In contemporary vehicular systems, data, signals, buses, and encoding play a pivotal role in ensuring effective communication and seamless operation of all systems. Data are transmitted as electrical signals among numerous electronic ECUs within the vehicle or within ECUs. This data encompasses a wide array of information, ranging from engine metrics and vehicle diagnostics to safety and entertainment system details. Signals utilise various custom data types, such as `boolean`, `enumeration`, or `integer` for transmitting discrete signals and `float` for conveying continuous signals.

Inter-component communication within the vehicle's E/E architecture is predominantly facilitated through bus systems. Bus systems act as conduits for data transmission amongst the various ECUs. Commonly utilised vehicular bus systems include the Controller Area Network (CAN) bus, Local Interconnect Network (LIN) bus, and Ethernet. These systems enable efficient and reliable communication across the vehicle's diverse components (cf. [4]). In the highlighted use cases (cf. Sect. 5), this becomes significant, as the root cause of a fault may originate from another ECU and become evident at the I/O layer of the fEPM's layered architecture. The encoding of data is crucial to ensure accurate interpretation of information. Specific standards and protocols are employed to guarantee uniform interpretation of data across the vehicle's systems.

Overall, data, signals, buses, and encoding are integral to the connectivity and smooth functioning of modern automotive electronic systems. The continuous evolution of these technologies contributes significantly to enhancing the efficiency, safety, and functionality of vehicles. The multitude of bus signals, as well as signals within the same ECU, significantly increases but also reflects the complexity of the embedded software architecture.

3.2   Diagnosis

The term "diagnosis" originally derives from the medical field, where a potential disease state is identified based on the presence of acute symptoms and historical indicators. This knowledge informs the initiation of appropriate treatment strategies. Extending from this general concept, Reif [13] defines classical vehicle diagnostics as follows: "Based on specific and vague symptoms reported by the driver, diagnostic systems in service centres are employed to create a precise error profile and initiate suitable repair actions."

This paper will rely on definitions established in the community [7]: An "error" refers to the deviation from correctness in a system's operation or output. A "fault" is the underlying cause in the system that has the potential to lead to an error. When an error affects the system's intended operation, leading to an inability to perform a required function, it is termed a "failure;" thus, a fault can lead to an error, which may result in a failure. Fundamentally, fault diagnosis in a system comprises three components. The first two stages are *fault detection* and *fault localisation*, for which algorithms are well-established in the literature. The final component, *fault evaluation*, will be a critical focus of this work [10].

Modern vehicles typically integrate a fault memory within one of the vehicle's control units, featuring non-volatile memory. This storage is used to record Diagnostic Trouble Codes (DTCs), diagnostic information, and event memory entries generated by various sensors and control units in the vehicle. Fault memory entries are made when specific faulty behaviours are detected, as well as other diagnostic-related events. In addition to DTCs, when faults occur, further environmental conditions such as *timestamp*, *frequency of occurrence*, current *speed*, *temperature*, and additional metrics like *voltage* and the *state of charge* are recorded to simplify troubleshooting in workshops.

3.3 Cause-Effect Chain

In scientific discourse, *"cause-effect chains"* refer to graphs or networks that visualise the cause-and-effect relationships between different elements, variables, or events within a system or process. These graphical representations illustrate how certain factors or events influence others, enabling researchers and analysts to investigate and comprehend the causal connections within systems and architectures. The forms of representation for cause-effect chains can vary depending on their application and context. These chains are employed across various research domains, providing insights into the causes and effects within systems composed of interdependent mechanical, biological, or socio-economic components (cf. [6]). In the example, b is dependent on a, i.e., $a \bullet\!\!-\!\!\!-\!\!\!-\!\!\!\to\!\!\bullet b$.

In the realm of automotive software architectures, the intricate interactions among *signals* are typically represented by extensive and complex graph structures. Here, nodes represent signals and edges visualise their dependencies.

3.4 Software Architecture

We demonstrate the practicality of our approach using a function from the energy supply system in modern vehicles. The fundamental software architecture adheres to the principle of a layered architecture with defined interfaces to the rest of the system. In this case, the entire functionality is deployed on a single ECU, providing both a software interface and a distinct technical interface to the vehicle's remaining system. The lowest layer is responsible for *input/output* (I/O) operations. Here, signals are processed via bus systems. A layer above isolates all subsequent layers from the I/O layer, ensuring functionality remains stable, even if the technical signals change, i.e., a *hardware abstraction layer* (HAL). The highest layer (fEPM) encapsulates the main functionality, implementing the idea of a *cybernetic system model* (cf. [3]). Signals are often

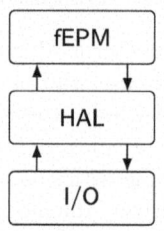

Fig. 1. Layered architecture

relayed or transformed, i.e., $in \bullet\!\!\xrightarrow{\ id\ }\!\!\bullet out$ or $in \bullet\!\!\xrightarrow{\ f\ }\!\!\bullet out$, where f is a *transformation* function. Note, in the first case, id is the *identity function* that only forwards a signal.

Further analysis reveals that the signal network is significantly more complex than these superficial layers suggest. Figure 1 illustrates the 3-layered software architecture. Due to confidentiality constraints, further details are withheld. Nonetheless, the complexity of the system is evident upon examining the network of signal dependencies presented in Fig. 2, which displays a pseudonymised overview of all signals and their inter-dependencies within the fEPM.

Fig. 2. fEPM's signal graph containing 4024 nodes, i.e., signals.

4   Approach

In this paper, we imple-
ment an approach based
on the extraction of
cause-effect chains from
a detailed model derived
from the requirements
specification. This me-
thodology proves to be
effective mainly because
the requirements are
highly formalised and
structured within the
responsible department.
The specifics of the

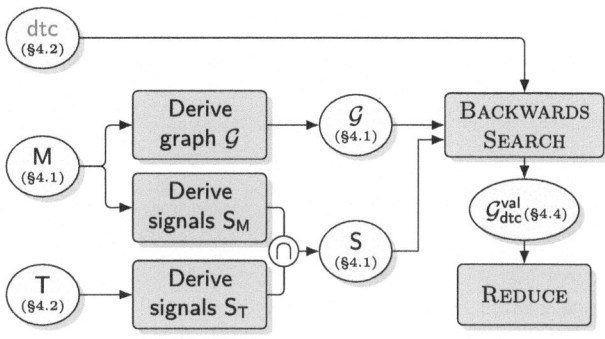

Fig. 3. Overall process

detailed model and structuring process are described next.

During the *simulation-based verification* (cf. Sect. 4.2) of fEPM's functionality, it is essential to generate suitable simulation stimuli. For this purpose,
we employ *traces* (T) in the simulation environment. These traces consist of
signal values, the valid ranges of which can primarily be determined by expert
knowledge. For example, within what voltage range can a lead-acid battery be
considered operational? This query is an illustrative example highlighting the
critical importance of whether corresponding signal values fall within a valid
range. Thus, a fully charged battery is unlikely to be considered a source of
under-voltage in the power network and its corresponding architecture. Figure 3
represents the process that will be explained further in the rest of the paper.

4.1   Detailed Model

The fEPM component embodies the whole logic through a sophisticated software architecture and implementation. It assumes functional responsibility for
detecting specific conditions within a vehicle's operation, such as identifying
under-voltage occurrences while driving and subsequently logging a fault record
entry. The instance of "under-voltage during driving," as mentioned, will be
repeatedly referenced and analysed throughout this paper.

This research significantly focuses on both the detailed model (M) and software tests related to the fEPM. The detailed model has been largely generated
automated based on the requirements specification and is available for further
analysis of the cause-effect chains. The subsequent approach is predicated on the
notion that within the department, the concept of "requirements engineering"
is not just perfunctorily regarded, but its benefits are actively embraced and
utilised. Through a consistently formalised and structured set of requirements,
which have been articulated at the signal level, it becomes feasible to model
causal relationships. The table below presents an excerpt (for clarity, additional
necessary columns such as ID, type of requirement, or chapter are not shown)

of a representative requirement. It includes source signal(s), a description, and target signal(s). The second example illustrates that limits (upper and lower bounds) are also employed in the description, thereby enabling their plausibility to be verified. From all these requirements, a graph ($\mathcal{G}$) can be derived, which is subsequently used for further analyses (cf. Fig. 3).

Source	Description	Target Signal
tsnPctSoc	The SOC of the currently installed battery [tsnPctSoc] is written to [ghgPctSoc]	ghgPctSoc

Source	Description	Target Signal
ghgBEcuValid, ghgPctSoc	If the current SOC [ghgPctSoc] is below a threshold of 30% and [ghgBEcuValid] = false, then [dfcBErrorOnBn] is set to true, otherwise to false	dfcBErrorOnBn

The second source/target signal pair would, for instance, yield the following sub-graph:

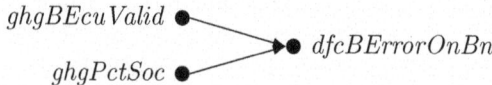

$$ghgBEcuValid \quad\quad\quad ghgPctSoc \quad\quad\quad dfcBErrorOnBn$$

In line with the high degree of formalisation, naming conventions are mandatory, e.g.,

$$\underbrace{\text{ghg}}_{\text{Prefix}}\ \underbrace{\text{B}}_{\text{Type}}\ \underbrace{\text{EcuValid}}_{\text{Name}}$$

In a structured naming convention for signals (S), the prefix identifies the originating interface, indicating the context or source from which the signal is derived. The type, which may comprise one or more characters, signifies the specific physical quantity represented by the signal. This aspect of the naming convention is crucial for understanding the nature of the data being conveyed. Additionally, the signal name is designated as free text, allowing for descriptive identification that facilitates intuitive understanding and reference. Within this naming framework, the character string denoting the type plays a pivotal role in classifying the signal's nature. Notably, the values B (Boolean), St (status), U (voltage), I (current), and Pct (percentage) emerge as critical identifiers, each representing distinct quantities.

4.2 Modelling Environment for Software and Hardware

At BMW, the Modelling Environment for Software and Hardware (MESH) is utilised in more complex testing procedures. The fEPM and the physical onboard power network, along with its physical properties, are implemented as Functional Mock-up Units (FMUs). For this purpose, each hardware component is simulated based on its physical and chemical characteristics to closely mirror reality.

For instance, the chemical ageing process significantly contributes to the lifespan of lead-acid batteries. Corrosion, as well as extreme temperatures or loads, cause a gradual decrease in maximum storage capacity over time. Furthermore, the battery's charge intake, discharge, and capacity are highly dependent on ambient temperature. Components such as the battery, alternator, and voltage converter are implemented as FMUs, among others. All these processes and software-hardware components can be modelled and their interactions examined using MESH. This approach allows for the precise description and comprehensive testing of test cases involving a simulated defect in any of these components. Test cases for the simulated entire onboard power network can be applied to this MESH network, enhancing the specificity and scope of testing capabilities.

The outcomes of the simulations conducted, are presented as traces. This approach involves the logging of all signal values from the simulation at predetermined discrete intervals (e. g., every 0.01 s). This process generates the entire test trajectory. The structure of a trace is depicted in (2): Each column represents the valuation μ of signal s_i, $1 \leq i \leq n$ where $s_i \in S$ and $n = |S|$ at a particular point in time (the time is monotonically increasing over the rows). The valuation function μ is defined in (1).

$$\mu \colon S \times \mathbb{N}^+ \to \mathbb{T}, \quad e.\,g.,\ \mu(s_1, 27) \mapsto \text{false}, \tag{1}$$

where S denotes the set of signals (cf. Fig. 3), $\mathbb{T}$ denotes the signal's type, e. g., $\mathbb{B}$ for Boolean values, or $\mathbb{R}$ for percentage values, i. e., $\mathbb{T} = \mathbb{N} \cup \mathbb{R} \cup \mathbb{B}$. The second argument of μ denotes the discrete-time tick in the trace. In the following, let $\mathsf{T} \in \mathbb{T}^{l \times |S|}$, be a *trace*, where l is it's length.

$$\mathsf{T} = \begin{bmatrix} \mu(s_1, 1) & \mu(s_2, 1) & \cdots & \mu(s_{|S|}, 1) \\ \mu(s_1, 2) & \mu(s_2, 2) & \cdots & \mu(s_{|S|}, 2) \\ \vdots & \vdots & & \vdots \\ \mu(s_1, t_{fault}) & \mu(s_2, t_{fault}) & \cdots & \mu(s_{|S|}, t_{fault}) \\ \vdots & \vdots & & \vdots \\ \mu(s_1, l) & \mu(s_2, l) & \cdots & \mu(s_{|S|}, l) \end{bmatrix} \tag{2}$$

In the context of the entry highlighted in red, specifically exemplified by the signal s_2, the evaluation of the signal value indicated a fault. Specifically, signal s_2 is associated with a corresponding DTC. From this point, we analyse all backward-linked signal values at this time, as they must have been crucial for triggering the fault memory entry. Consequently, only the cause-effect chain up to the point time t_{fault} is relevant. For this reason, the entire cause-effect chain graph can be reduced by analysing the cause-effect chain backwards from signal s_2 at time t_{fault}. This results in a significantly smaller cause-effect chain graph, which is relevant exclusively for the examined DTC/signal s_2. To deduce the original triggering cause, it is vital to identify which signals were outside the valid range. The determination of what constitutes a valid or invalid range will be further clarified subsequently.

4.3   Signal Value Range and Knowledge Base

The methodology of this work is predicated on several prerequisites that enable the implementation of an algorithmic backward chaining process. The segmentation of the value ranges of all signals into sections (sub-value ranges) characterises the state of a signal at a specific point in time t. For this purpose, the following sections for signals will be defined: NORMAL (green), NOTICEABLE (yellow), and FAULTY (red). However, what remains open is a mapping of signal values to the sections.

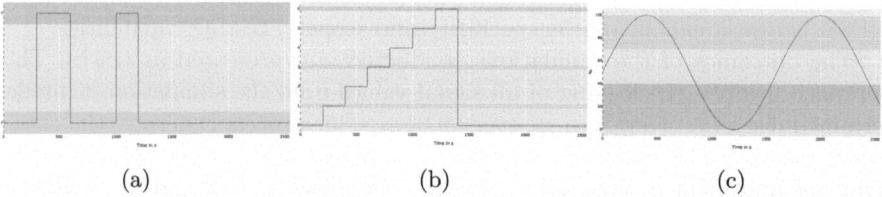

(a)   (b)   (c)

Fig. 4. (a) Boolean $\{$true (T), false $(F)\}$, (b) enumeration $\{0, 1, \ldots, 6\}$, and (c) continuous (percentage) $[0, \ldots, 100\%]$ signals. x-axis: continuous time. (Color figure online)

Expert knowledge plays a crucial role in acquiring necessary expertise. Experts can be consulted for specific threshold values, or data from signal traces of simulations or actual vehicles can be analysed, resulting in multiple sub-value ranges. Each approach has its pros and cons, requiring careful consideration. Generating diagnostic knowledge from existing data offers high automation but risks inaccurate or missing data. MESH traces may distort data as they only evaluate faulty traces; on the other hand, simulated traces could also lead to incorrect evaluations. No data source exists that offers high variance and represents the desired state of all relevant signals across many scenarios. Thus, this work utilises *expert knowledge*.

This expert knowledge is captured in a *knowledge base* κ that stores for each signal (set of signals S) and valuation μ one of the three sections, i.e.,

$$\kappa \colon \mathsf{S} \times \mu \to \{\mathsf{NORMAL}, \mathsf{NOTICEABLE}, \mathsf{FAULTY}\}$$

Expert knowledge represents a specialised and profound understanding in function development and vehicle diagnostics, acquired through years of experience, extensive research, and practical application. This knowledge, often extending beyond formal education, is enhanced through continuous learning, active community participation, and real-world application. Accessing this knowledge ranges from simple inquiries to integration into the fEPM's detailed model and the message catalogue, cataloguing all bus signals across vehicles. A significant part of signals in the fEPM is of the *Boolean* type, covering most signals through further specifications. Through literature review and detailed model study, conclusions can also be drawn about signals of the types *Pct* and *U*. For the two use cases detailed in Sect. 5, it was possible to determine the sub-value ranges

Algorithm 1. ANALYSEDTC $(\mathcal{G}, \mathsf{T}, \mathsf{dtc})$

1: Let j be dtc's column in T, i.e., signal s_j
2: **for** $1 \leq i \leq l$ **do**
3:     **if** $\kappa(\mathsf{s}_j, \mathsf{T}_{ij}) = \mathsf{FAULTY}$ **then**
4:         $\mathcal{G}_{\mathsf{dtc}} \leftarrow$ BACKWARDSSEARCH $(\mathcal{G}, \mathsf{s}_j)$
5:         $\mathcal{G}_{\mathsf{dtc}}^{\mathsf{val}} \leftarrow$ VALUATEGRAPH $(\mathcal{G}_{\mathsf{dtc}}, \kappa)$
6:         **return** $\mathcal{G}_{\mathsf{dtc}}^{\mathsf{val}}$
7:     **end if**
8: **end for**

for all involved signals. This was achieved through several methods: firstly, by utilising the detailed model, which already contained threshold values for some signals; secondly, by conducting an analysis or inquiry within a signal database that encompasses all vehicle signals along with their codifications; and thirdly, through targeted consultations with experts. Three exemplars of sub-value range identification are depicted in Fig. 4, showcasing the diverse signal types encountered in our analysis. Specifically, the illustrations cover (a) Boolean signals, (b) enumeration types, and (c) continuous signals, each representing distinct categories of data within our study framework.

4.4 Valuated Cause-Effect Chain Extraction

Having introduced the concepts of the detailed model, the simulation environment MESH, and the classification of signals into three categories, it is now possible to derive a valuated sub-graph $\mathcal{G}_{\mathsf{dtc}}^{\mathsf{val}}$ from the complete fEPM cause-effect chain graph. This sub-graph is specifically designed for analysing the root cause of a particular DTC dtc. The procedure is outlined in Algorithm 1. VALUATEGRAPH "colours" the graph according to the signal valuation and the knowledge base. Note, besides the three colours for NORMAL •, NOTICEABLE •, and FAULTY •, additional colours are used: • target signal, • bus signal, • not in trace, • not relevant, and • not specified.

4.5 Cause-Effect Chain Reduction

To enhance clarity by eliminating non-relevant signals and those within the normal range, along with all signals transitively leading into these nodes from the representation, the following methodology is applied. The rationale behind graph reduction is not only to gain clarity but also to limit the specification of sub-value ranges required. When multiple non-relevant, unspecified, or non-faulty signals converge into a signal that is specified as non-faulty, all nodes along this path can be removed from the representation without loss of information. Additionally, paths leading to inconspicuous or non-faulty nodes can also be removed if there are alternative paths to them. This approach streamlines the diagnostic visualisation by focusing only on potentially faulty or relevant paths, thereby simplifying the complexity of specifying sub-value ranges and enhancing the overall interpretability of the system's diagnostic data.

5   Use Cases

We illustrate the approach proposed in this work through two representative use cases, both of which address the issue of *under-voltage* within the vehicle's electrical power network: one during driving and the other during the transportation of a non-operating vehicle. Given that the fEPM software under investigation embodies the central logic with a corresponding sophisticated software architecture, the examination and analysis of the root cause of under-voltage is crucial. Assume, the two fault memory entries, dtc_1 and dtc_2 that trigger respective check control messages ccm_1 and ccm_2:

dtc_1 elineates the scenario of under-voltage during driving, which, is to be triggered if either the battery voltage drops below a certain threshold or if the alternator is faulty.

dtc_2 details the scenario of deep discharge during the transportation of the vehicle.

Starting from the complete signal graph of the fEPM, shown in Fig. 2, the evaluated graphs for the two use cases, dtc_1 and dtc_2, at the time of setting the fault memory entry are depicted in Figs. 5(a) and 5(b), respectively. Through further reduction, the final causality chains were calculated, depicted in Figs. 5(c) and 5(d). It becomes readily apparent that the graph reduction significantly enhanced clarity and understanding regarding the root cause of the fault.

The final step, the interpretation of the graph, yields promising results. In both instances, the fault induced by the test case was successfully identified. The signal s_{1243} (c) was set to false immediately after initialisation during the MESH test to provoke the fault memory entry. The signals s_{3508} and s_{3509} (c) both relate to the correct operation of cyclic and non-cyclic communication, implying that signal s_{1243}, which logically AND combines both signals of type bool, must have been externally overwritten. Both the current onboard voltage s_{128} and the voltage s_{523} at the control unit are directly from the bus. Additionally, the state of charge of the battery s_{140} (d) is directly from the LIN bus.

6   Evaluation

This interview-based study focuses on expert evaluations and experiences within BMW. Through targeted inquiries and analyses, the objective was to gain insights into the effectiveness of existing methods and tools for fault diagnosis. Particular attention was given to a newly developed vehicle diagnostic system, assessing its applicability, user-friendliness, and efficiency. Respondents provided feedback via text fields and Likert scales, a numerical method for expressing agreement or opinion on a statement, ranging abstractly from "very obstructive" to "very helpful" or "very inefficient" to "very efficient," i.e., from –2 to 2. The insights gained offer valuable perspectives on the current state of vehicle diagnostics, serving as a foundation for potential improvements and future developments. The following sections present the methodology, key evaluation questions, results, and conclusions.

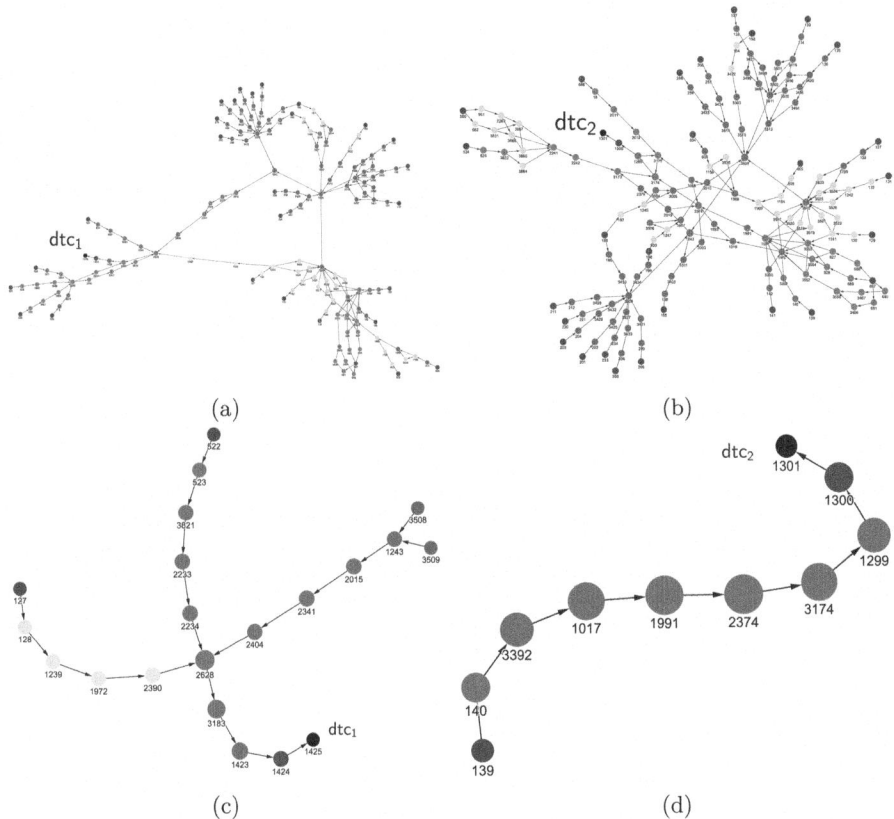

Fig. 5. Graph for (a) dtc$_1$ and (b) dtc$_2$ after applying Algorithm 1, and graph for dtc$_1$ (c) and dtc$_2$ (d) after graph reduction (Sect. 4.5). (Color figure online)

6.1   Participant Profile

The six participants varied in experience (cf. Fig. 6(a)) but shared similar roles in software development, test development, and function specification, all intersecting with diagnostics. Further expansion of participants was deemed unnecessary due to their expertise in the energy board network diagnostics.

6.2   Results

Evaluation of Fault Diagnosis and Environmental Conditions. The complexity of identifying fault root causes from vehicle traces was generally rated as *complex* and rather time consuming (cf. Fig. 6(c)). The tools used for fault identification ranged from consulting experts and commercial programs for trace extraction to in-house tools for signal correlation analysis. The current environmental conditions were rated as *moderately informative* on average.

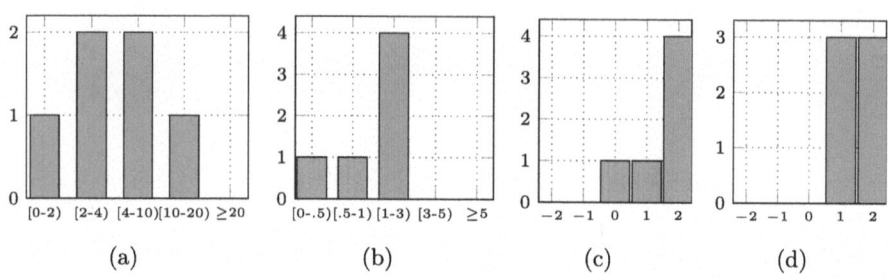

Fig. 6. (a) Interviewees' years of experience, (b) current time [h] for trace analysis, (c) usefulness, and (d) efficiency.

Algorithm Evaluation. To measure the impact on the industry, we asked, *"How much does the system facilitate fault diagnosis?"* The survey showed the distribution depicted in Fig. 6(c). Opinions on the user-friendliness of graphical representations, graphs, and legends were predominantly *positively high*. The key question, *"How do you compare the system's efficiency with previous fault diagnosis methods?"* received the responses shown in Fig. 6(d). Most respondents found the clarity gained from graph reduction *very good*, while the majority classified the process of graph reduction as *fairly reliable*.

7   Discussion

The use cases and results presented for the algorithm indicate its correct functionality. The most significant limitation of the system lies in the completeness of the range evaluations, as accurate results can only be achieved under this condition. If certain signals are not specified or do not appear in the trace, it becomes impossible to determine potential fault causes after graph reduction. If sub-value ranges for every fEPM signal were comprehensively specified, the algorithm would operate flawlessly and accurately. The challenge arises from sub-value ranges that are unspecified, not yet specified, or incorrectly specified, which can only be identified in a comprehensive view during development stages. Furthermore, any changes to the detailed model would require corresponding updates to the sub-value ranges, representing a significant additional effort in specifying vehicle functions.

The proposed method was assessed and found feasible for use in a single ECU, even with various integrated software components, some sourced from different suppliers and sub-departments. Therefore, this methodology is also applicable to diagnostics across ECU boundaries within complex networks. This requires a similarly consistent, formalised description of requirements so that a detailed model can be processed.

Utilising a layered architecture within the fEPM function further simplifies the diagnostic process. As demonstrated in Fig. 5(d), it is clear where signals, for example, in the I/O layer from buses or other control units are processed, and the internal processing of the signals is hardware-agnostic.

The methodology could be expanded to factor in a combination of different signal values for diagnostics. This aspect could be the subject of future study.

7.1   Threats to Validity

Internal Validity. One potential threat to internal validity in our study arises from the selective examination of only two DTCs within the fEPM system. While these DTCs have been identified as highly representative by both our assessment and expert opinions from the specialised department, focusing on a limited number of cases may not encompass the full spectrum of scenarios encountered in the fEPM system. This selection was justified by the central importance of the power supply within the vehicle's energy supply architecture, yet it remains a limitation that the findings may not be universally applicable to all DTCs within the system.

External Validity. A notable threat to external validity stems from the relatively small number of interview participants, which could limit the generalisability of our findings. To address this potential threat, we placed a strong emphasis on the selection process of participants, prioritising individuals with substantial expertise in the fEPM diagnosis. By incorporating participants with varying degrees of experience, we aimed to mitigate the impact of this limitation.

8   Conclusion

In this paper, we introduce a cause-effect chain-based methodology designed to enhance diagnostic processes within the core software component responsible for energy and power management across all BMW vehicles. Utilising a formalised detailed model alongside a comprehensive knowledge base, we pinpointed the root cause of a check control message through the analysis of simulation traces employing backward search techniques. Additionally, we employed graph reduction strategies to depict chains of effects in a manner that is readily comprehensible to test engineers, thereby supporting efficient and accessible diagnostics. The practicality of our approach was validated through two key use cases and further corroborated by an interview study. Participants' responses to the research question indicate a highly positive study outcome.

References

1. Abdelgayed, T.S., Morsi, W.G., Sidhu, T.S.: A new approach for fault classification in microgrids using optimal wavelet functions matching pursuit. IEEE Trans. Smart Grid **9**(5), 4838–4846 (2017). https://doi.org/10.1109/TSG.2017.2672881
2. Broy, M.: Challenges in automotive software engineering. In: Osterweil, L.J., Rombach, H.D., Soffa, M.L. (eds.) 28th International Conference on Software Engineering (ICSE 2006), Shanghai, China, 20–28 May 2006, pp. 33–42. ACM (2006). https://doi.org/10.1145/1134285.1134292

3. Fröschl, J.: Kybernetisches Energiemanagement elektrischer Energiewandlung in Kraftfahrzeugen. Ph.D. thesis, Technische Universität München (2020)
4. Fröschl, J., Sirch, O.: Bordnetze und E/E-Architektur Eine Einführung in die Zusammenhänge zwischen Elektrik/Elektronik-Architektur und Energiebordnetz im Automobil. Narr Francke Attempto Verlag GmbH + Co. KG (2023). https://doi.org/10.24053/9783816985327
5. Günzel, M., Chen, K., Ueter, N., von der Brüggen, G., Dürr, M., Chen, J.: Compositional timing analysis of asynchronized distributed cause-effect chains. ACM Trans. Embed. Comput. Syst. **22**(4), 63:1–63:34 (2023). https://doi.org/10.1145/3587036
6. Hira, S., Deshpande, P.S.: Mining precise cause and effect rules in large time series data of socio-economic indicators. Springerplus **5**(1), 1625 (2016). https://doi.org/10.1186/s40064-016-3292-0
7. ISO/IEC/IEEE: ISO/IEC/IEEE 24765: 2017(E): ISO/IEC/IEEE International Standard – Systems and Software Engineering – Vocabulary. IEEE Std, IEEE (2017)
8. Krieger, O.: Wahrscheinlichkeitsbasierte Fahrzeugdiagnose mit individueller Prüfstrategie. Ph.D. thesis, Technische Universität Carolo-Wilhelmina zu Braunschweig, September 2011. https://doi.org/10.24355/dbbs.084-201203280921-0
9. Kugele, S., Hettler, D., Peter, J.: Data-centric communication and containerization for future automotive software architectures. In: IEEE International Conference on Software Architecture, Seattle, WA, USA, April 30 - May 4, 2018, pp. 65–74. IEEE Computer Society (2018). https://doi.org/10.1109/ICSA.2018.00016
10. Lan, J., Patton, R.J.: A new strategy for integration of fault estimation within fault-tolerant control. Autom. **69**, 48–59 (2016). https://doi.org/10.1016/J.AUTOMATICA.2016.02.014
11. Min, H., Fang, Y., Wu, X., Lei, X., Chen, S., Teixeira, R., Zhu, B., Zhao, X., Xu, Z.: A fault diagnosis framework for autonomous vehicles with sensor self-diagnosis. Expert Syst. Appl. **224**, 120002 (2023). https://doi.org/10.1016/j.eswa.2023.120002
12. Nguyen, B.L.H., Vu, T.V., Nguyen, T.T., Panwar, M., Hovsapian, R.: Spatial-temporal recurrent graph neural networks for fault diagnostics in power distribution systems. IEEE Access **11**, 46039–46050 (2023). https://doi.org/10.1109/ACCESS.2023.3273292
13. Reif, K.: Automobilelektronik: Eine Einführung für Ingenieure. Springer Vieweg, Wiesbaden, 5 edn. (2014). https://doi.org/10.1007/978-3-658-05048-1
14. Song, Z., Wen, F., Zhang, H., Li, J.: A cooperative perception system robust to localization errors. In: IEEE Intelligent Vehicles Symposium, IV 2023, Anchorage, AK, USA, 4–7 June 2023, pp. 1–6. IEEE (2023). https://doi.org/10.1109/IV55152.2023.10186727
15. Zeller, A.: Isolating cause-effect chains from computer programs. In: Proceedings of the Tenth ACM SIGSOFT Symposium on Foundations of Software Engineering 2002, Charleston, South Carolina, USA, 18–22 November 2002, pp. 1–10. ACM (2002). https://doi.org/10.1145/587051.587053

Architecture-Based Issue Propagation Analysis

Sandro Speth[1]([✉])[iD], Niklas Krieger[1][iD], Robert Heinrich[2][iD],
and Steffen Becker[1][iD]

[1] Institute of Software Engineering, University of Stuttgart, Stuttgart, Germany
{sandro.speth,niklas.krieger,steffen.becker}@iste.uni-stuttgart.de
[2] Institute of Information Security and Dependability, Karlsruhe Institute
of Technology, Karlsruhe, Germany
heinrich@kit.edu

Abstract. Modern software systems usually consist of multiple components, e.g., libraries or (micro-)services, that can be defined through their contract. These components are often developed independently and use their own issue management systems (IMSs). Due to composing these components into a larger architecture, issues such as bugs can propagate along the call chains throughout the entire architecture. However, when experiencing an issue in one component, identifying that such an issue originates, e.g., from a component multiple calls deeper into the graph, is challenging and error-prone as the propagation path has to be identified and the components' IMS investigated for the root-causing issue. Therefore, this paper proposes a model-based issue propagation analysis concept by annotating the Gropius architecture graph with issue propagation rules. Based on the characteristics of the issue, the analysis follows the rules and creates a potential issue propagation graph. Developers can accept resulting issues and semi-automatically create them for the affected components. We conducted experiments on a reference architecture for 21 propagation root-causing issues and compared the results with manually identified propagation paths. The results show that such an analysis can uncover issue propagations well with a recall of 0.99. However, precision can differ among different issue characteristics.

Keywords: Issue Propagation · Software Architecture ·
Architecture-based Propagation Analysis · Cross-Component Issue
Management

1 Introduction

Modern software architectures usually consist of multiple components, e.g., libraries or (micro-)services. These components are often independently developed and defined through contracts. Even though these components might be independently developed and often not restricted to one application, dependencies arise by composing them into a larger architecture. Therefore, issues, e.g.,

M. Galster et al. (Eds.): ECSA 2024, LNCS 14889, pp. 121–137, 2024.
https://doi.org/10.1007/978-3-031-70797-1_8

bugs, can propagate along the call chains throughout the architecture. For example, if a library responsible for the API of a microservice has a bug due to an update, a downstream microservice's API might be affected, further breaking other microservices that call the affected service. Identifying such issue propagations is challenging and error-prone [16].

While the usual fault/failure propagation analysis might uncover such propagations (e.g., [15]), they often require access to the code base or instrumentalization such as tracing. Furthermore, these approaches often do not work on specific component types like libraries and are applied after noticing the effects of propagation. Instead, we require identifying issue propagation paths directly after an issue in a component's issue management system (IMS) is opened. However, as each component might use an independent IMS, downstream components might not directly notice a newly opened issue in their upstream component. Therefore, traditional issue management ends at the boundaries of one component [21], and the impact of an upstream component's issue on the downstream components might stay hidden until the propagated issues are noticed.

Architecture models are often used to predict potential propagations, e.g., for security attack path analysis [22,23], and deem good results. To enhance issue management for component-based systems, Speth et al. [18] developed a concept that manages issues with their architectural dependencies in a combined component model, i.e., annotating an architecture graph with issues and their dependencies. The Gropius metamodel [16] allows the modeling of component-based architectures. Through ontological modeling using templates, the metamodel supports different component types, e.g., libraries or microservices. Furthermore, developers can semantically link issues of different components, and Gropius synchronizes any change in the issues between Gropius and the components' original IMSs. Therefore, Gropius already combines issues with a component model, offering a means to extend it for propagation analysis, as done by Speth et al. [20] for an analysis of the impact of service-level objective violations.

This leads us to our research question: *Can an architecture-based analysis identify potential issue propagations accurately?* Therefore, we propose extending the Gropius metamodel with issue propagation rules that trigger depending on issue characteristics. Developers can model these rules and run analyses when new issues are opened to build potential issue propagation paths. We thereby analyzed three reference architectures on how issues propagate and which issue characteristics are relevant and created a taxonomy for that. Therefore, our contributions are (1) a taxonomy of issue characteristics and how they generally propagate, (2) an extension of the Gropius metamodel for issue propagation rules and issue characteristics, and (3) an analyzer to create potential issue propagation graphs and semi-automatically create arising issues.

We evaluated our concept by conducting experiments on the MiSArch microservice reference architecture[1]. To measure the quality of our results, we manually created a validation set with 21 issues and their propagation path and compared that with the results provided by the analyzer. Depending on the issue

---

[1] https://misarch.github.io/.

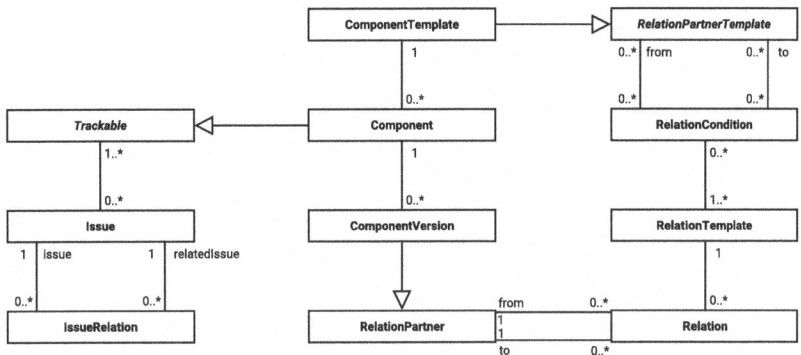

Fig. 1. An excerpt of the ontological Gropius metamodel.

characteristic and rules, we achieved an F_1 score for most issues between 0.5 and 0.7 with average recall of 0.99. The high difference of the F_1 score results of fluctuating precision as for some issue propagations our rules resulted in a large overestimation. This can be resolved by adding more specific rules instead of general ones. In total, the evaluation shows that our concept is applicable.

2 Background

In this section, we provide the required background regarding the Gropius metamodel. We focus only on parts that are necessary for this paper. Due to changing requirements, Speth et al. replaced the original metamodel [17] with an ontological metamodel. While the previous metamodel followed a linguistical approach that offered the means to model components primarily for (micro-)services and represent them via a graphical syntax similar to a UML component diagram, the new one allows developers to model to define custom component types at runtime using templates. Figure 1 depicts an excerpt of the Gropius metamodel, which we describe in the following concerning our requirements. Using a template mechanism, the Gropius metamodel defines component types, interfaces, and relations. It is noteworthy that the semantics of these templates can vary from project to project.

ComponentTemplate, Component, and ComponentVersion: Developers can create component templates for different component types. For example, a project could contain component templates that define the semantics for microservices, libraries, and databases. As issues often affect a specific version, each component can have multiple versions. For issue management, developers often want to distinguish between the general component blueprint, i.e., source code, and the concrete instantiation, which might differ in the config, environment, etc., for a specific project. This enables differentiating between issues that affect all instances and issues of a single instance that might be irrelevant to others. Such behavior can be achieved by creating component templates for a blueprint and instance separately. A typical project we focus on could define a component template for microservices that developers use to model a producer and consumer microservice with one version each.

RelationPartnerTemplate and RelationPartner: In the Gropius metamodel, component versions can have interfaces, and both entities can act as relation partners. This means they can be connected using relations. We can semantically define an interface through its interface specification template, i.e., templates that define the semantics of a component's interface, such as "REST API". Therefore, component templates and interface specification templates are specific relation partner templates. However, for the sake of simplicity, we focus primarily on relations between component versions directly instead of interfaces.

RelationTemplate, RelationCondition, and Relation: To model relations between the different components, the Gropius metamodel enables relation templates that define the semantics of relations at runtime. Each relation template contains a set of relation conditions to provide semantic constraints. A relation condition restricts the relation to possible relation partner templates as the start and end of the relation. For example, a project could contain a "microservice-calls-microservice" relation template or a "microservice-includes-library" relation template. Hence, using these templates, developers can model relations between specific versions of microservices and between a microservice version and a library version. The version restriction makes sense, as relations might change between different versions. Notably, developers could define relation templates that make semantically no sense, e.g., a relation template that defines a "hosted on" relation between a microservice and a library.

Trackable, IssueTemplate and Issue: As Gropius acts as a cross-component issue management system, it can synchronize between Gropius and the components' actual IMS. Therefore, in the metamodel, a component is a specific **Trackable** entity that is connected to an IMS. Even though Gropius projects are a **Trackable** too, we focus only on components for this paper. Furthermore, one issue can affect one or multiple trackables and, thus, multiple components. While this makes sense, especially for feature requests that span over multiple components, we assume for our propagation analysis that instead of directly propagating, the issue often results in a new one affecting the component the original issue propagated. The Gropius metamodel allows developers to state semantic relations between issues of the same or different components, e.g., a *results in* issue relation. An **IssueTemplate** further defines issue types, states, and other metadata, such as templated fields.

Example: Figure 2 depicts a small example webshop architecture consisting of two services and one library and a Gropius model representing this architecture. The model shown in the figure is divided into ontological levels, i.e., the elements are ontologically instantiated from top to bottom, and the lowest level represents the actual architecture. Ontological instance relations are shown in dashed arrows for components and dotted arrows for relations. To model the components of this architecture, we require two component templates, one for services and another

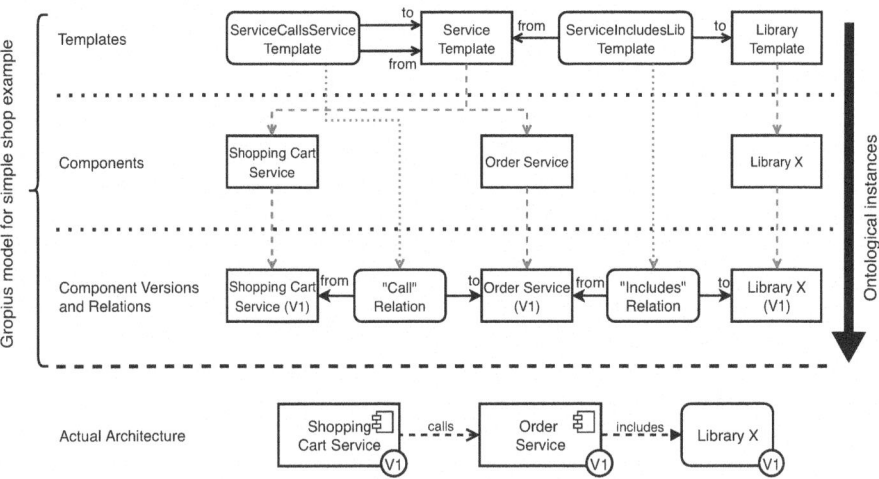

Fig. 2. Example of a small web shop architecture consisting of two services and one library and a Gropius model representing this architecture.

for libraries. Furthermore, on the highest level, we also define two templates for the relations, one for the "call" relation between the services and a "includes" relation for including the library in a service. For the sake of simplicity, Fig. 2 omits the relation conditions and shows the associations directly between the relation templates and component templates. Therefore, we add the service template as *from* and *to* for the "call" relation template, as well as the service template as *from* and the library template as *to* for the "includes" relation template. Using the two component templates, we can create the general components for the *shopping cart service*, the *order service*, and the *library X*. As the actual running architecture uses specific deployed component instances of a specific version, we must model component versions for the three components. In our example, all components have one version, but multiple versions would be possible in a more complex scenario. Using the two relation templates, we can model the "call" and "includes" relation. The "call" relation uses the *shopping cart service* component version as the starting point (*from*) and the *order service* component version as the endpoint (*to*). The "includes" relation acts between the *order service* component version and the *library X* component version. Please note that in this example, we could also create an "includes" relation between the *shopping cart service* component version and the *library X* component version or a "call" relation between a service's component version and itself.

3 Taxonomy

To get a better feeling of how issues generally propagate in component-based architectures, we analyzed the three microservice reference architectures

TeaStore [10], SockShop[2], and T2-Project [19]. We chose the TeaStore as it provides a simple starting point with a small architecture with only synchronous HTTP calls and is implemented in a single programming language. SockShop is more diverse in terms of architecture as different services are implemented in various programming languages. Still, the communication is primarily based on synchronous HTTP calls while using message queues for payment communication. The T2-Project is implemented in Java only but provides asynchronous communication via messaging queues and implements the Saga pattern for data consistency across multiple microservices. The Saga pattern is relevant to us as it might hide issues from the user due to compensation rollbacks. However, a model-based analysis still uncovers such issues.

Based on our analysis results, we created a taxonomy of typical issue characteristics relevant to issue propagation and how they propagate or whether a propagation does not make sense. Please note that we consider issue propagation from an issue management perspective, i.e., feature requests, bug reports, change notifications, etc., instead of the actual faults on an implementation level. For some issues, it is often not intended to report the issues resulting from propagation, even though it might propagate on the actual system. This might happen not accidentally providing confidential information to other issue management systems but also if an issue does not benefit the overall system or even spam the affected component's developers.

Typical Templates: By analyzing the reference architectures, we derived typical component and relation templates. When modeling the systems in Gropius, we would require (micro-)service, middleware, frontend, database, library, and infrastructure templates. As relation templates, we need a "call" relation template for calls between the frontend and services, services and services, services and middleware, as well as middleware and services, an "includes" relation template for services or frontend and libraries, an "hosted on" relation between services, frontend, or database and infrastructure, and an "connection" relation template to connect services, frontend or middleware to a database.

Taxonomy: We identified five relevant categories for an issue propagation: (1) API contract change or violation, (2) performance issues, (3) database issues, (4) middleware issues, and (5) infrastructure issues. We derived typical issue characteristics for each category and how these issues usually propagate. **API contract change/violation:** The first category affects components of library or service component templates and typically propagates from them to components of service or frontend component templates. For a *new endpoint or method*, a feature request propagates transitively from the component to all downstream components, while a bug should not have any propagation. The issue propagates transitively to all downstream components if there is a *removed endpoint or method*. For *DTO schema changes* in an endpoint, the issue transitively propagates to all downstream components. For an *API endpoint post-condition violation*, the bug propagates transitively to all downstream components. In the

---

[2] https://github.com/microservices-demo.

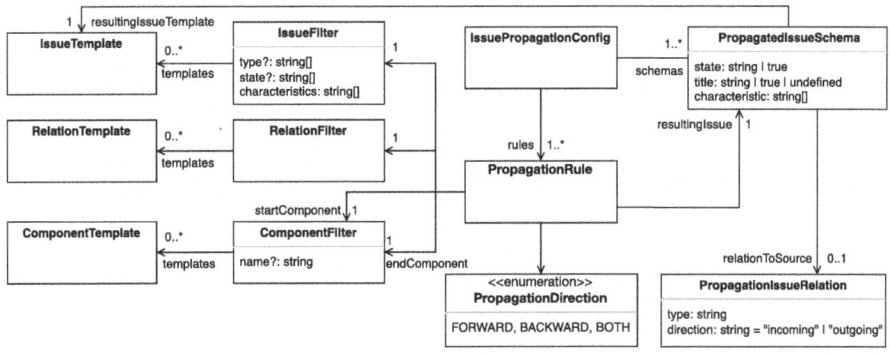

Fig. 3. Metamodel of the propagation rules and config to annotate Gropius models.

case of a *new version release*, all direct downstream components must adapt to the new version, and thus, the issue propagates one relation to each downstream component. If a component has an *availability change*, e.g., due to a change of license or payment model, the issue propagates to the direct downstream components.

Performance Issues: This category affects components of a service, database, or middleware component template. If there is a *service performance* issue, e.g., due to regular service-level objective violations, the bug transitively propagates against the call chains towards all downstream components. For *database performance* issues due to inefficient data schema or queries, the issue propagates to the connected downstream component. If the performance issue is due to a bug or inefficient implementation within the database, there should not be an issue propagation. Similarly, for *middleware performance* issues, they only propagate to calling downstream components if the issue is rooted in the configuration. Middleware implementation issues leading to performance problems should not propagate.

Database, Middleware, and Infrastructure Issues: These categories affect components of a database, middleware, or infrastructure component template. Implementation bugs should not propagate to connected, hosted, or calling components. If there is a *infrastructure crash*, e.g., due to lacking resources, the issue propagates to hosted service or frontend components.

4   Issue Propagation Rules and Characteristics

In this section, we elaborate on our concept of issue propagation rules and issue characteristics w.r.t. a Gropius model. We extend the Gropius metamodel to enable issue propagation rules and issue characteristics. Figure 3 depicts the metamodel extension and relevant classes significant to this extension.

PropagatedIssueSchema: An issue schema defines a placeholder example for resulting issues of one propagation step. It specifies the issue's type, state, issue template, and a placeholder for the title. The issue schema finally optionally defines the relation to the source issue of the propagation. Typical relations are "depends on", or "results in". A resulting issue can then be the source for the following issue propagation step. For example, we can define an issue schema for issues resulting from an issue propagation due to an accidental change of an API endpoint. Therefore, the schema would define "bug" as the issue type and "open" as the issue state. The schema would contain the resulting issue's type and state, as well as an example title describing the cause of the issue, i.e., the API endpoint bug. Furthermore, the schema defines a list of issue characteristics. For example, an issue characteristic could be "API endpoint bug". These characteristics are relevant later to selecting the correct propagation rule for further propagation. Based on the schema's definition, the issue characteristics may change between the source and the resulting issue.

PropagationRule and Target Filters: Each propagation rule defines a propagation direction, three filters for the propagation start, and one filter for the propagation end. We use these filters to identify the component affected by the rule. The rule further references the *propagated issue schema*, which defines the schema for the resulting issue. A propagation direction that can be *forward* (propagating along the relation direction from downstream to upstream component), *backward* (propagating against the relation direction from upstream to downstream component), or in *both* directions. The three filters for the propagation start restrict the issues for which the rule triggers (`IssueFilter`), the relation the propagation applies for (`RelationFilter`), and a filter for the starting component (`ComponentFilter`). The issue filter specifies a list of *characteristics*, i.e., a list of issue characteristics for which the propagation rule triggers, and optionally, a list of issue templates for which the propagation applies and a list of allowed issue types and states. The rule-triggering issue must match all properties specified by the issue filter. However, the issue characteristic filter itself is *or* combined, meaning that the rule triggers for any of the characteristics defined in the rule matching any of a causing issue's characteristics if the other properties match as well. The possible starting component of an issue propagation is restricted by a `component filter`. The component filter has an optional name to specify via a regex component names that are considered by the propagation and a list of component templates to constrain the possible component types. Propagations restrict the relations they follow by defining the type of relation this propagation rule applies with an *relation filter*, i.e., specifying a list of relation templates. Furthermore, the propagation rule can restrict the target component of the propagation by another component filter.

Example: As an example, we can model a propagation rule that should trigger for an "API bug" issue characteristic. Therefore, it has "API bug" as a characteristic filter. The rule further defines a *backward* propagation and restricts the propagation to all components with a component template for services as the starting point. To limit the propagation only to API calls, we could further define a relation filter for a "Service calls Service" relation and a "Frontend calls

```
1   function propagateIssues(initial, propagationRules):
2       issuesToPropagate = [initial issue]
3       issues = [initial issue]
4
5       function propagateIssue(issue, rule, relation):
6           if rule matches issue, relation, and start and end component based on the issue
                  and relation:
7               propagatedIssue = create new issue from the propagation
8               if issues contains an issue equal to propagatedIssue:
9                   update the existing issue with the new propagation
10              else:
11                  issues.append(propagatedIssue)
12                  issuesToPropagate.append(propagatedIssue)
13
14      while issuesToPropagate is not empty:
15          issue = issuesToPropagate.pop()
16          for rule in propagationRules:
17              for relation in issue.components.relations:
18                  propagateIssue(issue, rule, relation)
```

Listing 1.1. The pseudocode for the issue propagation algorithm.

Service" relation. Otherwise, the rule could also apply to relations that do not call the API but target the rule's starting component.

5 Model Annotation and Analysis

In this section, we discuss how we use the propagation rules and characteristics to analyze issue propagations based on a Gropius model. First, we describe how the annotation works, followed by the analysis concept. Finally, we provide an example to enhance understanding.

Annotating a Gropius Model: To annotate a Gropius model, we first define an issue propagation configuration for our analysis. The configuration consists of a record of propagated issue schemas and a list of issue propagation rules.

Analysis: The configuration is used in the analyzing step to compute potential issue propagation graphs. When a new issue is opened in one of the components modeled in Gropius, a developer gets notified to start the propagation analysis. He then adds a list of issue characteristics to the root causing issue based on the issue's content. Please note that the issue characteristics must match the ones defined in the propagation rules during the annotation step. The analyzing algorithm is described in Listing 1.1. The *analyzer* uses two lists, one for the resulting issues (`issues`) and one for storing issues for analyzing further propagation (`issuesToPropagate`). Both lists are initialized with the triggering issue. Afterwards, the *analyzer* starts the iterative process to create the propagation graph. While the `issuesToPropagate` list is not empty, it pops the first issue and calls for each propagation rule and component relation combination a function to calculate the propagation. If the rule matches the issue's characteristics, its affected component as the start component, the relation, and the relation's end component, it applies the rule. The algorithm creates a new propagated issue based on the propagation rule's issue schema. If the propagated issue is equal to an issue that already exists in the `issues` list, the algorithm takes the already existing issue and updates the issue with the new propagation. We consider the affected component, issue template, type, and state for equality. This might happen if multiple paths lead to the same propagated issue. Otherwise, the *analyzer* creates a new issue resulting in the propagation based on the schema and appends it to both lists for further processing. Note that at this point, the issues are not actually created in the issue management systems. The final result of the analyzing process is an issue propagation graph with the issues and their affected components as nodes, shown in the UI. Furthermore, as this is an iterative process, a developer can select/deselect relations and characteristics for the initial issues and see how this affects the propagation. Finally, the developer can reject paths that are removed from the graph and decide for each propagated issue whether Gropius should create it in the affected components' issue management systems. Thereby, the developer can override the title and must add the issue's body, etc., or take the data from the issue directly causing the propagated issue. By overriding or changing data, the algorithm takes the new issue and re-calculates the resulting issue propagations. This might lead to new propagation paths in the graph or other paths being removed.

Complete Example: Consider the example Gropius model of Fig. 2. We define two propagation rules and two propagated issue schemas. The first rule takes a component defined by a library component template as a starting point and triggers for a "REST API SDK Bug" characteristic. It applies to an "includes" relation and accepts any component as the endpoint. The propagated issue schema defines the type "bug", state "open", and characteristic "REST API Bug". The second rule takes a component defined by a service component template as a starting point and triggers for a "REST API Bug" characteristic. It applies to a "calls" relation and accepts any service component as the endpoint. The propagated issue schema defines the type "bug", state "open", and characteristic "REST API Bug". When an analysis is triggered for an issue in the library X with the characteristic "REST API SDK Bug", the analyzer applies the first rule and then twice the second rule to compute a propagation from the library until the "shopping cart service" component.

6   Evaluation

This section describes our evaluation. First, we describe how we collected the validation set. Next, we elaborate on our experiment setup. Finally, we present our results and discuss our limitations and threats to validity. We used the MiSArch reference architecture(see footnote 1) as the case for our experiments. MiSArch is a reference architecture consisting of multiple loosely coupled microservices that communicate via messaging and GraphQL. The services are written in different programming languages and thus include different dependencies. We selected MiSArch as a case for our evaluation as the architecture is actively maintained (at the time of this paper) with extensive documentation, which makes it easier for us to collect the data and conduct the experiment.

6.1   Data Collection

We require a validation set for issue propagations in MiSArch to conduct our evaluation. This validation set must contain initial issues triggering propagations and their actual propagation paths. Furthermore, we must define the relevant issue characteristics for each initial issue. Therefore, we asked the developers of the MiSArch project to provide us with a set of issues and how these issues propagate through the architecture. Please note that the collected set consists of actual issues the project faced and issues that might theoretically arise if they had not been covered by tests before. We require these artificial issues to gain a validation set that is large enough for our evaluation. The developers provided us in total with 21 different issues spanning across the entire architecture. These issues differ in their content, start components and propagation paths. As these issues were documented in an unstructured form, we manually created a JSON array based on them for automated processing. Each entry consists of the issue's title, type, state, initial component, and propagation as a list. We further added the initial characteristics based on the issues' titles. Our validation set is publicly and open source available in our replication package on Zenodo[3].

6.2   Experiment Setup

To conduct the experiment, we must model the MiSArch architecture in Gropius and define our propagation configuration. Therefore, based on the documentation, we defined component templates for libraries, databases, microservices, frontend, and infrastructure components. We further modeled respective relation templates and an issue template defining typical issue states, issue relations, and issue types. Using these templates and the MiSArch documentation, we modeled the architecture in Gropius. We further added library components to the model, which are described in the validation set's issues.

Independently of the validation set, we modeled the issue propagation rules and propagated issue schemas based on our taxonomy, understanding of the general architecture of MiSArch, and the templates we modeled in Gropius. In total, we modeled a propagation configuration with 18 propagation rules and their respective propagated issue schemas. We then ensured that the issue characteristics in the validation set and the issue filter of our propagation rules matched. The configuration is publicly and open source available on Zenodo(see footnote 3).

We conduct our experiments in an automated manner to ensure faster reproducibility. We trigger our analysis for each issue in the validation set and derive a potential propagation graph based on our analysis. We then compare each resulting propagation graph against the triggering issue's propagation based on the validation set and calculate *precision*, *recall*, and F_1 score.

6.3   Results

In total, the results of our analysis show a high recall with an overall average of 0.99. All issue propagations except for one achieve a recall of 1, but one issue

---

[3] https://zenodo.org/records/11069093.

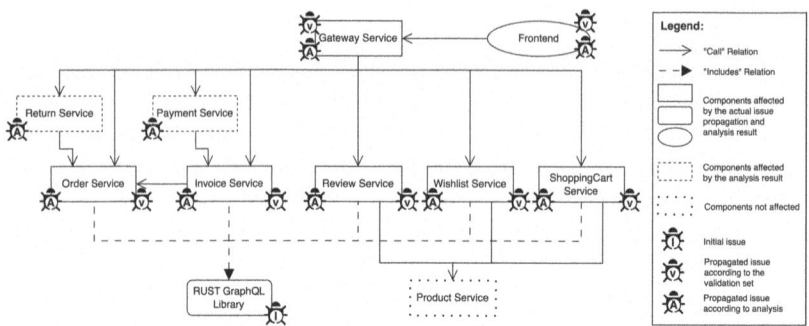

Fig. 4. Excerpt of the MiSArch architecture for one issue, its actual propagation, and the analysis result.

("Notifications should be sent for Wishlist items...") resulted in a recall of 0 (see Sect. 6.4 for an explanation). Regarding the F_1 score, our results vary between 0.22 and 1, except for the same issue having a recall of 0. This particular issue propagation also resulted in zero true positives. Our rules seem to cover the exact propagation on three occasions, leading to an F_1 score of 1 and 0.93 in one instance. Additionally, the analysis yielded an F_1 score between 0.5 and 0.8 for eight issue propagations. The resulting F_1 score was below 0.5 for eight issues. Low F_1 scores always resulted due to many false positives.

6.4   Discussion and Threats to Validity

Discussion: Considering the overall results and our research question, our concept enables an architecture-based analysis to identify potential issue propagations accurately. Nevertheless, our results highly depend on the modeled propagation rules and issue categories. The results show that our concept achieves an average recall of 0.99 with our validation data and rules. This is typical for such model-based impact analyses as these algorithms usually overestimate the results, such as issue propagations. This is often intended in order to identify more false positive propagations rather than overlooking potentially significant issue propagations. Nevertheless, recall values might drop if the modeled propagation rules are too specific or incomplete. It is, therefore, recommended that the rules be updated in case of a smaller recall, as this is a hint of missing propagation rules. However, one issue ("Notifications should be sent for Wishlist items...") yielded 0 true positives, resulting in a precision and recall of 0. The reason is that this issue represents a feature request for a new event between two services that have no direct or transitive relation in the current Gropius model. Therefore, our analysis cannot identify a propagation to the correct component.

While we achieved a high average recall for all issue propagations, our results differ greatly regarding the precision and, therefore, w.r.t. F_1 score. Worth mentioning is, for example, the issue "The invoice service does not send out its event for invoice creations /invoice/invoice/created" for the *Invoice service* as it achieved a very low F_1 score (0.23). According to the validation set, the issue only propagates to the *Payment service* and *Shipment Service* components, but our analysis results in 13 false positives. Here, our analyzer cannot distinguish semantically which "call" relations it should follow backward as the rules do not differentiate. Hence, our resulting issue propagation graph contains all transitive downstream components of the *Invoice service* instead of two specific ones. This could be fixed by introducing more specific rules with deeper domain focus. We identified the same reason for other results with an F_1 score lower than 0.5. In contrast, for example, the "Pods regularly crash in current K8 setup" resulted in an F_1 score of 1 as the issue propagation does not depend on domain knowledge (all components hosted on the affected infrastructure component are affected instead of a subset as for the issue described above). Nevertheless, our analysis yielded an F_1 score between 0.5 and 0.7 for many issues that we deem usable.

As mentioned, accuracy might differ between more general or more specific rules. On the one hand, more general rules increase recall but often decrease precision, and the resulting graphs are less helpful in pinpointing actual affected components. On the other hand, more specific rules might result in smaller recall and more precise and helpful graphs. However, if the rules are too strict, propagation paths might be missed. In our evaluation, we tried to model rules at an abstraction level that would be typical for such a system to gain valuable insights based on the propagation graph. Figure 4 shows an example for one issue ("Bug in GraphQL library for Rust introduces breaking changes"). The initial issue is for a "Rust GraphQL Library". According to our validation set, the issue propagates to the components, including this library, and transitively to the "API Gateway service" and "Frontend". However, our analysis resulted in an additional propagation to the "Return service" and "Payment service", as we modeled rules to follow "Call Relations" with "backward" direction. If we had defined the propagation rule with the "both" direction, the analysis would also have added additional propagations, e.g., to the "Product service".

Threats to Validity: There are three significant threats to the validity of our concept and evaluation. First, we analyzed only e-commerce microservice systems to obtain our taxonomy and conduct the evaluation. Therefore, results might not apply to other domains or architecture styles. Nevertheless, all issue characteristics of our taxonomy are independent of a domain, and most of them are applicable to other architectural styles as well. Second, the issue propagations in our validation set are based on the MiSArch developers' understanding of how these issues would propagate through the architecture. It could happen that they included propagation to components that would not actually happen in reality. Lastly, our evaluation uses only one case with 21 issues in the validation set. Especially the results based on a relatively small number of issues might not be generalizable. Nevertheless, the example issues cover typical issue propagations.

7 Related Work

Approaches to *architecture-based project planning* use the software architecture as an artefact in project planning. For example, the approach Architecture-Centered Software Project Planning (ACSPP) [12] estimates cost and schedule in a project using architectural knowledge. Carbon [4] proposes an approach to align software architecture and project planning. However, to the best of our knowledge no approach in this category supports change propagation analysis.

Examples of approaches to *architecture-based software evolution* are those by Garlan [5] and Naab [11]. The approach by Garlan [5] addresses the evolution of a given architecture and the reasoning about the quality of evolution paths. The approach by Naab [11] analyses the flexibility of an architecture for modification. However, these approaches do not support change propagation analysis.

Approaches to *scenario-based software architecture analysis* investigate the impact of a change based on change scenarios to a given architectural model. The Software Architecture Analysis Method (SAAM) [8] analyses the maintainability of the software architecture by investigating components and their connections as well as data flows. The Architecture Tradeoff Analysis Method (ATAM) [9] as a successor of SAAM allows making trade-offs between various quality properties like maintainability and performance. The approach Architecture-Level Prediction of Software Maintenance (ALPSM) [1] estimates the impact of a change based on the size of software components using weighted change scenarios. Similarly, Architecture-Level Modifiability Analysis (ALMA) [2] estimates the impact of change scenarios on a software architectural model. Change impact analysis in these approaches is limited to artefacts of the system structure and associated activities. Besides these structural artefacts, however, there are various technical and organisational artefacts, such as test cases and staff specifications, that may be affected by a change and thus cause activities, such as updating test cases and assigning tasks to responsible staff [13]. Consequently, they need to be considered in change impact analysis.

The Karlsruhe Architectural Maintainability Prediction (KAMP) approach [13] goes beyond aforementioned approaches by using comprehensive architectural models that allows to include technical and organisational artefacts, besides structural artefacts, in estimation. KAMP has been applied in various domains including business processes [14], PLC software [3] and automated production systems [6]. A general methodology for domain-spanning change impact analysis has been proposed in [7].

Building upon KAMP, Walter et al. [22] developed a metamodel and analysis for the architecture-based attack propagation to investigate attack paths. The approach has been extended in [23] for investigating multiple attack paths to a target element and support for filtering options. This enables architects to identify potential security risks to critical components. In contrast to related work the approach by Walter et al. considers fine-grained access control policies and vulnerabilities based on the software architecture for attack paths propagation. The derived attack paths can help software architects to harden the system.

There is also one previous work by us [20] which uses the previous Gropius metamodel [17] and extends it with BPMN, a modeling language for sagas, and trace links to model and analyze the impact of service-level objective violations for the architecture, saga, and business process.

8   Conclusion and Future Work

This paper presents a concept for model-based predicting potential issue propagations. We extend the Gropius metamodel to enable the annotation of the component model with propagation rules. Based on the annotated model and issue characteristics, e.g., a violation of an API contract, our analyzer creates a graph for potential issue propagations, which developers can manually check to reject paths or automatically create issues uncovered by the analysis. By inspecting multiple reference architectures, we identified how issues usually propagate and derived a taxonomy that builds the foundation of our concept. We evaluated our concept by conducting experiments on the MiSArch reference architecture, where we compared the results of our analyzer with manually obtained issue propagations. The results show that our concept can uncover such issue propagation, thus supporting developers in quickly identifying whether their components are affected by issues propagating from upstream components, e.g., libraries.

In future work, we plan to automatically map newly created issues to their characteristics and rules using natural-language processing, extending the taxonomy by analyzing further applications and conducting a more extensive study. Furthermore, we plan to compare the results of our analyzer with the results of traditional failure propagation analysis approaches.

Acknowledgements. This research has been partially funded by the German Federal Ministry of Education and Research (BMBF) through grant 01IS23072 Software Campus 3.0 (University of Stuttgart) as part of the Software Campus project "Gropius". This work was also supported by funding from the topic Engineering Secure Systems of the Helmholtz Association (HGF) and by KASTEL Security Research Labs.

Data Availability Statement. Our implementation and replication package with the results are publicly and open source available on GitHub (https://github.com/ ccims/propagation-analysis-frontend) (implementation) and Zenodo (https://zenodo. org/records/11069093) (complete replication package).

References

1. Bengtsson, P., Bosch, J.: Architecture level prediction of software maintenance. In: Third European Conference on Software Maintenance and Reengineering, pp. 139–147. IEEE (1999)
2. Bengtsson, P., et al.: Architecture-level modifiability analysis (alma). J. Syst. Softw. **69**(1–2), 129–147 (2004)

3. Busch, K., et al.: A model-based approach to calculate maintainability task lists of plc programs for factory automation. In: 44th Annual Conference of the IEEE Industrial Electronics Society, pp. 2949–2954 (2018)
4. Carbon, R.: Architecture-centric software producibility analysis. Ph.D. thesis, Fraunhofer IESE, Kaiserslautern; University of Kaiserslautern (2012)
5. Garlan, D., et al.: Evolution styles: Foundations and tool support for software architecture evolution. In: 2009 Joint Working IEEE/IFIP Conference on Software Architecture & European Conference on Software Architecture, pp. 131–140. IEEE (2009)
6. Heinrich, R., et al.: Architecture-based change impact analysis in cross-disciplinary automated production systems. J. Syst. Softw. **146**, 167–185 (2018)
7. Heinrich, R., et al.: A methodology for domain-spanning change impact analysis. In: 2018 44th Euromicro Conference on Software Engineering and Advanced Applications (SEAA), pp. 326–330 (2018)
8. Kazman, R., et al.: SAAM: a method for analyzing the properties of software architectures. In: 16th International Conference on Software Engineering, pp. 81–90. IEEE (1994)
9. Kazman, R., et al.: The architecture tradeoff analysis method. In: 4th International Conference on Engineering of Complex Computer Systems, pp. 68–78. IEEE (1998)
10. von Kistowski, J., et al.: TeaStore: a micro-service reference application for benchmarking, modeling and resource management research. In: IEEE 26th MASCOTS, pp. 223–236 (2018)
11. Naab, M.: Enhancing architecture design methods for improved flexibility in long-living information systems. Ph.D. thesis, Fraunhofer IESE, Kaiserslautern (2012)
12. Paulish, D.J., Bass, L.: Architecture-Centric Software Project Management: A Practical Guide. Addison-Wesley, Boston (2001)
13. Rostami, K., et al.: Architecture-based assessment and planning of change requests. In: 11th International ACM SIGSOFT Conference on Quality of Software Architectures, pp. 21–30. ACM (2015)
14. Rostami, K., et al.: Architecture-based change impact analysis in information systems and business processes. In: 2017 IEEE International Conference on Software Architecture (ICSA), pp. 179–188 (2017)
15. Soldani, J., Montesano, G., Brogi, A.: What went wrong? Explaining cascading failures in microservice-based applications. In: Barzen, J. (ed.) SummerSOC 2021. CCIS, vol. 1429, pp. 133–153. Springer, Cham (2021). https://doi.org/10.1007/978-3-030-87568-8_9
16. Speth, S.: Architecture-based cross-component issue management and propagation analysis. In: 2024 IEEE/ACM 46th International Conference on Software Engineering: Companion Proceedings (ICSE-Companion 2024), April 2024
17. Speth, S., Becker, S., Breitenbücher, U.: Cross-component issue metamodel and modelling language. In: Proceedings of the 11th International Conference on Cloud Computing and Services Science (CLOSER 2021), pp. 304–311. INSTICC, SciTePress, November 2021
18. Speth, S., Breitenbücher, U., Becker, S.: Gropius — a tool for managing cross-component issues. In: Muccini, H., et al. (eds.) ECSA 2020. CCIS, vol. 1269, pp. 82–94. Springer, Cham (2020). https://doi.org/10.1007/978-3-030-59155-7_7
19. Speth, S., et al.: A saga pattern microservice reference architecture for an elastic SLO violation analysis. In: 2022 IEEE 19th International Conference on Software Architecture Companion (ICSA-C), pp. 116–119 (2022)

20. Speth, S., et al.: Dromi: a tool for automatically reporting the impacts of sagas implemented in microservice architectures on the business processes. In: Sales, T.P., Proper, H.A., Guizzardi, G., Montali, M., Maggi, F.M., Fonseca, C.M. (eds.) Enterprise Design, Operations, and Computing. EDOC 2022 Workshops, EDOC 2022, LNBIP, vol. 466, pp. 326–331. Springer, Cham (2023). https://doi.org/10.1007/978-3-031-26886-1_20
21. Speth, S., Breitenbücher, U., Krieger, N., Wippermann, P., Becker, S.: Integrating issue management systems of independently developed software components. In: Stettina, C.J., Garbajosa, J., Kruchten, P. (eds.) Agile Processes in Software Engineering and Extreme Programming, XP 2023, LNBIP, vol. 475, pp. 3–19. Springer, Cham (2023). https://doi.org/10.1007/978-3-031-33976-9_1
22. Walter, M., et al.: Architectural attack propagation analysis for identifying confidentiality issues. In: 19th International Conference on Software Architecture, pp. 1–12. IEEE (2022)
23. Walter, M., Heinrich, R., Reussner, R.: Architecture-based attack path analysis for identifying potential security incidents. In: Tekinerdogan, B., Trubiani, C., Tibermacine, C., Scandurra, P., Cuesta, C.E. (eds.) Software Architecture, ECSA 2023, LNCS, vol. 14212, pp. 37–53. Springer, Cham (2023). https://doi.org/10.1007/978-3-031-42592-9_3

MDEPT: Microservices Design Evaluator and Performance Tester

Raghad Matar[1]([✉])[iD] and Jasmin Jahic[2]

[1] Fraunhofer Institute for Experimental Software Engineering, Kaiserslautern, Germany
raghad.matar@iese.fraunhofer.de
[2] University of Cambridge, Cambridge, UK
jj542@cam.ac.uk

Abstract. In microservices-based systems, architects find it hard to reason about the impact of their design decisions on performance before implementing them. While definitions of anti-patterns help to avoid inadequate design decisions, they are context-dependent. Static analysis of software design can identify constructs that conform to anti-patterns. However, this is not suitable for quantifying the extent to which these anti-patterns would affect system performance. Ideally, we should be able to predict the dynamic behavior of a system before it is implemented. However, existing approaches either cannot achieve this because they analyze the design statically or require complex and laborious modeling and simulation approaches. To address this challenge, we previously introduced a conceptual solution idea that facilitates rapid evaluation of high-level architectural models by combining both static and dynamic analysis. In this paper, we build upon our previous work and introduce the *Microservices Design Evaluator and Performance Tester (MDEPT)* approach. Mainly, we formalize modeling specifications for microservices systems, introduce a fully functional toolchain for our approach, and present the evaluation results.

Keywords: microservices · design decisions · anti-patterns · performance

1 Introduction

For microservices-based systems, there exist a number of design patterns and anti-patterns [5, 6, 16, 17, 19]. These are generally characterized by design outline rules and threshold parameters. A generic design outline often does not provide enough information to conclude whether or not a certain design constitutes an anti-pattern. To draw this conclusion, architects need to quantify thresholds at the design level that constitute an anti-pattern for a specific design outline [22]. For example, the design outline for the bottleneck service anti-pattern is defined as "a service used by too many other systems or external customers". While it is easy to spot potential bottleneck services in a design (those with more input/output requests than others), the challenge is how to define a threshold - what does "too many" mean in a concrete design? This challenge is well known in the design exploration domain, and it has been shown in the past that the perception of an anti-pattern varies from team to team and from system to system [4]. Defining anti-patterns in a concrete system using design outlines and their threshold parameters helps to identify anti-patterns automatically. However, architects still need to predict the scale of the negative impact that such anti-patterns have on their system.

Although there exist approaches that can automatically assess the conformance of microservices systems with known patterns and best practices and detect violations of good practices, including anti-patterns, they depend on the existence of source code [20]. Some of these approaches perform static analysis on the source code [17], while others perform static analysis on the design models abstracted from the source code [15]. Dynamic approaches [7] observe the behavior of microservices systems in production and trace performance observations to the design in order to detect potential anti-patterns. Finally, some approaches rely on simulation of architectural behavior [10]. Although they are able to answer many questions about the system upfront, they often tend to be effort-intensive and have a steep application curve.

The research question we are concerned with is: **How to assist architects in quickly determining whether design decisions that conform to anti-pattern definitions have a significant negative effect on microservices performance?** To answer this question, we build upon our previously introduced solution idea for evaluating the potential impact of anti-patterns on microservices performance [13] and introduce the *Microservices Design Evaluator and Performance Tester MDEPT* approach. Our approach first uses static analysis to detect potential anti-patterns in an architectural design and guides architects in creating alternative design variants (without the potential anti-patterns). Then it generates boilerplate source code for each of the variants based on the architectural design and executes the source code in several test scenarios. Architects compare the performance of the different variants and draw conclusions about the influence of potential anti-patterns on system performance. *MDEPT* is a lightweight approach positioned in the pre-implementation design space exploration phase and is aimed at enabling relative comparison of the performance of early architectural design ideas.

The remainder of this paper is structured as follows: In Sect. 2, we provide an overview of related work. In Sect. 3, we describe the concepts behind the performance evaluation of microservices using *MDEPT*, while in Sect. 4, we provide details about our prototype implementation of the approach. Section 5 presents details of the evaluation of *MDEPT*. We conclude with Sect. 6.

2 Related Work

Bogner et al. [5] conducted a systematic literature review to collect Service-Oriented Architecture (SOA) and microservices anti-patterns [1]. To identify anti-patterns specifically related to microservices systems, Taibi et al. [19] conducted interviews with 27 experienced practitioners from industry. In their paper, the authors extended upon the anti-patterns catalog in their previous work [18] and proposed a taxonomy for classifying microservices anti-patterns. Existing approaches do not provide insight into the correlation between patterns and anti-patterns on the one side and system performance on the other. Even though in [18] and [19], the authors reported on the harmfulness of microservices anti-patterns, they based their assessment on the experience of the interviewed practitioners. Furthermore, they did not specify whether the harmfulness assessment was based on the anti-patterns' impact on performance, or whether it was based on other aspects. To the best of our knowledge, no research has established a quantified connection between anti-patterns and performance.

Some studies focused on testing the performance of microservices and guiding architects through the design space exploration. Goncalves et al. [9] studied the impact on performance when migrating from monolithic architectures to microservices architectures. Avritzer et al. [3] introduced an approach for assessing the performance of microservices systems under different deployment alternatives. Palladio [10], an approach that is closest to our own, uses simulation and virtual prototyping to predict system performance. In particular, the work presented by [11], which relies on Palladio, focuses on achieving similar goals as our work. The authors present a heavy framework comprised of a combination of several tools, including instrumentation and tracing. They do not, however, provide any specification regarding the level of abstraction of the architectural model. In contrast to approaches focusing heavily on detailed and complex modeling of states in the architectural model [21], our approach, which we built on top of our previously introduced conceptual solution idea [13], is lightweight and targets the relative comparison of the performance of early architectural ideas.

3   MDEPT: Conceptual Solution

The following steps explain the *MDEPT* approach for assisting architects in evaluating their microservices system designs (see Fig. 1):

The architect starts by sketching a *Microservices Architectural Model* of the system under test (SUT) (step 1). *MDEPT*'s *Architectural Model Evaluator* component performs static analysis on this model to identify *potential* anti-patterns known to hinder system performance and generates an *Evaluation Report* that lists the detected potential anti-patterns along with suggestions for mitigating them (steps 2 and 3). Based on the *Evaluation Report*, the architect creates an alternative design variant (step 4). The architect can then repeat steps 1–4 to create several variants to compare. *MDEPT*'s *Code Generator* component then generates the source code for each variant (steps 5 and 6). The architect/developer can manually modify the code, should they wish to do

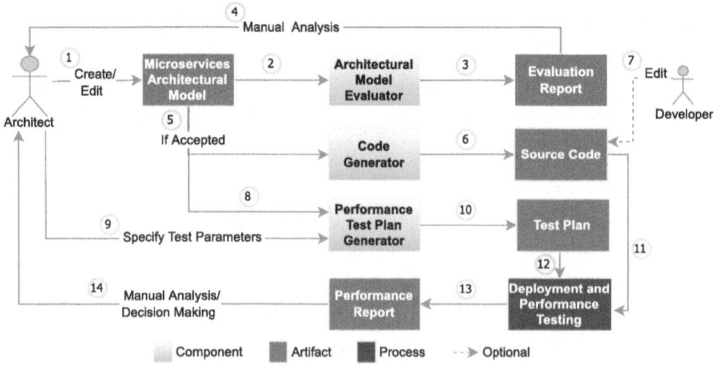

Fig. 1. MDEPT pipeline overview

so (step 7). The architect then provides the values of the parameters of the test plan(s) to the *Performance Test Plan Generator* component (step 9), which generates a test plan (step 10) for each *Microservices Architectural Model* using the provided parameters (input from steps 8 and 9). The architect then deploys the code of each design variant, imports the corresponding test plans into a performance testing tool, and triggers the execution of the tests (steps 11 and 12). Finally (step 14), the architect analyzes the performance measurements (from step 13) of each variant and compares the results. Analyzing and comparing the performance measurements of the different variants helps architects understand the potential impact of the detected anti-patterns on the performance of the system and supports them in making informed decisions about which variant to choose to meet the requirements with an increased level of confidence.

3.1 Formalization of MDEPT's Microservices Architectural Model

Our *MDEPT* approach currently models synchronous communication between microservices, since this is one of the most performance-relevant factors in microservices. Formally, we represent a microservices system as a weighted directed graph $G(N, E, W)$, such that:

- $N = \{MS_1, ..., MS_n\}$; $n \in \mathbb{N}$ *and* $n > 1$.
  Each microservice MS has a set of functionalities, such that: $MS = \{f_1, ..., f_m\}$; $m \in \mathbb{N}$ *and* $m > 0$. Thus, in *MDEPT*, a functionality represents a node in the graph, while a microservice represents a group of nodes.
- $E = \{c_1, ..., c_k\}$; $k \in \mathbb{N}$ *and* $k \geq 0$.
  A call c represents an inter-microservices *RESTful HTTP* direct call between two functionalities, so that: $c = \{(f_i, f_j)\}$; $f_i \in MS_x$, $f_j \in MS_y$ *and* $x \neq y$.
  Thus, in *MDEPT*, an inter-microservices call represents an edge in the graph.
- $W : E \to \mathbb{N}^+$.
  W is a weighting function for edges (i.e., calls). The weight of the edge represents the number of calls between two functionalities to process a request.

Based on this formal description, when using *MDEPT*, architects need to provide the following information about the microservices of a system:

- Microservice name: Identifier of the microservice.
- Functionality name: Identifier of the functionality.
- Functionality complexity/processing time: Architects can choose between specifying the estimated complexity of a function (using the "big O notation") or using an estimated processing time value (in milliseconds).
- Called by client: An indicator of whether the functionality would be called directly by the client or not.
- Inter-microservices call: Source functionality (caller); destination functionality (callee); number of calls.
- Functionality call method: *HTTP methods* (e.g., *GET, POST*).

From our point of view, this level of granularity for modeling microservices systems would be sufficient for capturing system properties needed to facilitate the static analysis of a system, the detection of design anti-patterns related to performance, and the generation of source code and performance testing plans.

3.2 Static Analysis: Potential Anti-Pattern Detection

The *Architectural Model Evaluator* component in *MDEPT* (Fig. 1) performs static analysis to draw the architect's attention to potentially problematic design decisions of the analyzed microservices system considering its performance. Furthermore, it provides a set of recommendations on how to adjust a design to avoid anti-patterns. This helps architects reason about alternative design variants. The results of the static analysis can also help architects design suitable performance test plans by focusing on the highlighted critical design parts. In the following, we present a summary of microservices design anti-patterns that can be detected by the *Architectural Model Evaluator*:

Bottleneck: A microservice that has high incoming and outgoing coupling. Such a microservice is considered an anti-pattern since it represents a single point of failure in the system [5, 14]. If a bottleneck microservice suffers from a huge workload (in production), this might affect its response time or, even worse, its availability. As a consequence, this also affects the performance of other microservices that depend on it.

Long Chain of Responsibilities: A chain of sequential microservices calls to fulfill one abstraction [2, 5]. This pattern is considered an ideal solution for functions that process shared data sequentially, since it allows autonomous scaling of resources based on the needs of each functionality [2]. However, according to [5], the existence of a chain of responsibilities could be due to wrong cuts. In such a case, it is considered an anti-pattern. Furthermore, the length of the chain of responsibilities can significantly affect the response time of the system. A long chain indicates a larger number of calls over the network, which, in turn, could lead to delays in sending responses.

Megaservice: A service that provides many functionalities. Such a service is considered an anti-pattern because it might suffer from the same problems as those associated with monolithic systems [19]. Different functionalities in a megaservice might have different scalability requirements; however, since they belong to the same microservice, the need to scale one of them will result in scaling the whole microservice, resulting in a potential waste of resources. Furthermore, a megaservice might suffer from being overloaded by requests during runtime, which might affect its performance and availability.

Nanoservice: A service with very few functionalities. Such a service is considered an anti-pattern because the effort needed for its maintainability and communication exceeds its benefits [5].

Scattered Functionalities: This is the case when one functionality in a microservice needs to perform multiple calls to another functionality belonging to a different microservice in order to perform a task. If such a case exists, it might be beneficial to place these two operations in the same microservice to reduce the number of network calls needed to respond to a request and thus enhance the performance of the system.

Infinite Cycle: Although not defined in the literature as an anti-pattern, the existence of an infinite cycle of inter-microservices calls is a logical error that might be introduced unconsciously into the architectural model and might be hard to detect manually. The existence of such cycles would lead to timeouts in the responses and thus affect the performance analysis.

To detect these anti-patterns in a microservices system, *MDEPT*'s *Architectural Model Evaluator* component relies on the following metrics:

Degree-in: Number of incoming calls to a microservice (client calls and inter-microservice calls). In the case of a high number of calls, overloading a microservice with requests will affect its performance and the performance of the calling microservices.

Degree-out: Number of outgoing calls from a microservice to other microservices in the system. If a microservice calls many other microservices, it will be busy formulating and sending requests and perhaps waiting for responses to continue its tasks. Thus, its performance might be affected by the performance of the called microservices as well.

Number of Inter-Microservice Calls: Number of calls between functionalities of two different microservices.

Number of Functionalities: Number of exposed functionalities of a microservice.

Chain Length: Number of inter-microservice calls required to respond to a single client request.

Complexity or Processing Time: Estimated complexity or processing time of the functionalities of a microservice (see Sect. 3.1). Overloading a microservice that has highly complex functionalities with many requests will impact its performance.

The *Architectural Model Evaluator* component relies on the *Degree-in, Degree-out,* and *Complexity or Processing Time* metrics to detect *Bottlenecks*; on the *Chain Length* metric to detect *Long Chains of Responsibilities*; on the *Number of Functionalities* metric to detect *Megaservices* and *Nanoservices*; and on the *Number of Inter-Microservice Calls* to detect *Scattered Functionalities*.

3.3 Dynamic Analysis: Code Generation and Performance Testing

MDEPT's *Code Generator* component generates low-fidelity boilerplate source code that is fully functional, based on the *Microservices Architectural Model* (Sect. 3.1). Furthermore, it generates the deployment configuration files for containerized microservices. Currently, *MDEPT* generates *Express.js*[1] code (as explained in Sect. 4). However, the approach can be extended to allow architects to choose a different language(s) for the generated code. To generate a test plan, architects need to provide the *Performance Test Plan Generator* component with the required test parameters, depending on the tool used for performance testing. Currently, *MDEPT* uses *JMeter*[2] for performance testing (see Sect. 4). The architect runs these test plans against the generated code in a test environment to collect performance measurements.

---

[1] https://expressjs.com/.

[2] https://jmeter.apache.org/.

4   MDEPT: Prototype Implementation Details

To demonstrate our approach, we implemented a prototype *MDEPT Tool* and composed a toolchain around it (Fig. 2). To model microservices systems, we chose *draw.io*[3], an open-source diagramming tool, because of its simplicity, online collaboration capabilities, and the fact that it enables the creation of modeling profiles and the export of designs as easy-to-parse XML files. We created a custom shape library, the *MDEPT-modeling library*, in *draw.io* to serve as a simple modeling profile for microservices systems. Architects can easily import the *MDEPT-modeling library* into *draw.io* and use its elements to create microservices system models that are compatible with the *MDEPT Tool*. The *MDEPT-modeling library* consists of the following elements:

- *Microservice*: A rectangle shape with the properties *Name, Type: microservice*.
- *Functionality*: A circle shape with the properties *Called By Client* (true or false), *Communication method* (GET or POST), *Name, Complexity* (in big O notation) or *Processing time* (in milliseconds), and *Type: functionality*.
- *Call*: A directional line with the properties *Number of calls* and *Type: call*.

Figure 3 illustrates an example of a microservices system architectural model created using the *MDEPT-modeling library* in *draw.io*. After creating the architectural model, the architect exports it as an *XML* file, which is the input for the *MDEPT Tool*.

We implemented the *MDEPT Tool* as a *Windows Forms Application*, using *C#*. The tool has a *Graphical User Interface (GUI)* to guide the architect during the evaluation process of a microservices system and get the required input. Taking the *XML*

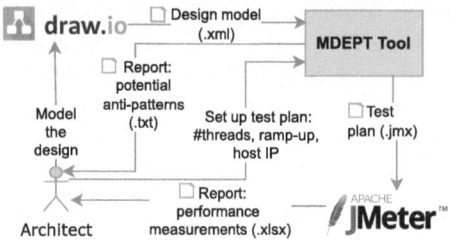

Fig. 2. MDEPT's toolchain

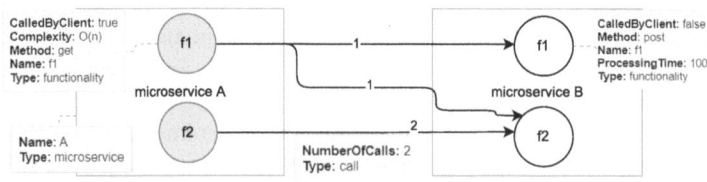

Fig. 3. Example of a microservices system architectural model using draw.io. Filled circles represent functionalities directly called by the client.

---

[3] https://www.drawio.com/.

file describing the architectural design as input, the *MDEPT Tool* extracts the graph model (as formalized in Sect. 3.1). After importing the architectural design model, the architect can select the types of anti-patterns that should be considered by the *MDEPT Tool*. The *Architectural Model Evaluator* component uses graph theory algorithms on the graph model to perform the static analysis. The component generates an overview report (.txt file), which serves as the starting point for the architect, as it lists the potential anti-patterns detected in the architectural model. It provides a definition for each anti-pattern, an explanation of its expected impact on performance, and suggestions for creating alternative design variants to remove the anti-patterns. Furthermore, for each type of anti-pattern, this report directs the architect to another dedicated report (.txt file) for a thorough overview.

For the sake of demonstrating our solution, we decided to use *Express.js*. The *Code Generator* component of the *MDEPT Tool* generates an *Express.js* application, along with the files needed to create a *docker image*[4], for each microservice in the system based on the *XML* architectural design model (taking all its elements and their defined properties into account). For each functionality in a microservice in the system, the *Code Generator* component generates a function that has the complexity or response time of the corresponding functionality and generates the inter-microservices calls of the functionality. For example, for $O(n)$ and $O(n^2)$ complexities, we generate dummy multiplication operations inside *for loop(s)* with n = 1000 to enable a relative comparison between all the functionalities in the system. Furthermore, to enable deploying and running the system, the *Code Generator* automatically generates the docker images and containers of the microservices along with the *docker-compose.yml*[5] file, which defines all the services in the system and the network between them.

The final step in this process is the generation of a test plan, which is used as input to the performance testing tool. We use *JMeter* in our toolchain, and therefore the following test plan parameters are required for this step:

- *Number of threads*: Number of users that send requests to the system. Each thread represents a user and simulates a concurrent request.
- *Ramp-up period*: The time that *JMeter* takes to start all the threads.
- *Host IP*: The *IP* address of the machine on which the SUT is deployed.

Using the test plan parameters provided by the architect, the *Performance Test Plan Generator* component creates a test plan (a *.jmx file*) that the architect can import into *JMeter* and execute it to test the performance of the deployed microservices system. Finally, the architect can import the test results as a *CSV* file and compare them with the results of another deployed variation for the microservices system under study. The results file contains measurements of essential performance metrics (e.g., average response time, 90th-99th percentiles of response time, percentage of failed requests).

---

[4] https://www.docker.com/.

[5] https://docs.docker.com/compose/.

5   Evaluation and Discussion

5.1   Experimental Study

Setup

We performed experiments that reflect the expected usage of *MDEPT* on three different microservices systems. We modeled these systems using *draw.io* and used the *MDEPT Tool* to detect potential anti-patterns in the models. Based on the results, we created new variants for each system under study, i.e., we applied fixes to remove the detected anti-patterns and thus created a new design model. For each system, we used the *MDEPT Tool* to generate the source code for each variant, i.e., the original system variant and the variant(s) created by removing anti-patterns.

For experiment (1) and experiment (2), we created two small demo microservices systems (Fig. 4-a and Fig. 5-a, respectively). We deliberately injected some anti-patterns into them to validate whether the *MDEPT Tool* would detect them during the static analysis. For experiment (3), which aimed to validate our approach on a real-world system, we conducted a use case study on *TeaStore*[6], a microservice reference and test application that consists of five microservices and 28 functionalities [12]. To achieve this, we worked with one of the *TeaStore* system engineers to model the *TeaStore* system (TeaStore-V1). Due to the lack of space, we do not show the system's model here, but it reflects the system's source code in accordance with the modeling rules we already presented. The purpose of this case study was to evaluate whether the *MDEPT* approach provides a sufficient level of abstraction for modeling real-world microservices systems, whether it can detect potential anti-patterns, whether it can provide valuable guidance to the architect in making architectural design decisions, and whether the limitations present in *MDEPT* are a limiting factor for application to real-world systems. During the case study, we observed the system engineer's experience while using *MDEPT* and collected his feedback.

To test the performance under different loads, we created two test plans. **Test Plan (1):** 300 users, 2 s ramp-up period, and **Test Plan (2):** 500 users, 2 s ramp-up period. We used three machines running Windows 10–64 bit, as follows: **Machine (1):** Intel(R)

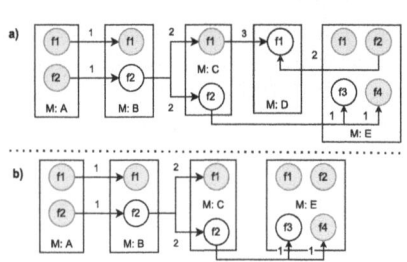

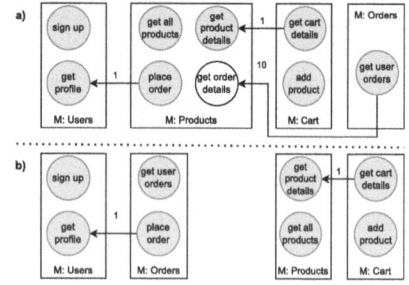

Fig. 4. Microservices system (1) architectural models: a) S1-V1; b) S1-V2

Fig. 5. Microservices system (2) architectural models: a) S2-V1; b) S2-V2

---

[6] https://github.com/DescartesResearch/TeaStore.

Core(TM) i5-1035G1 CPU, 8 GB RAM, 220 GB SSD, to host the system variants for experiments (1) and (2); **Machine (2):** AMD Ryzen 5 5500U CPU, 16 GB RAM, 512 GB SSD to host the system variants for experiment (3); **Machine (3):** Intel(R) Core(TM) i5-5300U CPU, 8 GB RAM, 120 GB SSD, to host *JMeter*. The machines were located in the same *Local Area Network (LAN)* and used the *WiFi 802.11* standard for communication. To evaluate the performance of the systems' variants, we measured both the *average response time* and the *average 90th percentile of the response time* of the functionalities of each variant for the two test plans.

Results

For experiment (1), the *MDEPT Tool* identified four potential anti-patterns after the static analysis of S1-V1 (Fig. 4-a): Microservice *D* corresponded to a nanoservice; microservice *C* was identified as a potential bottleneck; *C:f1* and *D:f1* were identified as scattered functionalities. Three potential long chains of responsibilities of length 3 were detected as well. After analyzing the reports of the *MDEPT Tool*, we concluded that microservice *D* could be contributing to too many issues in the design. In order to establish whether the identified anti-patterns did indeed cause performance degradation, we modified the system S1-V1 and created another design model variant, S1-V2 (Fig. 4-b). In S1-V2, we removed microservice *D* and moved its functionality *D:f1* inside both *C:f1* and *E:f2* (a sub-optimal design, as it causes the coupling of *C* and *E*, which we chose deliberately in order to analyze the impact of such a trade-off on performance) and applied the necessary updates to the properties (in this case, the processing times) of the affected functionalities. We then used the *MDEPT Tool* to generate the source code and the test plans for both S1-V1 and S1-V2, and measured the performance of the variants of the system. Figure 6 illustrates the results of the experiment.

For experiment (2), the *MDEPT Tool* identified three potential anti-patterns after the static analysis of S2-V1: The microservice *products* was a potential bottleneck; the microservice *orders* was a nanoservice; the functionalities *orders: get user orders* and *products: get order details* were scattered functionalities. After analyzing the reports, we concluded that the incoming calls from *orders: get user orders* to *products: get order details* contributed to too many issues. Based on *MDEPT's* recommendations, we modified the system S2-V1 and created S2-V2 (Fig. 5-b). In S2-V2, we moved the functionality *get order details* from the microservice *products* to the microservice *orders* and merged it into its functionality *get user orders*. We updated the processing time of the functionality *get user orders* to reflect this change (we increased its processing time by $10 \times$ the processing time of *product: get order details* from S2-V1). Furthermore, we also moved the functionality *products: place order* into the microservice *orders*. We then used the *MDEPT Tool* to generate the source code and the corresponding test plans for the two variants of the system. Finally, we measured the performance of the two variants of the system for the test plans. Figure 7 illustrates the results.

For experiment (3), we used the *MDEPT Tool* to carry out the static analysis on *Tea-Store* (TeaStore-V1). The *MDEPT Tool* identified the following potential anti-patterns: The microservice *WebUI* is a potential bottleneck with a total degree (sum of degree-in and degree-out values) of 61; the microservice *Persistence* is a megaservice with 10 functionalities; the microservice *Recommender* is a nanoservice with only one functionality. There were two pairs of scattered functionalities with 11 calls: *(WebUI: product,*

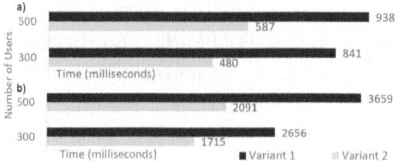
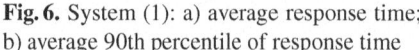

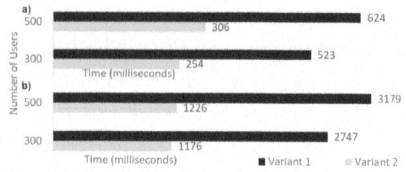

Fig. 6. System (1): a) average response time; b) average 90th percentile of response time

Fig. 7. System (2): a) average response time; b) average 90th percentile of response time

Persistence: prodFindEntityById) and *(WebUI: cart, Persistence: prodFindEntityById).* The first finding concerning the identified potential bottleneck (the microservice *WebUI*) is consistent with the results of a study conducted on *TeaStore* [8], which indicated that *WebUI* is the most CPU-intensive service, making it a bottleneck in the system. To validate the static analysis results, we interviewed one of the engineers of *TeaStore*. The engineer confirmed that the anti-patterns detected by *MDEPT* conform to their theoretical definitions and that most of them could lead to a degradation in system performance. The engineer stated that some of the detected anti-patterns were intentionally chosen to meet certain requirements, e.g., to simplify deployment, while some of them were present incautiously in the system design. He confirmed that the microservice *Persistence* is a megaservice that violates the microservices design principles. Nevertheless, the engineer and his team chose this design as it simplifies deployment. Regarding the *Recommender* microservice, the system engineer agreed that it can theoretically be identified as a nanoservice. However, he stated that this is also by design since this microservice "should be developed and maintained by the data science team, which has a different skill set than what is needed for the rest of the application". The engineer also confirmed that the detected pairs of scattered functionalities are anti-patterns and agreed that this is a "bad design".

During the interview with the *TeaStore* engineer, we worked on creating two other variants of the system (TeaStore-V2) and (TeaStore-V3) based on the recommendations of the *MDEPT Tool*. In (TeaStore-V2), to resolve the two pairs of scattered functionalities, the engineer decided to introduce a new functionality in the microservice *Persistence*: a bulk operation to find multiple products by their identifiers. He argued that this could be a simple solution that would not require much effort, but would probably lead to an improvement in performance. Furthermore, to improve the design (following domain-driven principles), he decided to move the login data from the *Persistence* microservice to the *Authentication* microservice. He achieved this by moving the *Persistence: userFindEntityByID* functionality to the *Authentication* microservice and moving the *Persistence: findByID* functionality to the *Authentication* microservice and merging it with its *login* functionality. The engineer estimated that this would require a medium amount of effort and might improve performance. As for the *WebUI* microservice, he decided to keep the current design, as introducing microfrontends would have been a major system redesign requiring considerable effort. To create TeaStore-V3, the engineer made some changes to TeaStore-V2. He moved the functionalities *Persistence: orderCreateEntity* and *Persistence: orderItemCreateEntity* to the *Authentication* microservice and merged them with its *placeOrder* functionality. He also decided

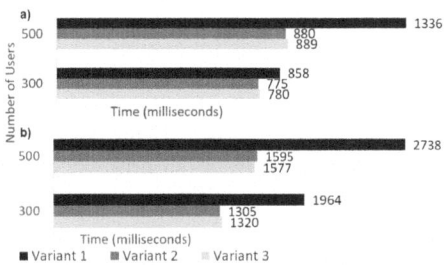

Fig. 8. System (3) - TeaStore: a) average response time; b) average 90th percentile of response time

to introduce a new microservice (*Order*) and moved the order-related functionalities from the *Authentication* microservices to this new microservice. According to the engineer, the variant (TeaStore-V3) would require more development effort than the variant (TeaStore-V2). After creating the design models, we used the *MDEPT Tool* to generate the source code and the test plans for the *TeaStore* variants (TeaStore-V1, TeaStore-V2, TeaStore-V3). Finally, we measured the performance of the variants for the test plans. Figure 8 illustrates the results.

Analyzing the results of experiments (1) and (2) revealed that both S1-V2 and S2-V2 had lower total *average response time* and *average 90th percentile* than S1-V1 and S2-V1, respectively. The system variants S1-V2 and S2-V2, which we created based on the suggestions from the static analysis of *MDEPT*, showed the following performance improvements: Considering the total *average response time*, S1-V2 showed improvements of 42.9% for 300 users (test plan 1) and 37.4% for 500 users (test plan 2) compared to S1-V1. S2-V2 showed improvements of 50.9% for 300 users and 51.4% for 500 users compared to S2-V1. The *average 90th percentile response time* was improved by 35.4% (300 users) and 42.8% (500 users) for S1-V2 compared to S1-V1. For S2-V2, this improvement was 57.1% (300 users) to 61.4% (500 users) compared to S2-V1.

For experiment (3), the results indicate that both TeaStore-V2 and TeaStore-V3 had lower *average response time* and *average 90th percentile* than TeaStore-V1. TeaStore-V2 and TeaStore-V3 resulted in improvements of 9.6% (for 300 users) to 34.1% (for 500 users) and 9% (for 300 users) to 33.4% (for 500 users), respectively, in *total average response time* compared with TeaStore-V1. Furthermore, the *average 90th percentile response time* of the system was improved by 33.5% (300 users) to 41.7% (500 users) in TeaStore-V2 and by 32.7% (300 users) to 42.4% (500 users) in TeaStore-V3. These results indicate that, for these concrete cases, implementing the suggestions of *MDEPT* would result in better performance of the system in terms of *average response time* and *average 90th percentile of the response time*. They also indicate that the statically identified anti-patterns in the original variants of the systems were indeed anti-patterns with respect to performance.

5.2 Survey Study

Setup

The goal of the survey was to understand how professionals perceive our approach and to get their feedback and suggestions on possible improvements. For this, we conducted interviews with twelve IT professionals from our professional network. We presented our approach and asked the interviewees to fill in a survey consisting of two parts: The first part was about the participants' domains, years of experience, and familiarity with microservices systems, while the second part was related to the *MDEPT* approach and contained both multiple-choice questions and open-ended questions.

Results

The results of the survey revealed that eight out of the twelve interviewees were working in the academic field, three in industry, and one in both fields. The majority of the interviewees (eight of them) had more than six years of experience in software engineering, and the other four had between three and six years of experience. In terms of knowledge of microservices systems, three interviewees identified themselves as experts, six reported having intermediate knowledge, one stated being familiar with microservices systems, and the remaining two indicated they were somewhat familiar with such systems. Regarding the level of experience in microservices anti-patterns, three interviewees reported being experts, while five had intermediate knowledge; the rest were less familiar with anti-patterns. Furthermore, three of the interviewees had participated in developing microservices benchmark systems. Table 1 summarizes the *multiple-choice questions* referring to our approach along with the answers. Regarding the *open-ended*

Table 1. The survey's multiple-choice questions and answers. Strongly agree (SA), Agree (A), Neutral (N), Disagree (D), Strongly disagree (SD)

Questions	SA	A	N	D	SD
Design properties expected by *MDEPT* are sufficient for modeling microservices systems.	0	11	0	1	0
Design properties considered by *MDEPT* are sufficient for detecting microservices anti-patterns.	0	7	5	0	0
MDEPT detects the most relevant anti-patterns for microservices performance.	1	4	7	0	0
The measurements considered by *MDEPT* are the most important ones for performance.	1	8	3	0	0
MDEPT provides a significant benefit/overhead gain.	1	8	3	0	0
The absence of database access emulation is not a significant missing factor.	2	2	7	1	0
The *MDEPT* approach is compatible with my toolchain.	0	6	5	1	0
Compared to my current approach for evaluating design decisions, *MDEPT* would give me more confidence.	1	8	2	1	0
I would like to use *MDEPT* in my development.	1	8	2	1	0

questions about *MDEPT*, we asked the interviewees about other design properties they would like to have as input to *MDEPT*. Suggestions included enabling modeling of the number of microservices instances and dependencies on external entities. When asked about the need to represent other types of calls between microservices, suggestions included support for asynchronous calls and protocols other than *HTTP*. Moreover, the interviewees expressed the need to detect as many anti-patterns as possible, since that would provide a more holistic evaluation of the architectural model. Furthermore, when asked about the approaches they are currently using to evaluate the expected performance of their microservices systems (multiple options could be selected here), eight interviewees stated that they are currently using *lab tests*, seven are using *simulation*, and five are using *prototyping*. As the numbers indicate, the majority of the interviewees are using more than one evaluation method.

Moreover, when asked about whether they would prefer having other programming languages supported by the *MDEPT Tool* for the generated source code, three interviewees mentioned Java, while two stated that supporting another language is not of importance. Finally, when asked about suggestions for improving the *MDEPT Tool*, seven interviewees suggested increasing the number of detected anti-patterns, six suggested measuring other performance parameters, and two suggested increasing the fidelity of the generated source code.

5.3   Discussion

The results of our experiments showed that using *MDEPT* helps architects choose a suitable design for their microservices systems during design space exploration, considering their performance requirements. In experiments (1) and (2), *MDEPT*'s static analysis identified all the anti-patterns that we deliberately injected into the demo systems. In experiment (3), one of TeaStore's engineers confirmed that the potential anti-patterns detected by *MDEPT*'s static analysis were consistent with their theoretical definitions, and that most of them could indeed cause performance degradation. The engineer also stated that the suggestions provided by *MDEPT* were reasonable and helpful in creating new variants of the system. The results of the dynamic analysis showed that by leveraging the recommendations of *MDEPT*'s static analysis to redesign a system under test, we created variants of the system that all had better overall performance (in all three experiments conducted). Although we observed a degradation in the performance of some of the functionalities of certain microservices, this degradation can be neglected compared to the overall performance gain for the system.

Our survey study showed that *MDEPT* was well perceived by the interviewed IT professionals. The interviewees expressed an interest in our approach, and most of them stated that they would like to try it. Neutral and negative feedback was mainly due to the fact that the interviewees had an existing toolchain where *MDEPT* would be difficult to integrate. 75% of the interviewees agreed that using *MDEPT* would provide them with a higher level of confidence when evaluating microservices design decisions than their current evaluation approach. One participant disagreed with this. Analyzing the answers further revealed that this participant had specified *"Self Controlled"*, *"Tested in Lab"*, and *"Tested in Field"* as their current evaluation approaches.

Threats to Validity. The systems S1 and S2 are small demo systems that we created. This is a potential threat to validity. However, we obtained similar results in experiment (3), which we conducted on *TeaStore*, a well-known reference microservices system for benchmarks and tests. The benefits of our approach were confirmed by one of the system's engineers. For the survey study, we interviewed only twelve IT professionals from our professional network. Although the surveyed group was small, it covered different domains, both in industry and academia. Furthermore, the interviewees had a significant number of years of experience in IT. Some of them are experts who regularly publish at the most important architectural conferences, while three of them had worked on developing microservices benchmarks.

6   Conclusion and Future Work

Although some design anti-patterns are known to lead to performance degradation, the impact that an anti-pattern would have on a system's performance is context-dependent. Thus, it is important to help architects evaluate microservices design models early during the design activity. In this paper, we presented the formalization and evaluation of our *MDEPT* approach, which combines both static and dynamic analysis to support architects in this process. To evaluate our approach, we carried out both experimental and survey studies. The experimental results show that following *MDEPT*'s recommendations can lead to better-performing variations of microservices systems. Furthermore, the results show that *MDEPT* can be easily applied to real-world systems. For a concrete case (*TeaStore*), *MDEPT* identified some design anti-patterns and helped to indicate how the design could be improved. For our qualitative evaluation, we conducted several interviews with IT professionals from both industry and academia. The results showed that *MDEPT* was well perceived by the professionals.

For future work, we aim to increase confidence in our experimental evaluation results by applying the design decisions suggested by *MDEPT* to the source code of *TeaStore* and measuring the performance of both the original system and the variants. We also plan to make conceptual improvements such as modeling more design properties for microservices systems and increasing the number of analyzed anti-patterns.

Acknowledgments. Parts of this work were carried out in the EnStadt:Pfaff project (grant no. 03SBE112D and 03SBE112G) of the German Federal Ministries BMWK and BMBF. We thank Simon Eismann for his support throughout our experiment using TeaStore.

Data Availibility Statement. The data that supports the findings of this study is available from the corresponding author upon reasonable request.

References

1. A knowledge base for Service-Based Antipatterns, created in a research project at the University of Stuttgart. https://xjreb.github.io/service-based-antipatterns. Accessed 04 Nov 2022
2. Akbulut, A., Perros, H.G.: Performance analysis of microservice design patterns. IEEE Internet Comput. **23**(6), 19–27 (2019)

3. Avritzer, A., Ferme, V., Janes, A., Russo, B., Schulz, H., van Hoorn, A.: A quantitative app-roach for the assessment of microservice architecture deployment alternatives by automated performance testing. In: ECSA, pp. 159–174 (2018)

4. Bass, L., Clements, P., Kazman, R.: Software Architecture in Practice, 3rd edn. Addison-Wesley Professional, Boston (2012)

5. Bogner, J., Boceck, T., Popp, M., Tschechlov, D., Wagner, S., Zimmermann, A.: Towards a collaborative repository for the documentation of service-based antipatterns and bad smells. In: 2019 International Conference on Software Architecture Companion (ICSA-C), pp. 95–101. IEEE (2019)

6. Brown, W.H., Malveau, R.C., McCormick, H.W.S., Mowbray, T.J.: AntiPatterns: Refactoring Software, Architectures, and Projects in Crisis. John Wiley & Sons, Inc., Hoboken (1998)

7. Cortellessa, V., Di Pompeo, D., Eramo, R., Tucci, M.: A model-driven approach for contin-uous performance engineering in microservice-based systems. J. Syst. Softw. **183**, 111084 (2022)

8. Eismann, S., Bezemer, C.P., Shang, W., Okanović, D., van Hoorn, A.: Microservices: a per-formance tester's dream or nightmare? In: International Conference on Performance Engi-neering, pp. 138–149 (2020)

9. Gonçalves, N., Faustino, D., Silva, A.R., Portela, M.: Monolith modularization towards microservices: refactoring and performance trade-offs. In: International Conference on Soft-ware Architecture Companion (ICSA-C), pp. 1–8. IEEE (2021)

10. Happe, J., Koziolek, H., Reussner, R.: Facilitating performance predictions using software components. IEEE Softw. **28**(3), 27–33 (2011)

11. Heinrich, R.: Architectural runtime models for integrating runtime observations and component-based models. J. Syst. Softw. **169**, 110722 (2020)

12. von Kistowski, J., Eismann, S., Schmitt, N., Bauer, A., Grohmann, J., Kounev, S.: TeaStore: a micro-service reference application for benchmarking, modeling and resource manage-ment research. In: Proceedings of the 26th IEEE International Symposium on the Modelling, Analysis, and Simulation of Computer and Telecommunication Systems. MASCOTS 2018 (2018)

13. Matar, R., Jahić, J.: An approach for evaluating the potential impact of anti-patterns on microservices performance. In: 2023 International Conference on Software Architecture Companion (ICSA-C), pp. 167–170. IEEE (2023)

14. Nayrolles, M., Moha, N., Valtchev, P.: Improving SOA antipatterns detection in service based systems by mining execution traces. In: 2013 20th Working Conference on Reverse Engi-neering (WCRE), pp. 321–330. IEEE (2013)

15. , Ntentos, E., Zdun, U., Plakidas, K., Geiger, S.: Semi-automatic feedback for improving architecture conformance to microservice patterns and practices. In: 2021 International Con-ference on Software Architecture (ICSA), pp. 36–46. IEEE (2021)

16. de Oliveira Rosa, T., Daniel, J.F.L., Guerra, E.M., Goldman, A.: A method for architec-tural trade-off analysis based on patterns: evaluating microservices structural attributes. In: Proceedings of the European Conference on Pattern Languages of Programs 2020, pp. 1–8 (2020)

17. Pigazzini, I., Fontana, F.A., Lenarduzzi, V., Taibi, D.: Towards microservice smells detection. In: International Conference on Technical Debt, pp. 92–97 (2020)

18. Taibi, D., Lenarduzzi, V.: On the definition of microservice bad smells. IEEE Softw. **35**(3), 56–62 (2018)

19. Taibi, D., Lenarduzzi, V., Pahl, C.: Microservices anti-patterns: a taxonomy. In: Microser-vices, pp. 111–128. Springer, Cham (2020). https://doi.org/10.1007/978-3-030-31646-4_5

20. Trubiani, C., Pinciroli, R., Biaggi, A., Fontana, F.A.: Automated detection of software per-formance antipatterns in java-based applications. IEEE Trans. Softw. Eng. **49**(4), 2873–2891 (2023)

21. Vögele, C., Hoorn, A., Schulz, E., Hasselbring, W., Krcmar, H.: Wessbas: extraction of prob-abilistic workload specifications for load testing and performance prediction-a model-driven approach for session-based application systems. Softw. Syst. Model. **17**(2), 443–477 (2018)
22. Wu, W., et al.: Software architecture measurement—experiences from a multinational com-pany. In: Cuesta, C.E., Garlan, D., Pérez, J. (eds.) ECSA 2018. LNCS, vol. 11048, pp. 303–319. Springer, Cham (2018). https://doi.org/10.1007/978-3-030-00761-4_20

Microservices Architecture

Exploring Architectural Evolution in Microservice Systems Using Repository Mining Techniques and Static Code Analysis

Patric Genfer[1,2]($\boxtimes$) and Uwe Zdun[1]

[1] Research Group Software Architecture, Faculty of Computer Science,
University of Vienna, Vienna, Austria
{`patric.genfer,uwe.zdun`}`@univie.ac.at`
[2] UniVie Doctoral School Computer Science DoCS, University of Vienna, Vienna,
Austria

Abstract. Microservices have gained popularity for isolating service functionality and mitigating issues such as architectural erosion and technical debt. However, their decentralized nature and rapid development often obscure the holistic view of the system and lead developers to lose sight of the overarching architecture. Our work addresses this challenge by proposing a novel approach to track and assess the evolution of microservice architectures through static source code analysis. We combine source code repository mining techniques with architectural reconstruction to measure various metrics throughout a system's development history. Our approach uses a formal API-based decomposition model that can easily be adapted for different scenarios by choosing various architectural metrics. We validated our method's scalability and robustness through a case study on an extensive open-source microservice reference system with more than 40 individual services written in different languages and more than 400 commits. Our research provides software architects with a powerful tool to identify and monitor problematic architectural trends before they become imminent threats, enabling the evolution of microservice-based systems while maintaining architectural coherence and integrity.

Keywords: Microservice API · evolution · source code repository mining · metrics · source code detectors

1 Introduction

The constant evolution of software systems is crucial for adapting to changing environments and meeting new requirements [1]. However, it also harbors the risk of architectural drift or erosion [2], leading to uncontrolled growth that may break previously defined conceptual boundaries. A software system that evolves without clear architectural guidance becomes more complex to maintain and

© The Author(s), under exclusive license to Springer Nature Switzerland AG 2024
M. Galster et al. (Eds.): ECSA 2024, LNCS 14889, pp. 157–173, 2024.
https://doi.org/10.1007/978-3-031-70797-1_10

more challenging to develop – its components can become too coupled, resulting in individual changes affecting too many parts of the system simultaneously [3]. Also, taking architectural shortcuts can often be necessary due to rough schedules and insufficient development resources [4]. Unfortunately, this approach may produce technical debt that has to be addressed later to prevent disproportionate delays in feature development [2].

The problem of architectural erosion is as old as software engineering itself and persists despite various prevention strategies [5]. Microservice architectures have emerged as one such strategy, with isolation as a core principle, facilitating localized changes and easier replacement of problematic services [6,7]. The idea is that a single service whose technical debt becomes worrisome can be easily replaced or rewritten [7]. These rewrites can also be valuable if business rules become better understood over time and allow for a leaner implementation [6].

However, while microservices mitigate technical debt within individual services, they exacerbate the challenge of maintaining architectural coherence across the system. Having isolated services means developers lose sight of the big picture, as each team works only on its small excerpt but misses a clear view of the overall architectural idea that underlies the entire system. The rapid development enabled by microservices can further accelerate architectural erosion, resulting in longer development times per feature with each new version [8].

To devise effective countermeasures against this erosion, identifying any problematic trends during development as early as possible would be necessary, but tracking the overall architectural integrity in a microservice system throughout its development is challenging: Many conventional metrics that work for monolithic applications, e.g., calculating coupling and cohesion, may not apply to highly distributed systems and loosely coupled architectures [9]. Also, existing approaches often focus only on single points in time, neglecting the continuous evolution of microservices [10–12]. Focusing only on an individual commit or snapshot bears the risk that erosion has already hit a threshold where implementing countermeasures may become too costly and bind too many resources that could be better spent on feature development [13].

Our work addresses this gap by proposing a novel approach to tracking and assessing the evolution of microservice architectures based on their APIs, which, as central points of their communication, play a significant role during their development. For this, we designed a multi-staged analysis process that, for the first time in this context, combines source code repository mining techniques with API-based architectural decomposition to reconstruct the evolution of a larger microservice architecture during the whole development process. We also derived a set of various metrics from our formal architectural model to analyze and assess the system's overall quality trends over time. We evaluated our solution in a case study on an extensive open-source microservice reference system with over 40 individual services and a history of 400 commits, verifying that our method works on a scale typical for mid-sized to large microservice architectures with a rich development history and polyglot nature. In contrast to research focusing on mining source code repositories, we do not use it as our central method to gather knowledge; instead, we consider it a tool to collect the information we need to reconstruct our architectural model. Accordingly, our study seeks to answer the following questions:

RQ1 *How can the evolution of microservice APIs be tracked efficiently?*
Creating an architectural model through source-code parsing is time-consuming, especially for larger microservice systems with a long commit history. A suitable approach must scale well, even for larger systems, and be robust enough to analyze and track large amounts of commits.

RQ2 *How can our approach be used to measure the different characteristics of microservice APIs over time?*
Microservices are highly dynamic systems tailored to specific domains, each with different requirements regarding the system's overall quality. Our analysis process must consider this variance by allowing software architects to adapt the analysis to their specific needs.

To our knowledge, this is the first study that uses source code repository mining to reconstruct the architectural evolution of large-scale microservice systems by using only static-analysis techniques. While we provide an initial set of metrics to measure architectural trends, we also demonstrate that our approach can easily be adapted to individual scenarios by using different metrics.

The remaining paper is organized as follows: Sect. 2 will look at the related research in this area, Sect. 3 will describe the approach we followed when implementing our work. The evaluation of our technique through a case study is the subject of Sect. 4. Section 5 presents and analyzes the findings we gathered through our case study. Potential validity threats affecting our results will be discussed in Sect. 6, while Sect. 7 concludes our work and contains an outlook on future work.

2   Related Work

The field of software architecture reconstruction [14] generally refers to works that reconstruct the software architecture or design from the source code, usually as a model or another intermediate representation. Our work aims to use repository mining techniques to reconstruct a microservice architecture during its evolution. Multiple works suggest approaches for microservice architecture reconstruction [15,16], but so far, none considers the evolution history in the source code repository systematically.

High-level architectural metrics can streamline the evaluation of software system quality, abstracting away implementation details and reducing information overload. Numerous studies offer metrics-based analyses of microservice systems: Walker et al. [11] use static source code analysis to detect code smells in microservice architectures, Ma et al. [12] propose an automated process to generate service-dependency graphs, and Zdun et al. [10] define a set of constraints to detect potential architectural erosions. Another study focusing on reconstructing the architecture of microservice systems through static analysis is the work of Bushong et al. [17]. They extract HTTP information from source code to create a communication diagram for visualizing inter-service communication.

When extending the analysis over the whole software evolution process, Barnes et al. [18] present a strategy for tracking the evolutional steps that affect software architectures over time. While their approach is not specific to microservice systems, they introduce the interesting concept of evolution operators to categorize architectural changes. In addition, there are also studies more microservice-related: Moreira and De França [19] show in their research how the cohesiveness of microservices changes over time. Similar research is performed by Tizzei et al. [20]: They track several size-oriented metrics, like lines of the code and the number of service operations throughout development. While both works go in the same direction as ours, they focus mainly on single-service metrics. In contrast, we concentrate on inter-service communication. Sampaio et al. [21] developed a service evolution model based on configuration files and by tracing runtime service communication. He et al. [22] are focused on runtime analysis, as they develop an online prediction system based on different runtime metrics to precalculate the costs and effects of microservice evolution. In contrast, our approach relies only on collecting static code artifacts, making the integration into development pipelines easier as expensive system execution can be avoided.

Stocker and Zimmermann [23] conducted a survey to analyze the reasons for microservice evolution. According to their results, new functional requirements are the primary reasons for API changes, followed by improving architectural quality. The main obstacle they identify that prevents developers from architectural refactoring is a need for more resources. Lercher et al. [24] further identified the tight organizational coupling that leads to additional communication overhead as a challenge to microservice evolution and consumer lock-in that requires long-term support for legacy APIs.

3 Approach

3.1 Source Code Repository Mining

Reconstructing a software system's architecture from the source code requires parsing and analyzing numerous source code artifacts. This is challenging as modern systems are complex and often use various programming languages and technologies. It becomes even more difficult for microservice systems, where the architecture is distributed across loosely coupled services. Tracking the evolution of such architectures further adds another level of difficulty to this already complex problem, as hundreds of commits, all containing several changes, must be analyzed. While exhaustively analyzing the entire development history of large microservice systems is feasible with sufficient resources, more efficient strategies exist. Considering the two primary aspects of a code repository – (1) the number and size of source code artifacts and (2) the number of commits containing changes over time – we present two strategies to reduce each.

Our first mining strategy capitalizes on the fact that modern source code repositories store information as sequential lists of patches [25], each containing atomic changes applied to individual files. This strategy is termed *Change-Based*

Repository Mining. In our approach, we suggest a novel approach using this strategy for reconstructing an architectural representation during evolution: This method starts with an empty architectural model or a previously generated snapshot (see below) and iteratively transforms and applies each commit to evolve it until we reach the final architecture's state. Our method updates only the parts of the model affected by a single commit (inspired by the approach taken in [26]), making it highly efficient, under the assumption that each commit is atomic and contains only a few changes [27]. While efficient, this approach has a downside: By focusing solely on changes instead of source code artifacts, every commit must be investigated, as missing even a single one can lead to incomplete models, rendering subsequent commits unprocessable.

Our second mining strategy tackles these problems with a different approach: Instead of reducing the source code to be analyzed by focusing only on the changes, we reduce the number of commits to be examined and include all artifacts of our codebase. We call this approach *Snapshot-Based Repository Mining*, as it operates only on a self-contained snapshot of the whole repository at a given time. Although seeming less efficient due to the need to parse more code artifacts, this approach offers greater flexibility and fault tolerance, as operating on the entire code base rather than incremental changes ensures a self-contained model independent of previous iterations. Also, since every commit can be processed in isolation, the analysis can be parallelized more easily.

We combined both strategies into a single mining process for our final approach, allowing us to benefit from the advantages of both methods (Fig. 1). In a first mining pass, we use the Snapshot-based approach on selected commits to create a macro-analysis of the overall system's architectural evolution. While this requires parsing more source code artifacts, it enables us to choose only specific commits we consider relevant. If we encounter any parsing errors, we skip the faulty commits and choose a nearby sample instead.

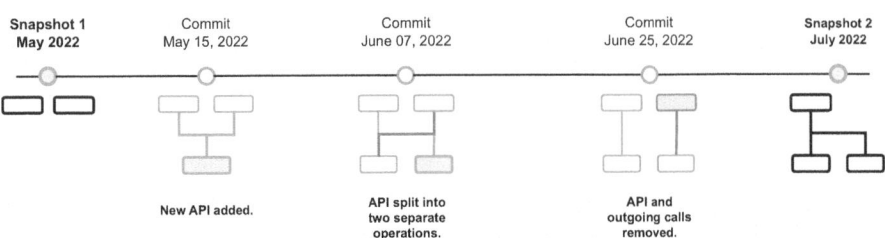

Fig. 1. The *Snapshot-based* mining approach creates the reference models for our start and endpoints. We then use the *Changed-based* mining approach to evolve our model iteratively between the snapshots to get a more in-depth view of the evolution.

After we pick enough commits to accurately recreate the architecture's development history, we then identify specific areas of interest between two commits and apply the change-based approach to get further a seamless visualization within the specific time range. With this approach, we can also detect any local minima or maxima we would otherwise not recognize.

Figure 2 shows a schematic view of our overall mining process. We begin by cloning the repository and creating local copies of all relevant commits, allowing further parallel execution. During the next phase, we use source code detectors, a lightweight parsing technology (see Sect. 3.2), to traverse each directory, identifying architectural patterns and capturing structural components and dependencies, resulting in our architectural decomposition model (also Sect. 3.2). Another set of detectors connects the identified services through API invocation call chains, forming a directed graph representing the system's static communication model. Using this structure, we calculate and store architectural metrics (see Sect. 3.3) to assess the system's health for each snapshot. After processing all selected commits, the resulting metrics are merged into a continuous trend diagram, showcasing the system's evolution over time. If desired, the change-based method can be run as a second pass between specific commits to fill any potential gaps in the analysis.

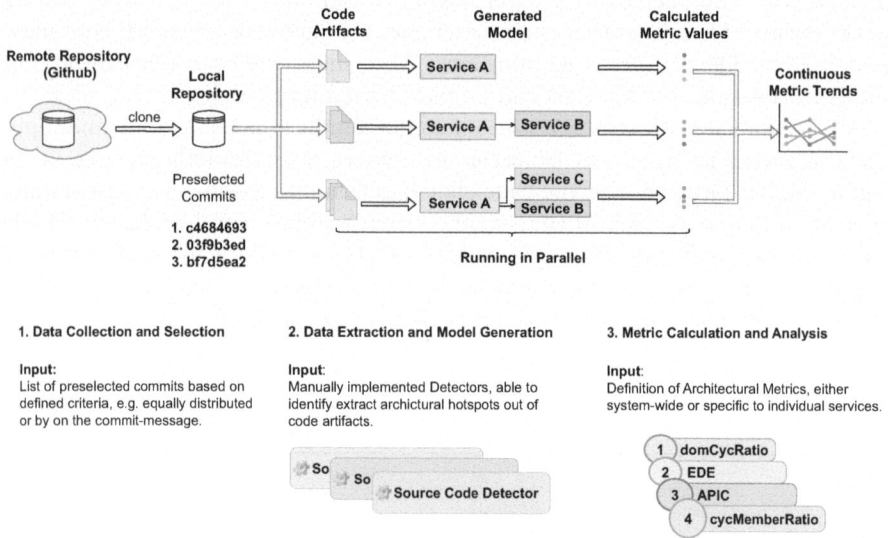

Fig. 2. The whole analysis process consists of three steps, where the second and parts of the third step can run in parallel to reduce the overall processing time.

3.2   Model Generation

Identifying architectural problems on the source code level can be challenging due to the lower level of abstraction, and even in well-documented systems, the documentation is not guaranteed to be precise and reflects the current architectural state [28]. To ensure our process accurately represents the overall system, we use the underlying code base as our single source of truth to reconstruct an abstract architectural model of each service and create a communication model showing the interconnection between these services. The model we decompose from the implementation focuses on API operations as the primary communication point within a microservice system. These include synchronous APIs and asynchronous message handlers, which we consider both part of a service's public interface. Besides API operations and their invocations, our model also contains optional elements like API Interfaces and Connection Hosts that help to create better traceability between model elements and code.

We use *Source Code Detectors*, lightweight Python-based parsers [29], to extract and generate our architectural model. These detectors look only for specific patterns, like a REST API method definition or asynchronous pipe subscriptions, and ignore all unrelated details, making their implementation more efficient than techniques that rely on full-fledged abstract-syntax-tree reconstruction. When identifying such a pattern, the detector can create a new architectural model element or enhance an existing one to reflect the newly gathered knowledge. Once the detectors traverse all artifacts, the resulting elements form an unlinked model of all architectural hotspots we consider relevant for our analysis [30]. In a second run, a different set of detectors identifies bridges between elements, such as local invocations or remote API calls, enhancing the model's coherence. Figure 3 illustrates this workflow.

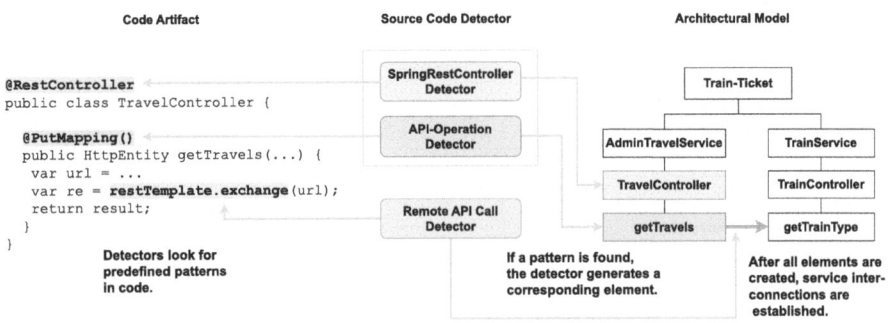

Fig. 3. Source Code Detectors are lightweight parsers that search the underlying code base for specific concepts. Whenever such a concept is found, a corresponding model element is generated.

Since our detectors are hand-crafted for every system we analyze, we can use heuristics based on project-specific coding conventions and style guides to simplify their implementation. However, our detection logic must remain robust enough to handle anomalies like typos or deviations, which may occur occasionally, especially when tracking a system's implementation over a longer time than we do in our research.

3.3 Architectural Metrics

To use our model to identify and track potential architectural trends, we give it a formal description that we then use to derive several architectural metrics. For this, we interpret our generated model as a directed graph $G = (V, E, F)$ [30]. Here, V denotes all architectural elements identified by our detectors, further categorized by their role within the system, e.g., V^{ms} for the set of all microservices and V^{api} for all API operations. E represents edges forming relations between elements, describing either a *has a* relation (e.g., between a microservice and its APIs) or an API or method invocation. F specifies predicates and utility functions we use for retrieving meta-data attached to the nodes. These can be annotations such as synchronous/asynchronous protocol usage or secure communication channels. For our case study, we collected a set of existing metrics from the literature that operate on higher abstraction levels and are thus implementation-independent. By purpose, we also chose metrics that measure different architectural and qualitative aspects to show that our approach is not restricted to any specific scenario. Table 1 shows the metrics we used for our case study and how we derived them from our formal model.

Table 1. A set of high-level architectural metrics we employ to obtain a comprehensive perspective of the overall structure of the microservice.

Name	Formula	Description
Average Service Interface Count (ASIC)	$\dfrac{\lvert V^{api} \rvert}{\lvert V^{ms} \rvert}$	Variant of *Weighted Service Interface Count* metric [31], with a uniform weight of 1. Quantifies the number of distinct API operations per microservice.
Average Path Length (APL)	$\dfrac{\sum \{p \in P(v) : \lvert p \rvert\}}{\lvert P(v) \rvert}$	Quantifies the consecutive API operations invoked when accessing any public API, with $P(v)$ being the set of all paths starting in v [30].
Cycle Ratio (cycRatio)	$\dfrac{\lvert \{p \in P(v) : cycle(p)\} \rvert}{\lvert P(v) \rvert}$	Tracks any cyclic paths when following the call chain of an API operation v. The predicate $cycle \rightarrow [0, 1]$ identifies paths containing at least one cycle [30].
Service Interaction via Intermediary Component (SIC)	$\dfrac{\lvert \{v \in V^{con} : is\_async(v)\} \rvert}{\lvert V^{con} \rvert}$	Ratio of service connections mediated through components like event buses or message brokers [32]. As this communication happens primarily asynchronously, we use the $is\_async \rightarrow [0, 1]$ predicate to identify such connectors.
Connection Anomaly Ratio (CAR)	$\dfrac{\lvert \{v \in V^{con} : is\_error(v)\} \rvert}{\lvert V^{con} \rvert}$	Ratio of service connections where the link to the target API could not be reconstructed. We also use this metric to verify that our detectors identified interservice calls correctly.

While the expressiveness of each metric may be limited when viewed in isolation, combining them can help provide a holistic view of the architectural trend of the system under observation, as we will show in our case study.

4   Case Study

We evaluated our approach on the *Train Ticket* repository[1], an open-source microservice benchmark system. It has more than 30 microservices, making it considerably larger than most other open-source implementations [33]. It was already part of various research studies [34,35], and according to the designers of the systems, they intended to provide a system with more public service APIs and deeper invocation chains than other open-source projects to better align with today's industry standards [33]. These architectural decisions and its extensive development history make it an ideal candidate for our analysis.

For our study, we focused on the backend architecture of the system and the API-based service interaction. i.e., we excluded the frontend-related *ts-ui-dashboard* microservice and the API Gateway as we do not consider these as parts of the backend. We also skipped three other services that were only connected to the UI or did not make any other inter-service calls: *ts-ticket-office*, *ts-avatar* (only called from UI without any other service interaction) and *ts-news-service* (does not contain any production code yet). This left us with 44 backend domain and infrastructure services, mostly implemented in Java, with the *voucher service* implemented in Python. This number should be large enough to represent a typical mid-size microservice system adequately.

Comparing the different branches and their commit frequency, we decided to run our analysis on the *Reconstruction* branch. From an evolutionary perspective, this branch provided the best source for reconstructing the system's evolution, as it contained the most commits and was frequently merged into the main branch. We selected every fourth entry from 440 commits in this branch, resulting in around 110 samples for our analysis process. Although our selection is not equally distributed throughout the repository's timeline, it should still cover enough of the development process to let us identify any possible trends (see Fig. 4).

| 9/22/17 | 4/10/18 | 10/27/18 | 5/15/19 | 12/1/19 | 6/18/20 | 1/4/21 | 7/23/21 | 2/8/22 | 8/27/22 | 3/15/23 |

Fig. 4. Distribution of selected commits throughout the repository's timeline.

As most service implementations use the Java-based Spring Framework, we only had to implement 13 different detectors to track all relevant features from the Java services and seven detectors for the Python-based service. The size of our detectors varies, with some containing only a few lines required to detect specific keywords and others having up to 50 lines of code needed to extract class methods and their annotations. Overall, the implementation effort for a system of such a size is manageable, especially since most detectors can be reused throughout different services and commits.

---

[1] https://github.com/FudanSELab/train-ticket.

4.1   Connection Anomaly Ratio

We first focused on the *Connection Anomaly Ratio* metric of every sample to verify our approach's correctness. All spikes we encountered in the trendline were cross-checked with our detector implementation to determine whether the reason for each missed connection could be a flawed detector. We continued this process iteratively until our detectors could cover most edge cases, and the remaining detector-related connection faults accounted for less than 1%, which we consider a value good enough. Aiming to reach an even lower fault rate would be challenging, as there were cases where target API addresses were constructed through various conditions depending on the input variables. These scenarios are hard to solve solely through pattern detection and require more complex parsing techniques, like reconstructing abstract syntax trees.

We identified only a short time range between 2018 and 2019, where the CAR metric peaked at 0.15, meaning that roughly 15% of interservice connections could not be resolved (see Fig. 5). Further investigation showed that it was due to a wrongly placed section in the configuration file of the *ts-config-service*. Whether this would manifest in real connection problems at runtime depends on how tolerant the system's configuration loader is against this type of error. However, we would still clearly flag this as an anomaly as it diverges significantly from all other services' configurations.

Despite that, we found only minor errors, e.g., isolated API calls using wrong target addresses. Overall, the system's interconnectivity implementation is very correct and mostly free of errors or issues.

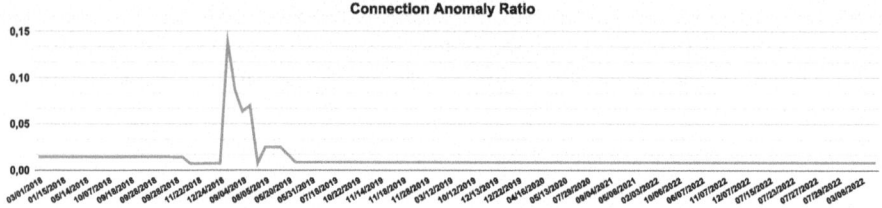

Fig. 5. *Connection Anomaly Ratio* throughout the *Reconstruction* branch. Despite a service misconfiguration at the beginning of 2019, which was later corrected, service interconnection remains highly error-free throughout the development history.

4.2   Average Service Interface Count and Path Length

The *ASIC* and *APL* metrics can both provide meaningful insights when assessing a system's architecture. A combination of high *ASIC* and *APL* values could indicate many small and atomic API operations that must be composed to implement more complex use cases, which in turn can result in longer invocation chains. Looking at the *train-ticket* system, we can see a similar pattern beginning in 2018 (Fig. 6): The number of available operations per service (blue line) increases slightly during development, which could indicate that the service interfaces have become more use-case-specific and less generic. However, verifying this assumption would require a deeper analysis of the APIs' semantics, which was out of the scope of our study.

Regarding the average number of API invocations (red line), we also see an increasing metric value till July 2022, i.e., calling a single API results in a growing number of follow-up calls. This trend is not unproblematic, as longer call paths create stronger interservice coupling, especially when these calls are synchronous and directed, as is the case here. Nevertheless, it seems the system's authors were also aware of this trend and restructured the architecture towards reduced invocation chains as the value dropped significantly in July 2022.

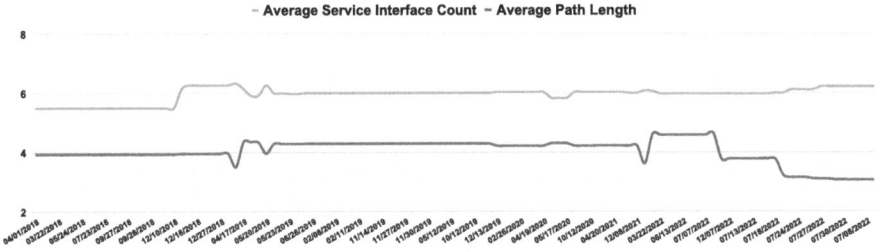

Fig. 6. While increasing *Average Service Interface Count* (blue) and *Average Path Length* (red) metrics indicate a stronger service coupling, several system redesigns were applied to constantly reduce the path length starting in the middle of 2022. (Color figure online)

4.3   Cycle Ratio and Asynchronous Service Communication

As Fig. 7 shows, there were times when between seven and up to 15% of API call chains in the system had at least one *weak* cycle, meaning the same service was addressed more than once during a call sequence [12]. These cyclic relations increase service coupling and can lead to higher network traffic, as calls must travel to the same service several times.

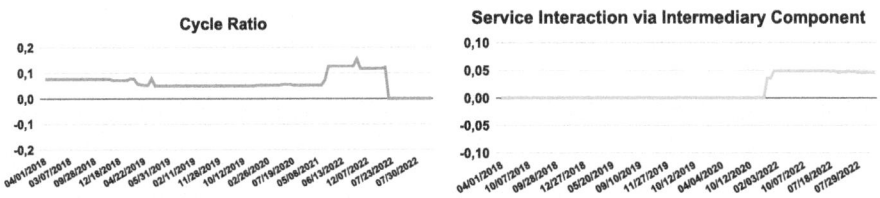

Fig. 7. *Ratio of cyclic connections* (left) and *Service Interaction via Intermediary Component Ratio* and other (right). While the amount of cycles decreases over time, the asynchronous communication increases. Both measures can help reduce the service coupling.

However, the system developers were probably aware of this problem and introduced a redesign as both metrics changed significantly by the end of the development history. Comparing two snapshots of the same API call from April and July 2022 (Fig. 8), the earlier one indicates a cyclic relation between the *Travel* and the *Seat Service*, creating a strong dependency and also additional overhead as the *TravelService* has to be visited twice. Moving further in the commit history, the system designers resolved this cycle: The second call to the *Travel Service* has been moved up in the hierarchy, making the actual caller – the *Route Plan Service* – responsible for orchestrating the API calls. Albeit this reduces the coupling, the problem of additional communication overhead remains. Here, the architecture could, for instance, be further improved by introducing a new API in the *TravelService* that bundles both requests.

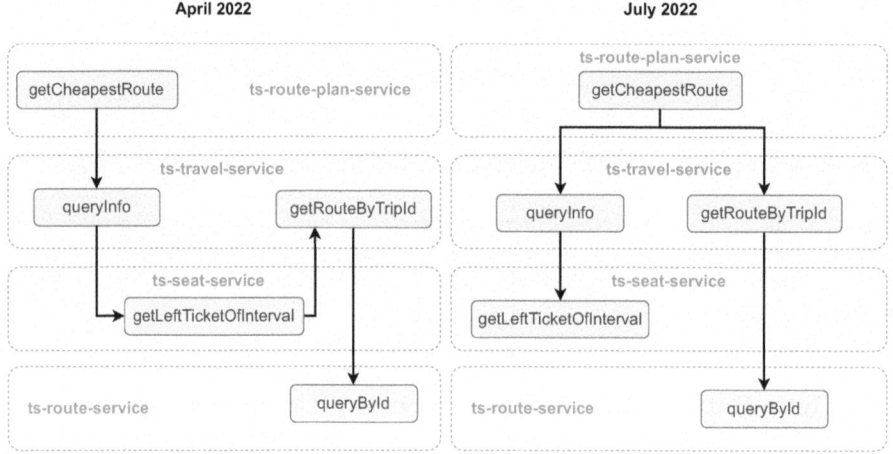

Fig. 8. Redesign of an API call chain. While the version from April 2022 contains a cyclic reference between two services, this was later resolved by letting the caller orchestrate the API calls instead.

Besides removing cycles, in 2021, the system's authors added asynchronous communication channels, mainly used for notification mechanisms. Compared to earlier versions of the system, this is another step towards a less coupled and more scalable solution, as it allows the services to be changed independently as long as they conform to the same message protocol and payload. Whether this approach could also be used to decouple additional parts of the business logic may depend on the individual scenario.

4.4   Summary

Overall, our static commit analysis revealed that the system is very stable. While there were some misconfigurations in the past, these were all corrected within a relatively short time frame. Following the trend of our metrics throughout *train ticket's* development history, we further see evidence suggesting the maintainers were constantly improving their architecture towards less coupling and better maintenance. One of the most vital indicators is the reduced length of API invocation chains and removing all cyclic paths in the system. Asynchronous communication patterns, added during the last third of the commit, further improve this trend towards decoupling.

In contrast, the number of API operations per service increases, which gives room for interpretation: Either existing functionality was split up into more fine-grained API operations, allowing for better API composition, or over time, more features were added to the services, increasing the overall number of available APIs. The simultaneous reduction of the average invocation path length suggests the latter, but a more semantic-based API analysis would be necessary here to verify this assumption.

5   Discussion

Considering **RQ1**, our case study demonstrated that our approach makes reconstructing a microservice's architectural evolution manageable, even for larger systems with several highly interconnected services. We developed a mining process consisting of two different strategies, one based on the analysis of individual snapshots while the other one focusing on applying the changes of every commit iteratively. Combining both methods makes our approach robust against errors while allowing for detailed and seamless in-depth analysis for specific points of interest. For this, we first analyzed a selection of snapshots to get an overall impression of the system's architectural trend while using change-based mining later to fill the gaps between relevant commits. We could also show that implementing the detectors to parse the code artifacts and decompose the architectural model is manageable when the system's code base follows coding guidelines and best patterns or practices. In that case, most detectors can be reused throughout different services, reducing the overall implementation effort.

Still, we also recognized that our implementation leaves room for improvement, e.g., using more optimized pattern matching.

When assessing and analyzing a microservice system's evolution (**RQ2**), we could also verify that the formal graph-based model generated by our detectors provides a solid ground for deriving various architectural metrics. By combining different metrics, we tracked various architectural qualities of our benchmarking system throughout its development history and identified significant architectural trends. In the case of our reference system, we could verify that some of our measured quality indicators, like cyclic dependencies or asynchronous communication, improved over time. We consider both valuable strategies to reduce the overall coupling within the system.

6 Threats to Validity

Construct Validity: Our detectors create an architectural model from source code, and with all models, there is always the risk of missing crucial details. We manually reviewed and cross-checked our model with the underlying implementation and documentation[2] to ensure all service interactions were tracked. We also added error checks during the reconstruction process to catch anomalies like missing target operations. However, implementing automated verification processes, such as additional runtime tests, would improve our work, a path we are exploring for future research.

External Validity discusses whether our results are generalizable to other systems or on a larger scale. While we doubt that there is such a thing as a general microservice architecture – every system is tailored to a specific domain and has to deal with unique requirements – the example application we chose for our case study is a widely accepted benchmark system with a relatively large scale compared to most other open-source implementations[3], making it well comparable to real-world systems. Besides, our approach could easily be applied to larger systems, considering the additional effort required to implement the detectors.

Internal Validity: Our whole approach to reconstructing the architecture relies on the source code and configuration files that evolve. While these artifacts are undoubtedly the most essential source of knowledge when investigating such systems, other aspects, like non-functional requirements or specific domain knowledge, may also affect the overall design of the architecture.

7 Conclusions and Future Directions

In this paper, we propose a method for monitoring microservice architecture evolution by reconstructing an API-based communication model through static source code repository analysis. Our mining process creates an architectural trend and allows seamless evolution analysis between selected commits. We employ lightweight, project-specific source code detectors to extract relevant architectural hotspots. The initial effort to implement these detectors is manageable even for large systems as long as recurring coding patterns and guidelines are applied. Our generated graph-based architectural model enables the derivation of various metrics to assess and identify quality trends. We validate our method with a case study on an extensive benchmark system.

---

[2] https://github.com/FudanSELab/train-ticket/wiki/Service-Guide-and-API-Reference.

[3] see, for instance, this curated list of various microservice open source systems: https://github.com/davidetaibi/Microservices_Project_List.

To our knowledge, this is the first study to demonstrate such an approach across the entire development history of a complex microservice system, providing software architects with a tool to monitor evolution and detect negative quality trends early.

Acknowledgments. This work was supported by: FWF (Austrian Science Fund) projects API-ACE: I 4268.

Data Availibility Statement. We offer the whole source code and data of our study in a data set published on Zenodo: https://doi.org/10.5281/zenodo.10961768.

References

1. Godfrey, M.W., German, D.M.: The past, present, and future of software evolution. In: 2008 Frontiers of Software Maintenance, pp. 129–138. IEEE (2008)
2. Whiting, E., Andrews, S.: Drift and erosion in software architecture: summary and prevention strategies. In: Proceedings of the 2020 the 4th International Conference on Information System and Data Mining, pp. 132–138 (2020)
3. Baum, D., Dietrich, J., Anslow, C., Müller, R.: Visualizing design erosion: how big balls of mud are made. In: 2018 IEEE Working Conference on Software Visualization (VISSOFT), pp. 122–126. IEEE (2018)
4. Rios, N., Spínola, R.O., Mendonça, M., Seaman, C.: The most common causes and effects of technical debt: first results from a global family of industrial surveys. In: Proceedings of the 12th ACM/IEEE International Symposium on Empirical Software Engineering and Measurement, pp. 1–10 (2018)
5. De Silva, L., Balasubramaniam, D.: Controlling software architecture erosion: a survey. J. Syst. Softw. **85**(1), 132–151 (2012)
6. Dragoni, N., Giallorenzo, S., Lafuente, A.L., Mazzara, M., Montesi, F., Mustafin, R., Safina, L.: Microservices: yesterday, today, and tomorrow. In: Present and Ulterior Software Engineering, pp. 195–216. Springer, Cham (2017). https://doi.org/10.1007/978-3-319-67425-4_12
7. Newman, S.: Building Microservices. O'Reilly Media, Inc., Newton (2021)
8. Bogner, J., Fritzsch, J., Wagner, S., Zimmermann, A.: Limiting technical debt with maintainability assurance: an industry survey on used techniques and differences with service-and microservice-based systems. In: Proceedings of the 2018 International Conference on Technical Debt, pp. 125–133 (2018)
9. Bogner, J., Wagner, S., Zimmermann, A.: Automatically measuring the maintainability of service-and microservice-based systems: a literature review. In: Proceedings of the 27th International Workshop on Software Measurement and 12th International Conference on Software Process and Product Measurement, pp. 107–115 (2017)
10. Zdun, U., Navarro, E., Leymann, F.: Ensuring and assessing architecture conformance to microservice decomposition patterns. In: Maximilien, M., Vallecillo, A., Wang, J., Oriol, M. (eds.) ICSOC 2017. LNCS, vol. 10601, pp. 411–429. Springer, Cham (2017). https://doi.org/10.1007/978-3-319-69035-3_29
11. Walker, A., Das, D., Cerny, T.: Automated code-smell detection in microservices through static analysis: a case study. Appl. Sci. **10**(21), 7800 (2020)

12. Ma, S.-P., Fan, C.-Y., Chuang, Y., Liu, I.-H., Lan, C.-W.: Graph-based and scenario-driven microservice analysis, retrieval, and testing. Futur. Gener. Comput. Syst. **100**, 724–735 (2019)
13. de Toledo, S.S., Martini, A., Przybyszewska, A., Sjøberg, D.I.: Architectural technical debt in microservices: a case study in a large company. In: 2019 IEEE/ACM International Conference on Technical Debt, pp. 78–87. IEEE (2019)
14. Ducasse, S., Pollet, D.: Software architecture reconstruction: a process-oriented taxonomy. IEEE Trans. Softw. Eng. **35**(4), 573–591 (2009)
15. Cerny, T., Abdelfattah, A.S., Bushong, V., Al Maruf, A., Taibi, D.: Microservice architecture reconstruction and visualization techniques: a review. In: 2022 IEEE International Conference on Service-Oriented System Engineering (SOSE), pp. 39–48. IEEE (2022)
16. Bushong, V., Das, D., Al Maruf, A., Cerny, T.: Using static analysis to address microservice architecture reconstruction. In: 2021 36th IEEE/ACM International Conference on Automated Software Engineering (ASE), pp. 1199–1201. IEEE (2021)
17. Bushong, V., Das, D., Cernỳ, T.: Reconstructing the holistic architecture of microservice systems using static analysis. In: CLOSER, pp. 149–157 (2022)
18. Barnes, J.M., Garlan, D., Schmerl, B.: Evolution styles: foundations and models for software architecture evolution. Softw. Syst. Model. **13**, 649–678 (2014)
19. Moreira, M.G., De França, B.B.N.: Analysis of microservice evolution using cohesion metrics. In: Proceedings of the 16th Brazilian Symposium on Software Components, Architectures, and Reuse, pp. 40–49 (2022)
20. Tizzei, L.P., Azevedo, L., Soares, E., Thiago, R., Costa, R.: On the maintenance of a scientific application based on microservices: an experience report. In: 2020 IEEE International Conference on Web Services (ICWS), pp. 102–109. IEEE (2020)
21. Sampaio, A.R., et al.: Supporting microservice evolution. In: 2017 IEEE International Conference on Software Maintenance and Evolution (ICSME), pp. 539–543. IEEE (2017)
22. He, X., Shao, Z., Wang, T., Shi, H., Chen, Y., Wang, Z.: Predicting effect and cost of microservice system evolution using graph neural network. In: Monti, F., Rinderle-Ma, S., Ruiz Cortes, A., Zheng, Z., Mecella, M. (eds.) International Conference on Service-Oriented Computing, pp. 103–118. Springer, Heidelberg (2023). https://doi.org/10.1007/978-3-031-48421-6_8
23. Stocker, M., Zimmermann, O.: From code refactoring to API refactoring: agile service design and evolution. In: Barzen, J. (ed.) SummerSOC 2021. CCIS, vol. 1429, pp. 174–193. Springer, Cham (2021). https://doi.org/10.1007/978-3-030-87568-8_11
24. Lercher, A., Glock, J., Macho, C., Pinzger, M.: Microservice api evolution in practice: a study on strategies and challenges. arXiv:2311.08175 (2023)
25. Assunção, W.K., Krüger, J., Mosser, S., Selaoui, S.: How do microservices evolve? an empirical analysis of changes in open-source microservice repositories. J. Syst. Softw. **204**, 111788 (2023)
26. Heseding, F., Scheibel, W., Döllner, J.: Tooling for time-and space-efficient git repository mining. In: Proceedings of the 19th International Conference on Mining Software Repositories, pp. 413–417 (2022)
27. Kolassa, C., Riehle, D., Salim, M.A.: A model of the commit size distribution of open source. In: van Emde Boas, P., Groen, F.C.A., Italiano, G.F., Nawrocki, J., Sack, H. (eds.) SOFSEM 2013. LNCS, vol. 7741, pp. 52–66. Springer, Heidelberg (2013). https://doi.org/10.1007/978-3-642-35843-2_6

28. Rosik, J., Le Gear, A., Buckley, J., Babar, M. A., Connolly, D.: Assessing architectural drift in commercial software development: a case study. Softw. Pract. Exp. **41**(1), 63–86 (2011)
29. Ntentos, E.: Detector-based component model abstraction for microservice-based systems. Computing **103**(11), 2521–2551 (2021)
30. Genfer, P., Zdun, U.: Identifying domain-based cyclic dependencies in microservice APIs using source code detectors. In: Biffl, S., Navarro, E., Löwe, W., Sirjani, M., Mirandola, R., Weyns, D. (eds.) ECSA 2021. LNCS, vol. 12857, pp. 207–222. Springer, Cham (2021). https://doi.org/10.1007/978-3-030-86044-8_15
31. Hirzalla, M., Cleland-Huang, J., Arsanjani, A.: A metrics suite for evaluating flexibility and complexity in service oriented architectures. In: Feuerlicht, G., Lamersdorf, W. (eds.) ICSOC 2008. LNCS, vol. 5472, pp. 41–52. Springer, Heidelberg (2009). https://doi.org/10.1007/978-3-642-01247-1_5
32. Ntentos, E., Zdun, U., Plakidas, K., Meixner, S., Geiger, S.: Assessing architecture conformance to coupling-related patterns and practices in microservices. In: Jansen, A., Malavolta, I., Muccini, H., Ozkaya, I., Zimmermann, O. (eds.) ECSA 2020. LNCS, vol. 12292, pp. 3–20. Springer, Cham (2020). https://doi.org/10.1007/978-3-030-58923-3_1
33. p Zhou, X., et al.: Benchmarking microservice systems for software engineering research. In: Proceedings of the 40th International Conference on Software Engineering: Companion Proceeedings, ser. ICSE 2018, pp. 323–324. Association for Computing Machinery, New York (2018). https://doi.org/10.1145/3183440.3194991
34. Li, B., Peng, X., Xiang, Q., Wang, H., Xie, T., Sun, J., Liu, X.: Enjoy your observability: an industrial survey of microservice tracing and analysis. Empir. Softw. Eng. **27**, 1–28 (2022)
35. Zhang, C., et al.: Deeptralog: trace-log combined microservice anomaly detection through graph-based deep learning. In: Proceedings of the 44th International Conference on Software Engineering, pp. 623–634 (2022)

Temporal Community Detection in Developer Collaboration Networks of Microservice Projects

Alexander Bakhtin$^{(\boxtimes)}$ (ID), Xiaozhou Li (ID), and Davide Taibi (ID)

Univeristy of Oulu, Oulu, Finland
{alexander.bakhtin,xiaozhou.li,davide.taibi}@oulu.fi

Abstract. Analysis of developer collaboration in software projects can provide meaningful insights into the development culture of the team as well as enforce good and prevent bad development practices, such as observing the *one service - one team* rule for microservice projects. In this paper, we combine two previous works from different domains to perform temporal community detection in two networks of developer collaboration. We perform a case study of an open-source microservice benchmark project. We find several communities and their activity trends across time. The findings show that only one connected group of developers was working on the project at any given time, with several core developers consistently contributing and many developers entering and leaving the project. To our knowledge, it is the first attempt to apply temporal network analysis to developer collaboration networks.

Keywords: microservices · temporal networks · community detection · developer collaboration

1 Introduction

Microservice has been one of the most popular software architecture styles for cloud-native applications in the last decade. Many studies have been conducted in academia on maintaining and enhancing the overall quality of microservices by reconstructing the architecture on different layers [4]. However, besides the analysis conducted on the code level, the potential benefits from the microservice architecture maintenance on the organizational level have drawn increasing attention lately [2].

One approach towards understanding the organizational structure of microservice teams is to investigate it based on the collaboration of developers. Wolf et al. found in their study that developer collaboration plays a crucial role in determining the quality of software integration [17]. In the industry, the practitioners claim that "one microservice per developer/team" [10] is the best practice for microservice projects, while, for example, in OSS, the reality is hardly the case [1].

M. Galster et al. (Eds.): ECSA 2024, LNCS 14889, pp. 174–182, 2024.
https://doi.org/10.1007/978-3-031-70797-1_11

Li et al. [7] propose to use network modularity as the metric for monitoring the changes in the communities of developer collaboration networks through different releases of the project. In particular, they aim to *monitor the changes in the organization structure ... during system evolution* [7]. However, the authors determine the communities for each version separately and only comment on the changes in the structure by observing the change in the modularity metric. They do not comment on the composition of the communities or their evolution through time.

About such approaches, Gauvin et al. say that they *prove useful in specific cases but fail in the case of discontinuous activity patterns, abrupt structure formation or dissolution* [5].

Hence, to extend the work of Li et al. [7], we conducted a case study [12] on the same open-source project, `eShopOnContainers` [9], and considered the following research questions:

RQ$_1$ *Can we identify community structure evolution using the temporal network analysis?*

RQ$_2$ *Can we observe whether one service - one team rule was followed during the development, and at which point?*

By utilizing the selected temporal community detection method [5], we detected the network communities of the project and observed their activity trends over time. The contributions of this study are: **1)** demonstrate the usefulness of the temporal network perspective for understanding the evolution of developer collaboration structure; **2)** investigate the *one service - one team* principle via a case study with temporal network analysis.

The remainder of the article is organized as follows. Section 2 presents the previous works related to this study. Section 3 presents the steps, and Sect. 4 the results of the case study. Section 5 discusses the Threats to Validity while Sect. 6 concludes the article.

2   Related Works

Many studies have adopted methods to detect the latent social structures in the developer communities of software projects. The collaboration network is one of the commonly adopted methods for such analysis. Via collaboration network analysis, Singh finds a statistically significant connection between the technical and commercial success of software projects and the small-world properties of its developer community [14]. Surian et al. extract sub-graph patterns from large graphs among collaborating developers via leveraging network-level properties [16]. Meneely et al. examine the collaboration structure with the developer network using code churn information and predict the failures [8].

Though these studies adopted network analysis on developer collaboration, such approaches are not extended to microservice-related projects. Li et al. propose to use the modularity of developer collaboration network communities as the metric for monitoring the changes in developer collaboration and activity through different releases of microservice-based projects [7]. The study analyzes each snapshot network independently and then comments on the general trends in the community structure and centrality scores. On the other hand, regarding developer collaboration in microservice-based projects, d'Aragona et al. investigate the application of the "one microservice per developer" principle in OSS microservice projects as well as the technical roles of different developer groups using exploratory factor analysis (EFA) [1].

Saarimäki et al. discuss the temporal factor as a novel but critical perspective in software engineering, especially with time-dependent data involved [13]. However, they consider temporal series data and not temporal graphs. In Software Engineering, Rotschke and Schurr applied temporal graph queries to provide a natural means to express trend-oriented metrics and consistency rules to support software evolution [11]. However, to our knowledge, no studies have used temporal graph analysis for the analysis of microservice architecture or its organizational structure.

3   Case Study: eShopOnContainers

In this section, we describe the case study: the selected benchmark project, the data collection process adapted from [7], and the community detection method adapted from [5].

The used benchmark is eShopOnContainers, a project developed by Microsoft to showcase their .NET framework and Azure cloud infrastructure for developing and deploying microservices [9].

3.1   Data Collection

Data collection consists of mining the commits of the projects, and constructing the developer collaboration network based on the amount of co-modified files and microservices.

Commit Mining. We create our own tool MisON[1], which is a wrapper around the pydriller library [15], to enable running different mining jobs from the command line or script.

We use the same releases as [7] to split the data and adopt the same method of splitting the commits by the release date. We consider all commits from all branches. Li et al. in [7] use release 5.0.0 as one of the networks. However, the release was prepared by squashing the commits of a feature branch (i.e., all commits of a branch represented as one commit in the main branch). So, we remove the release 5.0.0 from the final analysis, because pydriller cannot identify the original commits.

---

[1] https://github.com/M3SOulu/mison.

Developer Collaboration Network Construction. After we collect the commits for each release, we construct the developer collaboration networks. For each release, two developers are nodes in the network joined by a link whose weight is the number of files (or microservices) that the developers have in common. The links are undirected.

We identify unique developers based on the author email field. After listing all the available emails, we noticed that several people seemed to have several emails, so we manually grouped emails evidently belonging to the same person.

Table 1 summarizes the network snapshots for each release. Data collection and replication scripts are available in our repository.[2]

Table 1. Summary of the mined releases and the corresponding networks

Release	From date	To date	File collaboration		MS Collaboration	
			Nodes	Edges	Nodes	Edges
2.0.0	–	2017-10-27	48	484	29	441
2.0.5	2017-10-27	2018-04-05	20	61	16	106
2.0.8	2018-04-05	2018-11-12	23	49	15	95
2.2.0	2018-11-12	2019-03-21	12	23	12	57
3.0.0	2019-03-21	2019-11-26	22	37	14	83
5.0.0	2019-11-26	2021-10-27	–	–	–	–

3.2 Data Analysis

The data analysis consists of performing temporal community detection in the mined networks and examining and describing the output to answer our research questions.

Community Detection with PARAFAC Decomposition. In studies about temporal networks, the temporal network with N nodes and T timestamps is usually represented as a third-order tensor $T \in R^{N \times N \times T}$, so that the element $T[i, j, t]$ is the weight of the link connecting node i to node j at time t.

Gauvin et al. [5] propose to use the PARARAC decomposition of the tensor [6] to obtain the community structure of the temporal network.

Given the temporal network tensor T and the desired amount of communities R as input, the algorithm outputs three matrices: $A, B \in R^{N \times R}$ show the strengths of each node n belonging to each community r, and $C \in R^{T \times R}$ shows the relative (raw) activity score of each community r in each timestamp t, according to Gauvin et al. in [5].

The authors, however, propose to scale the temporal activity of communities. They do not comment on the differences to using the raw scores. We will only consider raw scores in our results.

---

[2] See Data Availability Statement at the end.

Selecting the Number of Communities. We follow the approach of [5] when selecting R (the number of communities to detect) - we run the community detection algorithm several times for increasing R, and for each outcome, determine the quality of the fit PARAFAC model by computing the core consistency metric [3]. The metric is scaled to have a maximum of 100, and values above 50 are considered good [3]. We compute the core consistency metric for different choices of $R = \{2, 3, 4, 5, 6\}$. The values are $\{99.94, -86042.67, -101250.98, -127061.42, -1655778.94\}$ and $\{99.44, 99.29, 97.73, 96.22, 77.12\}$ for file and service collaboration networks, respectively. We then determine the best choice of R using the *elbow method* - we select the first value of R after which the value of the metric drops sharply. Judging by the numbers, $R = 2$ is the best choice for the file network and $R = 5$ for the service network.

4   Results

After performing the community detection with the selected amount of communities, we are ready to answer the research questions. The output of the decomposition is available with the rest of the replication scripts in our repository.[3]

RQ$_1$ - Can we identify community structure evolution using the temporal network analysis?

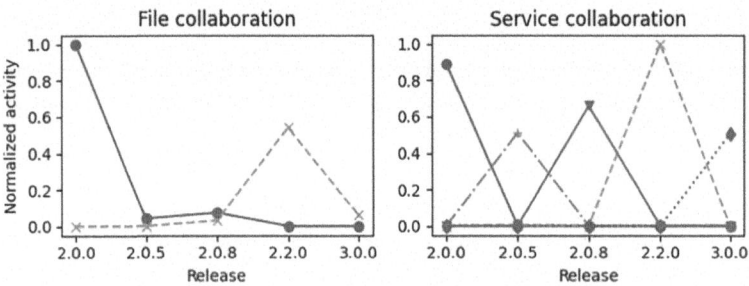

Fig. 1. Temporal activities of the communities detected in file and service collaboration networks. Values are normalized with respect to the maximum.

As shown in Fig. 1, we can observe certain aspects of the temporal community structure that would be hard to find by purely static methods:

When it comes to file collaboration network, we see that there are only two communities in the data. The activity plot indicates that one of the communities (blue solid line with dots) is active in earlier releases, with the other one (orange dashed line with crosses) gaining traction after release 2.0.5 and being the more

---

[3] See Data Availability Statement at the end.

important community in release 2.2.0. This was not observed by Li et al. [7] - their static method did not allow for overlapping or evolving communities, so they found a fit of 2 or 3 communities to each snapshot with a low modularity score, which implies a poor fit.

As for the service collaboration network, the activity results show something surprising - the five identified communities are completely disjoint in their activity, with each of the five releases being dominated by only one community of the developers that are highly connected, so no teams are working in parallel - there is only one team with several core members that persist throughout the development and other developers contributing different releases of the project. Such structure could not have been observed with static methods - Li et al. [7] could only fit 2 communities to all snapshots in the service network, and they comment on the low modularity score for all those fits, which is now also supported by our results - we see that at each release, there is essentially only one strongly connected group of people.

By considering the strengths of belonging to communities provided by the decomposition[4], we see that some developers have significant degrees of participation in several communities.[5] A total of 11 developers participate in more than one community. Notably, one developer participates actively in all five communities, so this must be a core developer overseeing the project. They and three other developers have a near maximum strength of participation before release 3.0.0. One developer participates in four out of five communities (joined after release 2.0.0). Five developers belong to three communities, and four developers - to two different communities. On the other hand, we notice that the majority (52) of the developers participate in only one community - 22, 9, 9, 7, and 5 developers participate in only one of the five communities, respectively.

Li et al. also list *core contributors* that they identified through centrality analysis, and the list is almost the same for all releases (Table III in [7]). While it could be deduced that the core contributors are the same for all snapshots and that the rest join and leave the project at different times by carefully observing the centrality scores of developers in each snapshot, our temporal method provided such an insight automatically.

RQ2 - Can we observe whether *one service - one team* rule was followed during the development, and at which point? The temporal community detection results indicate that the *one service - one team* rule was not followed. The detected community structure and activities in the service network show that there was always just one strongly connected group of developers contributing to the project, with several core members and many people joining and leaving the group at different stages of development.

---

[4] We do not demonstrate the strength values here due to space limitations.

[5] We remind that the same node in the network can belong to several communities with different strengths.

5   Threats to Validity

We acknowledge the possible threats and challenges to the validity of this work:

Construct Validity. The correct mapping of commits to different releases, files to corresponding microservices, and identifying unique contributors all play a crucial role in constructing a valid network for analysis.

Internal Validity. Since we lack the ground truth of developers' intended division into teams and the distribution of tasks, we can only make preliminary and general conclusions about the identified community structure. We only use 5 snapshots of a network, while the original study has over 100. The choice of core consistency as the guiding metric to select the best fit of communities is motivated in the original work. However, different choices guided by other principles are possible.

External Validity. Community detection is always exploratory in nature, and the results of community detection are only as good as the interpretation of communities that is possible to derive from observation. In this work, we lack the validation by the developers of the project of whether or not the identified communities (and lack of separate teams) provide any meaningful insight into their operations and workflow. Additionally, *one service - one team* is a desired trend in the Industry, while the current case study used an OSS project.

6   Conclusion

In this work, we extended the approach by Li et al. [7] to construct developer collaboration networks based on file and microservice co-editing and detect communities by applying a temporal community detection algorithm based on PARAFAC decomposition of the tensor representing the network, originally proposed by Gauvin et al. in [5], and performing a case study on eShopOnContainers project.

We determined the activity patterns of communities during development. The results indicate that the *one service - one team* rule was not followed: at every point during development, there was only one connected team of developers contributing to the project, with 1 core maintainer, 11 developers contributing code to several releases of the project and other 52 developers joining and leaving the project at different stages.

Future development of this work could include industrial validation in a microservice company, and comparing against other temporal community detection methods.

Acknowledgments. This material is based upon work supported by grants from the Research Council of Finland (grants n. 349487 and 349488 - MuFAno).

Data Availibility Statement. We provide the replication package of the work consisting of the tools and data replication scripts in our repositories https://doi.org/10.5281/zenodo.12190571 https://github.com/M3SOulu/ECSA2024-eShopOnContainers.

References

1. Amoroso d'Aragona, D., Li, X., Cerny, T., Janes, A., Lenarduzzi, V., Taibi, D.: One microservice per developer: is this the trend in OSS? In: Papadopoulos, G.A., Rademacher, F., Soldani, J. (eds.) European Conference on Service-Oriented and Cloud Computing, vol. 14183, pp. 19–34. Springer, Heidelberg (2023). https://doi.org/10.1007/978-3-031-46235-1_2

2. Baškarada, S., Nguyen, V., Koronios, A.: Architecting microservices: practical opportunities and challenges. J. Comput. Inf. Syst. (2020)

3. Bro, R., Kiers, H.A.L.: A new efficient method for determining the number of components in parafac models. J. Chemom. **17**(5), 274–286 (2003)

4. Cerny, T., Abdelfattah, A.S., Bushong, V., Al Maruf, A., Taibi, D.: Microservice architecture reconstruction and visualization techniques: a review. In: 2022 IEEE SOSE, pp. 39–48. IEEE (2022)

5. Gauvin, L., Panisson, A., Cattuto, C.: Detecting the community structure and activity patterns of temporal networks: a non-negative tensor factorization approach. PLOS ONE **9**(1), 1–13 (2014)

6. Kolda, T.G., Bader, B.W.: Tensor decompositions and applications. SIAM Rev. **51**(3), 455–500 (2009)

7. Li, X., Abdelfattah, A.S., Yero, J., d'Aragona, D.A., Cerny, T., Taibi, D.: Analyzing organizational structure of microservice projects based on contributor collaboration. In: 2023 IEEE SOS, pp. 1–8 (2023)

8. Meneely, A., Williams, L., Snipes, W., Osborne, J.: Predicting failures with developer networks and social network analysis. In: 2008 ACM SIGSOFT FSE, pp. 13–23 (2008)

9. Microsoft: Introduce eshoponcontainers reference application (2024). https://learn.microsoft.com/en-us/dotnet/architecture/cloud-native/introduce-eshoponcontainers-reference-app

10. Richardson, C.: Microservice patterns: Service per team (2023). https://microservices.io/patterns/decomposition/service-per-team.html. Accessed 17 Apr 2024

11. Rötschke, T., Schürr, A.: Temporal graph queries to support software evolution. In: Corradini, A., Ehrig, H., Montanari, U., Ribeiro, L., Rozenberg, G. (eds.) ICGT 2006. LNCS, vol. 4178, pp. 291–305. Springer, Heidelberg (2006). https://doi.org/10.1007/11841883_21

12. Runeson, P., Höst, M.: Guidelines for conducting and reporting case study research in software engineering. Empir. Softw. Eng. **14**(2), 131–164 (2009)

13. Saarimäki, N., Moreschini, S., Lomio, F., Penaloza, R., Lenarduzzi, V.: Towards a robust approach to analyze time-dependent data in software engineering. In: 2022 SANER, pp. 36–40. IEEE (2022)

14. Singh, P.V.: The small-world effect: the influence of macro-level properties of developer collaboration networks on open-source project success. ACM Trans. Softw. Eng. Methodol. (TOSEM) **20**(2), 1–27 (2010)

15. Spadini, D., Aniche, M., Bacchelli, A.: PyDriller: python framework for mining software repositories (2018)

16. Surian, D., Lo, D., Lim, E.P.: Mining collaboration patterns from a large developer network. In: 2010 17th Working Conference on Reverse Engineering, pp. 269–273. IEEE (2010)
17. Wolf, T., Schroter, A., Damian, D., Nguyen, T.: Predicting build failures using social network analysis on developer communication. In: 2009 ICSE, pp. 1–11. IEEE (2009)

Uncertainty Calculation-as-a-Service: Microservice-Based Metrology Applications

Anil Cetinkaya[1,2]([✉])[iD], M. Cagri Kaya[3][iD], Teklie Belay Bzuneh[3], and Halit Oguztuzun[1][iD]

[1] Department of Computer Engineering, Middle East Technical University, 06800 Ankara, Türkiye
{cetinkaya,oguztuzn}@ceng.metu.edu.tr
[2] Department of Computer Engineering, Iskenderun Technical University, 31200 Hatay, Türkiye
[3] Department of Computer Science and Engineering, Chalmers University of Technology and University of Gothenburg, 41756 Gothenburg, Sweden
cagri.kaya@chalmers.se, gusbzute@student.gu.se

Abstract. The calibration industry faces significant challenges due to its diverse and sophisticated equipment and complex traditional processes. Rapidly advancing technology highlights existing challenges while inspiring the adoption of innovative solutions to meet industry demands. This paper introduces a microservice-based cloud architecture that addresses these difficulties by managing the inherent heterogeneity in the industry. The presented architecture combines various equipment types, communication technologies, and diverse stakeholder expectations into a cohesive system that ensures efficiency and accuracy in calibration processes by utilizing different methods of uncertainty calculation and facilitating the generation of digital calibration certificates (DCCs). Our solution provides a holistic approach to managing data flow from the calibration equipment to the final generation of DCCs, employing cloud-based services in between to process data.

Keywords: digital metrology · microservices · software architecture · uncertainty calculation

1 Introduction

The continuous development of technology reshapes every sector. Demands for reducing costs, accelerating business processes, and efficiently using human resources are increasing. The metrology and calibration industry is no exception to this as it undergoes a digital transformation [18] to adapt to the new age. This transformation is characterized by the formulation and dissemination of data standards and the adoption of automated and network-based solutions [14]. In this way, efficiency and data accuracy can be improved, allowing stakeholders to remain competitive in a rapidly evolving global marketplace.

Any measurement equipment has a margin of error in its measurements and may deviate over time due to various reasons, including misuse and external factors such as temperature and humidity. To keep this shift in the results of the equipment within acceptable limits, devices need to be calibrated regularly. In the calibration process [6], the device under test (DUT) is evaluated at specific test points, and the measurement results are recorded. A trusted "standard device", typically calibrated by a higher authority [10], is used for comparison. After calibration, the DUT receives a calibration certificate that includes device details, calibration lab information, and uncertainty values.

The calibration process roughly involves three phases: data collection, uncertainty calculation, and calibration certificate generation. Manual data handling is still common throughout the process. Although there are tools for automation, they mostly focus on specific parts of the process, such as device communication and uncertainty calculation [13], lacking an end-to-end perspective. Additionally, traditional calibration certificates are often paper-based or digitally signed PDFs, which are considered the cheapest and safest options. However, machine-readable digital calibration certificate (DCC) standards [9] are now being proposed to address these concerns. Therefore, there is a need for an application that handles the calibration process holistically: starting with data collection from the calibration equipment, performing uncertainty calculations, and producing the calibration certificate. Our application meets these expectations by adopting state-of-the-art cloud-based development to better contribute to the digital transformation of the metrology and calibration industry.

This paper demonstrates how microservice-based solutions improve the digitalization of the metrology and calibration industry through cloud-based, scalable, and maintainable applications, ensuring data accuracy, integrity, and standardization. We developed a cloud-based application capable of handling different stakeholders' needs and data, performing uncertainty calculations, and producing calibration certificates on the cloud. Two specialized microservices were developed to perform uncertainty calculations, each based on a specific technique and implemented in different technologies. Additionally, we developed microservices for the generation and management of DCCs, designed with the flexibility to accommodate future extensions. Google Kubernetes Engine (GKE) [8] was utilized for orchestration and other benefits including automated roll-outs and rollbacks, easier management of containerized services, and automatic scaling and load balancing capabilities.

In previous work, Internet of Measurement Things (IoMT) was introduced as a layered architecture with physical, cloud services, and application layers specifically designed for the metrology and calibration industry [17] [12]. The AutoRFPower application was initially developed as a desktop tool to automatically calculate uncertainties in RF power measurement devices [2]. Following the IoMT architecture, the concept and initial steps of migrating the uncertainty modules of AutoRFPower to a cloud environment were outlined [3]. **In this paper**, we enhance the IoMT architecture with a microservices perspective and demonstrate its applicability by migrating specific modules of AutoRF-Power to the cloud. These modules were re-implemented as microservices while maintaining the original business logic. Additionally, new functionalities were added as microservices including DCC generation modules.

2   Related Work

To the best of our knowledge, our approach is one of the pioneering efforts to address the calibration process holistically. Existing studies, tend to focus on specific aspects or phases of calibration. As a result, there are only a limited number of studies available for comparison.

Zet et al. [20] describe an automated process for calibration and DCC generation using blockchain technology to improve accessibility among stakeholders. In contrast, our system utilizes cloud technologies, offering enhanced flexibility and scalability. Our approach focuses on RF Power measurements and incorporates microservices to integrate various measurement types, reducing interoperability issues. This architecture supports the easy integration of different standards and measurements, without being constrained by the inherent limitations of blockchain technology.

Oppermann et al. [15] introduce the "operation layer" at PTB (Physikalisch-Technische Bundesanstalt) to enhance domain workflows using a cloud-native, distributed microservices architecture. This layer streamlines processes, breaks down data silos, and automates the creation of DCCs by connecting laboratory workflows with administrative data. While their approach provides a broad solution for managing and improving workflows in the domain, including the DCC workflows, we propose a more focused tool for the calibration process providing details at the implementation level.

The NIST Uncertainty Machine (NUM) [13] offers uncertainty calculations through an online tool with a client-server architecture. Unlike our approach, which integrates measurements from physical equipment and generates DCCs, NUM focuses solely on server-side calculations without any addressing hardware-level measurements or DCC production. Another example is Metas.UncLib [19], a stand-alone desktop application designed to solve complex metrology problems, but it lacks a cloud-based perspective.

3   Proposed Approach

IoMT was extended by adopting microservices architecture and related technologies on the cloud layer. Figure 1 represents the overall architecture detailing the entities residing in each layer.

The physical layer comprises measurement setups that include a PC configured to control and process data generated by the calibration equipment which typically includes power signal generator, power meter, power sensor, and attenuator. All components in this layer work together to enable precise measurement and calibration. Device-specific drivers and communication libraries are also part of this layer.

The cloud services layer contains uncertainty calculation, DCC generation, and cloud storage services. Both measurement and user data are hosted on

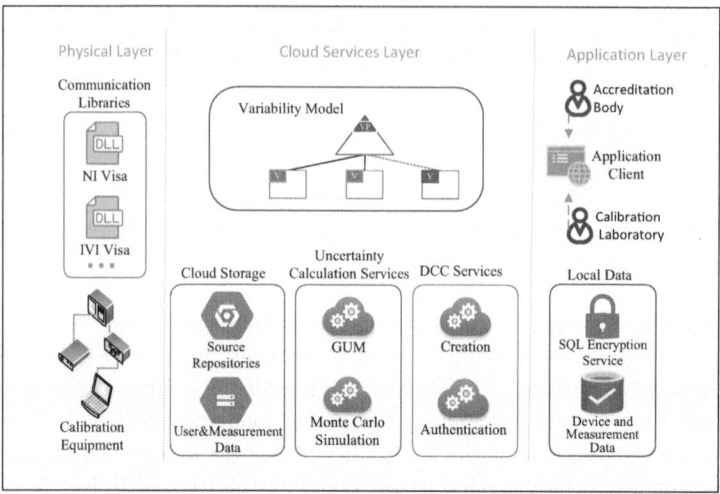

Fig. 1. An abstract representation of the realized architecture and its components.

Microsoft SQL Server in a container, ensuring high availability and secure data handling. Our architecture employes advanced encryption and access control mechanisms to secure data transmission and storage, facilitating rapid data retrieval and processing, which is critical for real-time uncertainty calculations and DCC generation. This layer supports high data throughput via purpose-built microservice containers and interacts seamlessly with lower layers, enabling effective product configuration using variability models.

The application layer hosts the User Interface (UI) along with essential components that facilitate communication with the hardware in the physical layer. Furthermore, the certificate information for the registered devices is collected from users and safely stored in an encrypted database. Communication with the measurement devices in the lower level is handled using the drivers and libraries located in this layer, then measurement results are kept in the mentioned storage.

3.1   Utilizing a CI/CD Pipeline

Figure 2 illustrates the system's overall architecture which uses Google Cloud. We have adopted and enhanced the recommended workflow for Google Cloud applications by integrating automated triggers and variability resolution for the development and deployment of containerized applications, utilizing a Continuous Integration and Continuous Delivery/Deployment (CI/CD) pipeline [7]. This architecture supports a full-cycle development workflow, starting with the collection of measurements from the hardware at the lowest level, transferring this data to the cloud environment, performing uncertainty calculations, and ultimately generating DCCs.

Our CI/CD pipeline relies on three key Google Cloud services: Cloud Build, Artifact Registry, and Cloud Deploy. Cloud Build compiles source code, runs

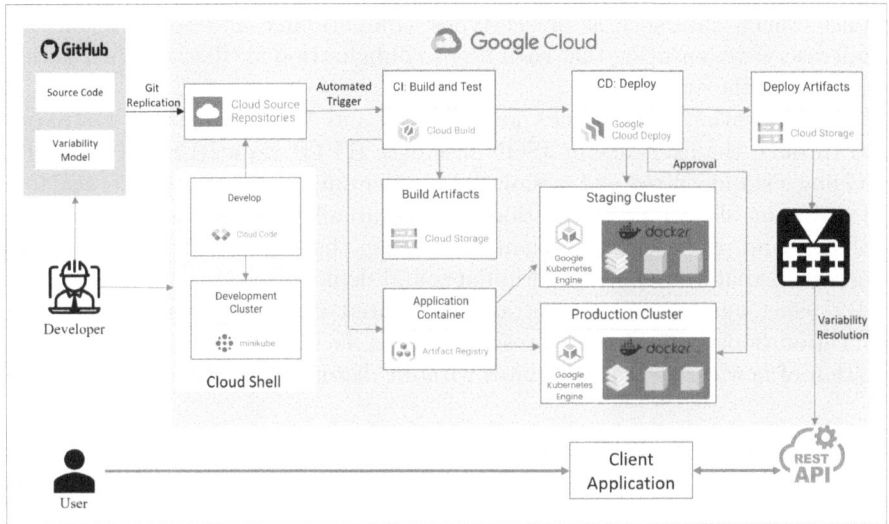

Fig. 2. Overall architecture and CI/CD pipeline (Adapted from [7]).

tests, and creates deployable software packages. Artifact Registry securely manages container images and libraries, and Cloud Deploy automates application delivery to Google Cloud environments. These services collectively streamline software deployment, improving the system's development lifecycle.

In addition to the conventional workflow, an automated trigger has been added to the Cloud Build Service using YAML configuration files. It activates when a Git repository is replicated to the Cloud Source Repository. This triggers a new build for any updates, which is then staged for deployment without interrupting the running instance.

Moreover, our system architecture utilizes Docker containers to host various microservices within the cloud, orchestrated through GKE. Each microservice is securely isolated in its own container, which boosts development speed, scalability, and fault tolerance while ensuring consistent environments across all phases of development, testing, and production. The use of GKE enhances our system's resource management, enables automatic scaling, and improves load balancing, which is vital for computationally demanding tasks. Overall, this setup increases operational efficiency and boosts system reliability and adaptability in response to the rapid technological changes and industry demands.

3.2 Variability Management for Product Configuration

Users of AutoRFPower are calibration laboratories that use different families of equipment with diverse communication protocols and libraries. To manage this diversity and the resulting code variability, we have systematically integrated a Textual Variability Model (TVM) into the source code. This model handles crucial data such as device types, compatibility, and access tokens for all microservices, ensuring that each user's configuration file dictates their access rights within the system.

Communication between the Cloud Services and Application layers is streamlined through the exchange of JSON files over HTTP via a RESTful API [16], providing a standardized and session-independent method that supports scalable and maintainable system integration. This setup allows for dynamic variability resolution and efficient data exchange, ensuring that any changes in the TVM trigger immediate, uninterrupted updates and deployments.

By using automated triggers and integrated variability management our application boosts the system's responsiveness, facilitating the seamless incorporation of new services and updates without disrupting existing operations.

3.3   Uncertainy Calculation and DCC Generation Microservices

In this research, we implemented two uncertainty calculation methods: Guide to the Expression of Uncertainty in Measurement (GUM) and its Monte Carlo Simulation (MCS) description provided by the International Bureau of Weights and Measures [11]. While there are no specific constraints on the tools and technologies used for these calculations, the need for diverse programming environments and libraries has led us to adopt a microservices architecture with containerization for flexibility. The GUM calculations are executed in C#, utilizing its native libraries, and MCS calculations are performed in Python 3 with the Pandas library, catering to complex mathematical requirements. Each set of equipment runs in its tailored containerized environment.

We use an XML format for DCCs to ensure compliance with PTB standards and facilitate easy data exchange. The uncertainty calculation results from GUM and MCS are integrated into this XML format via our DCC microservice, leveraging the REST API and Flask framework [5]. The data is then converted into a human-readable PDF accessible through a secure cloud service, where users can retrieve their certificates via a link or QR code.

4   Discussion

Digitalization in metrology is an important yet emerging field, characterized by the sector's diversity and the complexity of its devices, which often suffer from a lack of interoperability due to proprietary standards, technical challenges, and business decisions. The sensitive nature of metrological data requires high computational power and strict security measures. Adopting cloud computing and microservices effectively addresses these challenges. Our work utilizes dockerized microservices to enhance system flexibility and extendibility while reducing dependencies.

Our solution emphasizes the benefits of microservice architecture in digital metrology and calibration applications, including scalability and maintainability [1,4]. We aim to drive digital transformation in the metrology and calibration industry by using advanced microservice-based tools and technologies, as traditional monolithic architectures are costly and less effective due to inherent diversity and scalability needs.

This research may face threats to its validity, such as the upfront cost of cloud technologies posing a barrier to adoption. Additionally, stakeholders, especially in sensitive fields like defense industry, may have reservations about storing data on the cloud due to security concerns.

5   Conclusion

A microservice-based enhancement of the IoMT architecture is introduced, presenting cutting-edge cloud technologies to the metrology and calibration industry. An automated power measurement application, AutoRFPower, is implemented based on the proposed architecture. Google Cloud has been chosen as the deployment environment to manage containerized assets and microservices, with GKE orchestrating them. Additionally, RESTful API is incorporated to facilitate interactions between the client software running on the user's local environment and the cloud services, ensuring efficiency and security. The successful implementation of the AutoRFPower application demonstrates the proposed architecture's effectiveness for metrology and calibration applications.

Future work will be on increasing the utilization of the advantages of the microservice-based solution, other calibration processes, such as the scope of accreditation, will also be implemented by adhering to the proposed architecture.

Acknowledgment. The authors thank Dr. Erkan Danaci for his guidance in the metrology and calibration field. This work is supported by the European Partnership on Metrology through the Metrology Partnership Programme under grant agreement No. 22RPT04, entitled "Development of RF and microwave metrology capability II (RFMicrowave2)".

Data Availability Statement. Data sharing does not apply to this paper.

References

1. Baškarada, S., Nguyen, V., Koronios, A.: Architecting microservices: Practical opportunities and challenges. J. Comput. Inf. Syst. (2020)
2. Cetinkaya, A., Dogan, A.K., Danaci, E., Oguztuzun, H.: Autorfpower: automatic rf power measurement software for metrological applications. In: 2021 2nd International Informatics and Software Engineering Conference, pp. 1–4. IEEE (2021)
3. Cetinkaya, A., Kaya, M.C., Danaci, E., Oguztuzun, H.: Uncertainty calculation-as-a-service: an iiot application for automated RF power sensor calibration. In: Proceedings of the IMEKO TC6 International Conference on Metrology and Digital Transformation. IMEKO, Berlin, Germany (2022)
4. De Lauretis, L.: From monolithic architecture to microservices architecture. In: 2019 IEEE International Symposium on Software Reliability Engineering Workshops (ISSREW), pp. 93–96 (2019). https://doi.org/10.1109/ISSREW.2019.00050

5. Flask: Flask framework (2024). https://flask.palletsprojects.com/en/3.0.x/. Accessed 25 Apr 2024
6. Gadelrab, M.S., Abouhogail, R.A.: Towards a new generation of digital calibration certificate: analysis and survey. Measurement **181**, 109611 (2021)
7. Google: Develop and deploy containerized apps using a CI/CD pipeline (2024). https://cloud.google.com/architecture/app-development-and-delivery-with-cloud-code-gcb-cd-and-gke/deployment. Accessed 25 Apr 2024
8. Google Cloud: Google kubernetes engine (GKE) (2024). https://cloud.google.com/kubernetes-engine. Accessed 25 Apr 2024
9. Hackel, S., Härtig, F., Hornig, J., Wiedenhöfer, T.: The digital calibration certificate. PTB-Mitteilungen **127**(4), 75–81 (2017)
10. Hackel, S., Schönhals, S., Doering, L., Engel, T., Baumfalk, R.: The digital calibration certificate (DCC) for an end-to-end digital quality infrastructure for industry 4.0. Sci **5**(1), 11 (2023)
11. JCGM 101:2008: Evaluation of measurement data - Supplement 1 to the 'Guide to the expression of uncertainty in measurement' - Propagation of distributions using a Monte Carlo method (2008). https://www.bipm.org/documents/20126/2071204/JCGM_101_2008_E.pdf. Accessed 19 Apr 2024
12. Kaya, M.C., Saeedi Nikoo, M., Schwartz, M.L., Oguztuzun, H.: Internet of measurement things architecture: proof of concept with scope of accreditation. Sensors **20**(2) (2020)
13. Lafarge, T., Possolo, A.: The NIST uncertainty machine. NCSLI Meas. **10**(3), 20–27 (2015)
14. Nikoo, M.S., Kaya, M.C., Schwartz, M.L., Oguztuzun, H.: An MII-aware SoA editor for the industrial internet of things. In: 2019 II Workshop on Metrology for Industry 4.0 and IoT (MetroInd4. 0&IoT), pp. 213–218. IEEE (2019)
15. Oppermann, A., Eickelberg, S., Meiborg, M.: Digital transformation: towards process automation in a cloud native architecture. Acta IMEKO **12**(1), 1–6 (2023)
16. RedHat: What is a REST api? (2020). https://www.redhat.com/en/topics/api/what-is-a-rest-api. Accessed 25 Apr 2024
17. Saeedi Nikoo, M., Kaya, M.C., Schwartz, M.L., Oguztuzun, H.: Internet of measurement things: toward an architectural framework for the calibration industry. In: Mahmood, Z. (ed.) The Internet of Things in the Industrial Sector. CCN, pp. 81–102. Springer, Cham (2019). https://doi.org/10.1007/978-3-030-24892-5_4
18. Varshney, A., et al.: Challenges in sensors technology for industry 4.0 for futuristic metrological applications. Mapan **36**(2), 215–226 (2021)
19. Zeier, M., Hoffmann, J., Wollensack, M.: Metas.unclib-a measurement uncertainty calculator for advanced problems. Metrologia **49**(6), 809 (2012)
20. Zet, C., Dumitriu, G., Fosalau, C., Sarbu, G.C.: Automated calibration and DCC generation system with storage in private permissioned blockchain network. Acta IMEKO **12**(1), 1–7 (2023)

Performance Impact of Microservice Granularity Decisions: An Empirical Evaluation Using the Service Weaver Framework

Ricardo César Mendonça Filho[1] and Nabor C. Mendonça[2]([✉]) [ID]

[1] Center of Technological Sciences, Fortaleza, Ceará, Brazil
ricardocesarfilho@unifor.br
[2] Post Graduate Program in Applied InformaticsUniversity of Fortaleza,
Fortaleza, Ceará, Brazil
nabor@unifor.br

Abstract. The Service Weaver framework enables the development of distributed applications in Go as modular monoliths, with the flexibility to deploy monolith components in different environments and at various levels of granularity without code changes. This work evaluates the performance of an open-source microservice application using Service Weaver, considering multiple service granularity decisions in environments of one and two virtual machines under various workloads. The results indicate that service decoupling, while beneficial to modularity and maintenance, can significantly increase communication overhead between processes and virtual machines, negatively affecting the application's performance and scalability. These findings highlight the importance of balancing service granularity and communication costs in the design and deployment of microservice applications.

Keywords: Microservice granularity · Performance analysis

1 Introduction

One of the crucial challenges in designing microservices-based applications is determining the ideal level of granularity for each service [10,27]. If the services are too large, the benefits of loose coupling and independence are lost. If they are too small, there may be unnecessary complexity and communication overhead, affecting the performance and scalability of the application [15]. Traditional solutions used in microservices development are not suitable to face these challenges [5]. The main reason is that they all expect the developer to manually decompose the application into several independent services at design time. This means that the communication topology of the application is predetermined by the application developer. In practice, this constraint reduces the granularity options that can be explored by microservices developers, forcing them to make sub-optimal design decisions prematurely [4].

© The Author(s), under exclusive license to Springer Nature Switzerland AG 2024
M. Galster et al. (Eds.): ECSA 2024, LNCS 14889, pp. 191–206, 2024.
https://doi.org/10.1007/978-3-031-70797-1_13

The Service Weaver framework, launched by Google in early 2023, is an open-source solution with the potential to address the granularity challenges discussed above [6]. With Service Weaver, developers can write their distributed applications in Go as a modular monolith [25]. Later, at deployment time, the framework splits this monolith into multiple distributed services, according to the communication topology defined by the developer, without the need for changes in the monolith's code. This allows the application's services to be easily deployed locally or in the cloud, under different communication topologies, with minimal configuration effort [5]. In this sense, Service Weaver innovates by combining the advantages of a modular monolith with the flexibility and scalability of microservices [11].

Our work leverages the Service Weaver framework to explore key issues in microservice architecture design and deployment concerning possible service granularity decisions. To this end, we aim to answer the following research questions:

RQ1 What is the impact on the performance of a microservice application when the application's services are deployed with different levels of granularity using Service Weaver?

RQ2 How is the observed granularity impact influenced by factors such as the expected workload and the allocation of services to the resources of the underlying infrastructure?

To answer these two questions, we: (i) selected a widely adopted open-source microservice demo application as a research *benchmark*; and (ii) utilized the flexible service deployment features originally introduced by the Service Weaver framework to test the performance of the target microservice application considering different service deployment configurations, under different workloads, in scenarios with one and two virtual machines (VMs).

Our experimental results reveal that the choice of service granularity in microservices architectures has a significant impact on performance, especially regarding communication strategies between processes located on the same VM or different VMs. The deployment configurations tested on a single VM indicated that, although service decoupling can offer initial benefits in terms of modularity and maintenance, the overhead of inter-process communication can become a limiting factor under high workloads. In contrast, configurations distributed across two VMs showed that VM-to-VM communication introduces an additional layer of complexity and latency, particularly under high load conditions. These findings demonstrate that architectural decisions regarding service granularity and the allocation of services to the underlying computing infrastructure are critical aspects to be considered in the design and deployment of microservices applications, directly influencing their performance and scalability.

The rest of this work is organized as follows: Sect. 2 provides an overview of the Service Weaver framework; Sect. 3 describes the experimental evaluation method used; Sect. 4 presents and analyzes the obtained results; Sect. 5 discusses the limitations and broader implications of our findings; Sect. 6 compares our work with other related studies; finally, Sect. 7 offers some concluding remarks and suggests topics for future research.

2   Service Weaver

Service Weaver is a framework developed to facilitate the development, deployment, and management of distributed applications in Go [6]. The two main innovations introduced by the framework are (i) a programming model that allows developers to write modular distributed applications focused solely on business rules, compiled into a single binary artifact, and (ii) a runtime environment for the optimized distribution and management of these applications [5].

The proposed programming model allows the programmer to write a distributed application as a single program, where the code is divided by modular units, called *components*. The framework decouples the implementation of business rules from the configuration and management of infrastructure resources: the components are centered around logical boundaries based on the application's business rules, and the framework's runtime environment is centered around physical boundaries based on application performance (for example, two components may be co-located to improve performance). This decoupling, along with the fact that physical boundaries can be dynamically changed, helps to alleviate the challenges of modularity and granularity caused by architecturally rigid APIs, common in modern distributed architectures [5].

At deployment time, the Service Weaver's runtime environment analyzes the application's source code, generating the necessary code to handle remote calls between services, allowing them to be executed on different processes or machines without the need for explicit implementation of remote calls by the developer. Additionally, the framework manages the scheduling, replication, and co-location of the created services, without requiring developer intervention [6].

2.1   Components

A Service Weaver application is composed of a set of components. A component is described as an interface implemented in Go, and components can communicate with each other through method calls defined in these interfaces. This feature allows for the abstraction of remote calling mechanisms and data serialization, which do not need to be explicitly implemented by the application developer [6].

Components can be hosted on different operating system processes and possibly on multiple machines. Method invocations of components turn into remote procedure calls when necessary, but remain as local procedure calls if the calling and called components are in the same process [5].

2.2   Deployment Modes

Service Weaver does not include specific code for a particular deployment environment. Instead, the framework uses a special API, called *Deployer*, to integrate the application logic with the details of the execution environment, such as the physical distribution in multiple virtual machines in a public cloud. This deployment API allows developers to write Go applications that are independent of the deployment environment. The latest version of Service Weaver offers various types of Deployers that allow deploying application components in different execution environments, including local, Kubernetes and Google Cloud Deployers [6].

2.3   Deployment Configuration

Service Weaver provides a high-level notation for developers to explicitly define aspects of the application's infrastructure and communication to be considered at deployment time, such as which components should run locally, in a single process, or be distributed in multiple processes, possibly on multiple machines. Moreover, developers can define more complex deployment configurations, such as deploying components in different virtual machines, by customizing the configuration files Service Weaver automatically generates for each specific deployment environment. For example, a developer can customize the configuration files for a Kubernetes deployment by specifying which Deployment should run in which Kubernetes node.[1] These features give application developers more flexibility to experiment with different levels of service granularity as well as to assess the performance impact of each granularity decision under multiple deployment environments. Further details on the deployment configuration features of Service Weaver can be found in the official documentation [6].

3   Evaluation Method

This section presents the evaluation method adopted to answer the two research questions introduced in Sect. 1, including the target microservices application, the testing environment, and the performance metric and load generation tool used.

3.1   Target Application

The target application chosen was Online Boutique [7]. This application implements a virtual store where users can search for products, add products to the cart, and purchase them. It was originally developed by Google as a way to demonstrate a set of cloud-native technologies, such as Kubernetes, GKE, Istio, and gRPC. Since then, it has been used as a microservice evaluation benchmark in several research studies reported in the literature (e.g., [3, 17, 20, 23]).

The architecture of Online Boutique consists of 11 microservices, responsible for different aspects of the virtual store's operation, as illustrated in Fig. 1. In the original version of the application, these services are implemented in different programming languages, such as Go, Java, C#, Python, and Node.js. To enable

---

[1] When using the Kubernetes deployer, Service Weaver creates a separate Deployment specification for each group of co-located components specified in the application's configuration file [6].

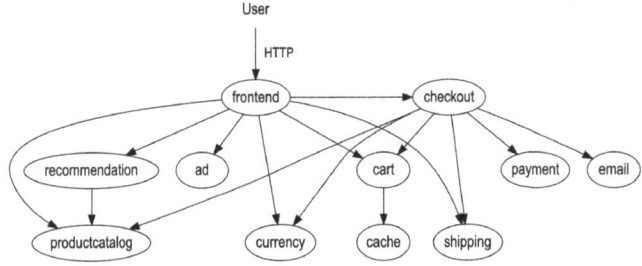

Fig. 1. Architecture of the Online Boutique application (source: [7]).

the application to be executed using Service Weaver, the framework's developers ported the remaining (non-Go) Online Boutique services to the Go language, in addition to replacing the use of the Redis cache service with a native Go version [5]. This new version of the application, written entirely in Go, was used in the experiments conducted in this work.

From the architectural diagram shown in Fig. 1, it is evident that the *frontend* service serves as the entry point of the application. This service plays the dual role of implementing the virtual store's web pages, as well as orchestrating the other services.

3.2   Testing Environment

All tests were conducted on a single host machine with the following characteristics: AMD Ryzen 7 5800H 3.2GHz processor, 8 processing cores, and 16GB of DDR4 RAM. To better represent potential real-world deployment scenarios, we deployed the application in the host machine in configurations involving both a single VM and a dual VM setup. To this end, we created a Kubernetes cluster using Minikube [26] containing two virtual machines created with VirtualBox [16]. Each VM was standardized with 4 CPUs and 4GB of RAM to maintain consistency across tests, allowing us to isolate the impact of service distribution from hardware variability.

The deployment process of the target application in this cluster was carried out using the configuration files automatically generated by the Kubernetes deployer provided by Service Weaver. These files are dynamically modified to adapt to the desired architectures, allowing for the creation of specific infrastructure conditions for each test scenario. All performance measurements were obtained from the actual execution of the application under different configurations on the local Kubernetes cluster. No artificial communication delays were introduced to differentiate local calls from remote calls between services.

The decision to use a single host machine with only two virtual machines as the testing infrastructure was motivated by three factors. First, the target application is relatively small, with only 11 non-persistent services, making it possible to deploy all services in a single process. Second, the use of a single dedicated host machine allows for more controlled performance testing, enabling a more accurate analysis of the performance impact resulting from the deployment granularity of the application's services. Finally, the use of only two virtual machines simplifies the configuration and execution of tests, reducing the complexity and time required for the experiments.

3.3 Load Generation and Performance Metric

The Locust [13] load generation tool was used to simulate different levels of workload on the target application. In addition to being widely used in web application load testing, the choice of Locust was a natural decision, as the original Online Boutique application comes with a Locust load testing script. This script invokes a series of operations offered by the application's *frontend* service, simulating the behavior of a typical store user:

- opening the store's homepage;
- performing two currency exchanges;
- conducting ten product searches;
- adding two products to the shopping cart;
- visualizing the cart items three times; and
- completing the purchases (checkout).

Each load test involved running Locust with the above script for about 5 min, with a growing number of virtual users, starting with 500 users up to the limit of 2250 users, in increments of 250 users. These load values were defined after conducting a set of initial tests, where we determined the maximum number of virtual users supported by the application when deployed on the previously described physical infrastructure.

At the end of each test, Locust produces a set of performance metrics. These metrics include a variety of statistics calculated on the response times measured for each operation requested from the target application, such as minimum, maximum, average, median, and various percentiles, as well as the throughput and error rate of each operation. Locust also calculates consolidated statistics, considering the set of operations performed during the test in an aggregated manner.

For the experiments conducted in this work, the performance metric used was the 95th percentile of response times, in milliseconds, calculated in an aggregated manner over the set of operations requested from the application by each user. The choice of a high percentile (>90%) as the performance metric is strategically advantageous due to its ability to realistically capture the application's behavior under stress conditions [14]. Unlike more traditional metrics such as average and median, high percentile metrics avoid distortions caused by outliers, which tend to occur more frequently in overload scenarios.

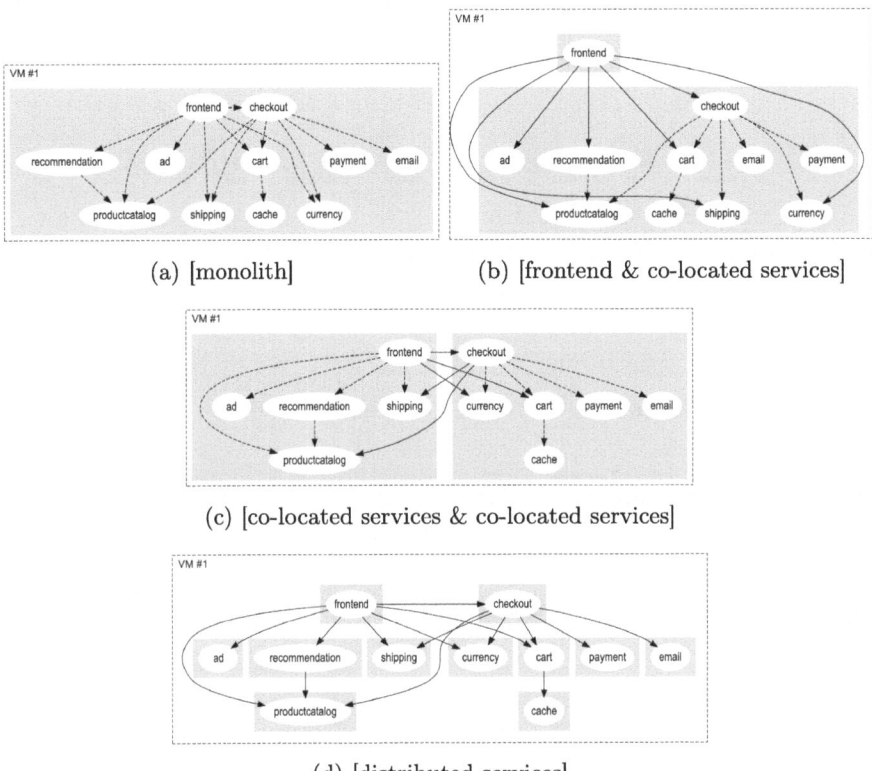

(a) [monolith]   (b) [frontend & co-located services]

(c) [co-located services & co-located services]

(d) [distributed services]

Fig. 2. Deployment configurations of the Online Boutique application in one VM. Dashed rectangles represent virtual machines, filled grey rectangles represent processes, white ellipses represent application components, dashed arrows represent in-process local calls, and solid arrows represent remote calls.

3.4   Selected Deployment Configurations

To investigate how service granularity can affect the performance of a microservice application, we used the deployment features of Service Weaver to define eight deployment configurations out of the eleven Online Boutique microservices, considering scenarios with one and two virtual machines.

Figures 2 and 3 illustrate these eight configurations, with four deployed on a single virtual machine (Fig. 2) and four on two virtual machines (Fig. 3). In all configuration diagrams shown in Figs. 2 and 3, unfilled rectangles with dashed borders represent virtual machines, filled grey rectangles represent processes, and white ellipses represent application components. Dashed arrows represent local calls between components executed within the same process, while solid arrows represent remote calls between components executed in different processes.

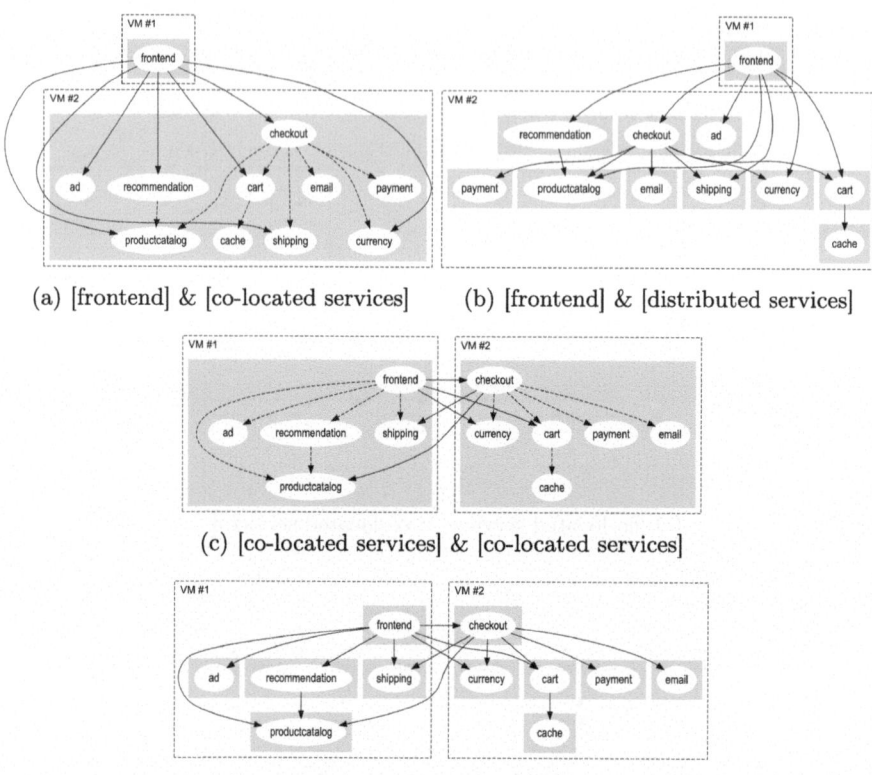

(a) [frontend] & [co-located services]   (b) [frontend] & [distributed services]

(c) [co-located services] & [co-located services]

(d) [distributed services] & [distributed services]

Fig. 3. Deployment configurations of the Online Boutique application in two VMs.

The chosen configurations range from a monolithic configuration (Fig. 3(a)), executed on a single virtual machine, to fully distributed configurations (Figs. 4(b) and 4(d)), executed on two virtual machines. Moreover, some configurations were designed to isolate a specific service of the application from the other services, such as the configurations that isolate the *frontend* service, illustrated in Figs. 3(b), 4(a), and 4(b). These three configurations were selected given the importance of the *frontend* service as the application's entry point. The configuration names were defined in a way to represent the strategy used to deploy the eleven application services, with brackets used to distinguish the services deployed in each VM. For example, the name [frontend & co-located services] (Fig. 3(b)) indicates that the *frontend* service is deployed isolated in a single process, while the remaining services are co-located in a second process, with both processes on a single VM. Similarly, the name [distributed services] & [distributed services] (Fig. 4(d)) indicates that all services are fully distributed across the two VMs, with each VM containing a subset of the application's services.

3.5   Considerations on the Number of Possible Configurations

The eight deployment configurations selected in our study comprise only a small subset of all possible configurations that could be used to deploy the eleven microservices of the target application to the two VMs. The problem of calculating the total number of possible deployment configurations of a given number of services in a single VM can be solved using combinatorics, as the number of ways to partition a set of n elements into non-empty subsets. This number is given by the Bell numbers, named after Eric Temple Bell [19]. A Bell number, denoted as B_n, represents the number of different ways to partition a set of n elements. The sequence of Bell numbers begins $B_0, B_1, B_2, \ldots$ with values $1, 1, 2, 5, 15, \ldots$. For example, the set $\{1, 2, 3\}$ can be partitioned in non-empty subsets in five different ways ($B_3 = 5$): $\{\{1\}, \{2\}, \{3\}\}$, $\{\{1, 2\}, \{3\}\}$, $\{\{1, 3\}, \{2\}\}$, $\{\{2, 3\}, \{1\}\}$, and $\{\{1, 2, 3\}\}$.

Bell numbers can be calculated, among other methods, using the recursive formula [19]:

$$B_{n+1} = \sum_{k=0}^{n} \binom{n}{k} B_k$$

where $B_0 = 1$.

For the case of the Online Boutique application, with eleven services, the number of possible deployment configurations in a single VM is given by the Bell number B_{11}, which is equal to 678,570. This number represents the total number of ways to partition the eleven services into non-empty subsets, considering all possible combinations of service granularity decisions. The number of possible deployment configurations in multiple VMs is much greater, as it involves partitioning the eleven services into two non-empty subsets, with each subset deployed in either one of the VMs. For example, if a partition consists of k subsets, the number of ways to allocate these k subsets across two VMs is $2^{k-1} - 1$, reflecting the binary decision for each subset to join one of the two VMs, excluding the case where all subsets end up in one VM. This means that there are $2^{11-1} - 1 = 1023$ possible ways to deploy the eleven Online Boutique microservices across two VMs in a fully distributed fashion, i.e., with each service in a separate process, which is only one out of the 678,570 possible ways the application's microservices can be deployed in terms of granularity choices.

The magnitude of these numbers highlights the complexity of choosing an optimal deployment strategy even for a small number of services and virtual machines. Finding the optimal deployment configuration for a given microservice application across a cluster of VMs is a challenging problem, as it involves balancing the trade-offs between service granularity, communication overhead, and resource allocation, among other factors, in an exponentially large deployment space. The experimental evaluation conducted in this work aims to shed an early light on these trade-offs by analyzing the performance of the Online Boutique application under eight representative deployment configurations, considering scenarios with only one or two VMs.

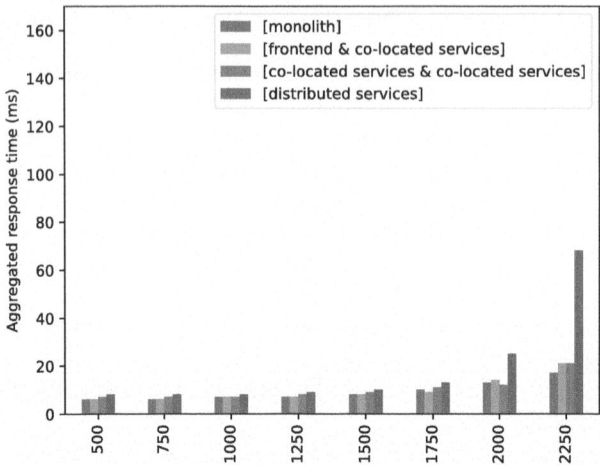

Fig. 4. Performance of the four deployment configurations in one VM.

4   Results

Figures 4 and 5 show the performance of the eight deployment configurations of the target application for the scenarios with one and two VMs, respectively. Each figure displays a plot for the 95th percentile of the application's aggregated response times measured in milliseconds for each configuration, under different workloads.

4.1   Configurations in One VM

Figure 4 displays the performance of the four deployment configurations on a single VM. The monolithic configuration showed a progressive increase in response time, from 6 ms for 500 users to 17 ms for 2250 users, indicating scalability limitations under larger loads due to its centralized nature. The [frontend & co-located services] configuration, with response times increasing from 6 ms to 21 ms, had a similar initial performance but suffered a slightly more pronounced increase under the higher load, likely caused by the greater number of remote invocations resulting from the decoupling of the *frontend* service from the other services. The [co-located services & co-located services] configuration showed a similar trend to the previous ones, with an initial response time of 7 ms and an increase to 21 ms, indicating that distributing the services into two coupled groups offers some benefit in terms of load distribution, but still faces similar scalability challenges due to remote invocations between the two groups of services. Finally, the [distributed services] configuration, with response times ranging from 8 ms to 68 ms, showed performance similar to the other configurations initially but experienced a significant increase at the highest load, the largest among the four single-VM configurations. The fully distributed nature of the services in this configuration,

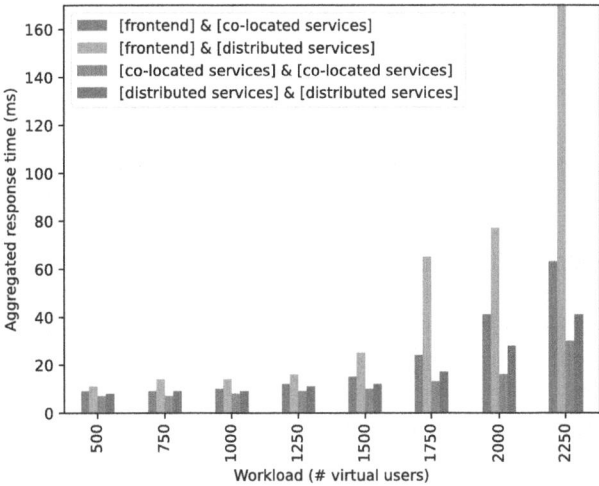

Fig. 5. Performance of the four deployment configurations in two VMs.

although offering greater flexibility and initial scalability potential, ultimately deteriorates its performance due to excessive communication overhead among the remote services.

4.2 Configurations in Two VMs

The configurations in two VMs (Fig. 5) showed greater variability in results, reflecting the greater complexity of distributing services in this deployment scenario. The [frontend] & [co-located services] configuration displayed an increase in response time from 9 ms to 63 ms, indicating that isolating the *frontend* on a separate VM became a bottleneck at high loads. The [frontend] & [distributed services] configuration started with a response time of 11 ms, increasing dramatically to 170 ms under maximum load, by far the worst performance among all the evaluated configurations. This result suggests that the communication overhead among fully distributed services on separate VMs is a critical factor at high loads. The [co-located services] & [co-located services] configuration offered the best scalability among the two VM configurations, with an increase from 7 ms to 30 ms, similar to the scalability of the best one VM configurations. Distributing the services into two coupled groups on separate VMs appears to offer a good balance between service granularity and communication overhead between them, resulting in a gradual and controlled increase in response time. Finally, the [distributed services] & [distributed services] configuration performed similarly to the [frontend] & [co-located services] configuration, with response times increasing from 8 ms to 41 ms, revealing that the initial flexibility offered by total decoupling between services on the two VMs faces significant communication challenges between VMs under high loads, although with a lesser impact than the distributed configurations with the *frontend* service isolated on the first VM.

4.3   Joint Analysis of the Two Deployment Scenarios

Comparing the performance of all eight configurations, it is possible to identify some general trends about the impact of service granularity and communication overhead between machines. In the one VM configurations, there is a progressive increase in response times with increasing load, particularly evident in the fully distributed configurations, where the communication overhead among processes becomes a limiting factor. In contrast, the two VM configurations demonstrate that the physical distribution of services across different machines can introduce an additional cost of communication between services. Particularly, configurations with services fully distributed on two VMs suffer a more pronounced increase in response times under high loads, indicating that communication among processes located on the same VM may be more critical than communication among processes located on different VMs. On the other hand, configurations with co-located services on two VMs performed similarly to configurations with co-located services on a single VM, indicating that communication between VMs may be less critical when services are co-located within each VM.

These results suggest that while service granularity offers scalability benefits at lower loads, the complexity and overhead associated with remote invocations, especially in configurations with services distributed across multiple VMs, become dominant factors under high loads. Thus, the choice between different microservice granularity decisions should consider a balance between the flexibility offered by the physical distribution of services and the performance implications associated with communication strategies among processes of the same VM or different VMs.

5   Discussion

Our work is an early step toward understanding the performance impact of service granularity when designing and deploying microservice applications. As such, it has several limitations that should be considered when interpreting our results. First, the results were derived from experiments conducted on a single demo application, the Online Boutique, with a single performance metric, aggregated response time. While this application and metric were selected to be representative of a real-world microservices architecture, different applications measured using different performance metrics could yield different results. In that direction, a more fine-grained analysis of the response time and resource consumption of individual services could provide further insights into the performance bottlenecks and their root causes. Second, the experiments were conducted in controlled environments with a limited number of virtual machines, which may not fully capture the performance dynamics of more complex cloud environments with elastic scaling capabilities. Finally, the experiments were limited to a predefined set of communication patterns established by the Service Weaver framework. Future research could explore the impact of deploying the application in more realistic cloud environments using alternative communication strategies, such as event-driven approaches, which might affect the interservice communication dynamics.

Despite these limitations, our work contributes to a deeper understanding of several broader implications for microservices architecture design. In particular, our results suggest that the choice of service granularity can significantly impact the performance and scalability of a microservice application, with tradeoffs between the benefits of decoupling and the performance penalties due to increased communication overhead. Architects must carefully consider this balance, especially when designing systems expected to operate under high-load conditions. The results also highlight the importance of strategic deployment planning, as the performance of the application can vary significantly across different deployment configurations. Finally, the study underscores the need for tools and methods that can assist architects in making informed decisions about service granularity, as the optimal granularity may vary depending on specific application characteristics and performance objectives.

6 Related Work

The impact of microservice granularity decisions has been the focus of several studies in the literature, e.g., [1,2,8,9,11,18,21,22,25,27]. According to Vera-Rivera et al. [27], most studies on this topic focus on the decomposition of monolithic systems into microservices, with less attention given to the empirical evaluation of granularity decisions in existing systems. In this regard, the studies most related to our work are those of Shadija *et al.* [21] and Costa & Ribeiro [2], as well as the recent investigations on modular monoliths by Su and Li [25] and Johnson et al. [11] discussed below.

The study by Shadija *et al.* [21] investigates the effect of service co-location on the latency of an academic admission application, simulating two deployment approaches: one with all services in a single container and another with services distributed across two containers. The results showed a negligible increase in latency for the two-container deployment compared to a single container. The study highlights that, although finer granularity of microservices may lead to an increase in the number of method invocations, this is manageable when microservices are within the same container. However, communication between containers introduces additional variables that can impact performance and reliability.

The work by Costa & Ribeiro [2] compares the performance of an e-commerce application developed following a traditional monolithic approach and two equivalent distributed versions based on different microservices communication patterns: direct communication and event-driven communication. The results showed that the monolithic approach had the lowest latency at all evaluated load levels, while the distributed approaches with direct communication and event-driven communication performed worst, in that order. According to the authors, the latency associated with distributed systems, the eventual nature of responses to asynchronous operations, and the cost of data synchronization imposed a significant negative impact on the performance of the distributed solutions based on microservices compared to the monolithic solution.

Our work differs from those of Shadija *et al.* and Costa & Ribeiro in several aspects. First, while the work of Shadija *et al.* compares the performance of two deployment configurations of the target application, in one and two containers, our work compares the performance of eight deployment configurations of the target application, in one and two virtual machines. Second, while the work of Costa & Ribeiro compares the performance of a monolithic application with that of two equivalent distributed versions using different architectural patterns, our work compares the performance of a monolithic version of the target application with that of seven equivalent distributed versions, using the same communication topology among microservices. Finally, both works limited themselves to evaluating service granularity levels that were predetermined at the time of the applications' design, while our work benefits from the Service Weaver framework to evaluate multiple levels of granularity, defined at deployment time from the in-process coupling of different subsets of application components.

In terms of modular monoliths, Su and Li [25] conducted a systematic grey literature review, finding that modular monoliths combine the advantages of monoliths and microservices. They identified three frameworks for building modular monoliths: Service Weaver, Spring Modulith [24], and Light-hybrid-4j [12]. They also found that modular monoliths are increasingly adopted as an alternative to microservice decomposition, offering a balance between development velocity and operational complexity. Johnson et al. [11] examined Service Weaver through the lens of cloud-native principles, comparing its capabilities to traditional microservice architectures. They found that Service Weaver simplifies development and deployment processes but may lack certain features like polyglot support and fine-grained control over the system. Their work highlights the trade-offs involved in choosing between a modular monolith and microservices, emphasizing the importance of considering specific project requirements and constraints.

Our work builds upon the concept of a modular monolith by leveraging Service Weaver's unique capabilities to define service granularity at deployment time. By harnessing this flexibility, we can explore a wider range of deployment configurations and gain insights into the performance trade-offs associated with different levels of service granularity. This approach offers a more nuanced understanding of the relationship between service granularity and performance, contributing to the development of more effective strategies for designing and deploying microservice applications.

7   Conclusion and Future Work

This work investigated the impact of microservices deployment granularity on the performance of the distributed application Online Boutique, using the Service Weaver operational framework. The application performance was analyzed under various deployment configurations and workloads, in environments with one and two virtual machines. Our findings demonstrated that both the granularity of deployment and the communication topology among microservices significantly influence application performance and scalability. These results underscore the importance of strategically balancing these factors to enhance application efficiency and maintainability.

Based on the insights gained from this study, several specific areas for future research can be identified, including the development of predictive models for granularity optimization, the exploration of more diverse application domains and granularity choices, and the assessment of the impact of different cloud-native platforms on microservice granularity design. Additionally, real-world application assessments, advanced deployment strategies, and resiliency aspects of service granularity should be further investigated to provide more concrete guidelines for architects and developers.

Acknowledgments. The generative AI tools ChatGPT-4 Turbo and GitHub Copilot were used to assist in structuring and revising this paper.

Data Availibility Statement. To facilitate reproduction and potential extension of this work by the research community, all experimental artifacts and data generated are publicly available at https://github.com/mendoncas/granularity-evaluation/.

References

1. Andrade, B., Santos, S., Silva, A.R.: A comparison of static and dynamic analysis to identify microservices in monolith systems. In: Tekinerdogan, B., Trubiani, C., Tibermacine, C., Scandurra, P., Cuesta, C.E. (eds.) ECSA 2023. LNCS, vol. 14212, pp. 354–361. Springer, Cham (2023). https://doi.org/10.1007/978-3-031-42592-9_25
2. Costa, L., Ribeiro, A.N.: Performance evaluation of microservices featuring different implementation patterns. In: Abraham, A., Gandhi, N., Hanne, T., Hong, T.-P., Nogueira Rios, T., Ding, W. (eds.) ISDA 2021. LNNS, vol. 418, pp. 165–176. Springer, Cham (2022). https://doi.org/10.1007/978-3-030-96308-8_15
3. Cui, J., Chen, P., Yu, G.: A learning-based dynamic load balancing approach for microservice systems in multi-cloud environment. In: IEEE 26th International Conference on Parallel and Distributed Systems (ICPADS), pp. 334–341. IEEE (2020)
4. Dragoni, N., Giallorenzo, S., Lafuente, A.L., Mazzara, M., Montesi, F., Mustafin, R., Safina, L.: Microservices: yesterday, today, and tomorrow. In: Mazzara, M., Meyer, B. (eds.) Present and Ulterior Software Engineering, pp. 195–216. Springer, Cham (2017). https://doi.org/10.1007/978-3-319-67425-4_12
5. Ghemawat, S., et al.: Towards modern development of cloud applications. In: Proceedings of the 19th Workshop on Hot Topics in Operating Systems, pp. 110–117 (2023)
6. Google: Service Weaver: A Programming Framework for Writing and Deploying Cloud Applications (2024). https://serviceweaver.dev. Accessed 15 Aug 2024
7. Google Cloud: Online Boutique (2023). https://github.com/GoogleCloudPlatform/microservices-demo. Accessed 15 Aug 2024

8. Hassan, S., Bahsoon, R., Kazman, R.: Microservice transition and its granularity problem: a systematic mapping study. Softw. Pract. Exp. **50**(9), 1651–1681 (2020)
9. Homay, A., Zoitl, A., de Sousa, M., Wollschlaeger, M., Chrysoulas, C.: Granularity cost analysis for function block as a service. In: 2019 IEEE 17th International Conference on Industrial Informatics (INDIN), vol. 1, pp. 1199–1204. IEEE (2019)
10. Jamshidi, P., Pahl, C., Mendonça, N.C., Lewis, J., Tilkov, S.: Microservices: the journey so far and challenges ahead. IEEE Softw. **35**(3), 24–35 (2018)
11. Johnson, J., Kharel, S., Mannamplackal, A., Abdelfattah, A.S., Cerny, T.: Service Weaver: A Promising Direction for Cloud-native Systems? arXiv preprint arXiv:2404.09357 (2024)
12. Light: Light Hybrid 4j (2024). https://www.networknt.com/getting-started/light-hybrid-4j/. Accessed 15 Aug 2024
13. Locust.io: Locust: an open source load testing tool (2024). https://locust.io/. Accessed 15 Aug 2024
14. Molyneaux, I.: The Art of Application Performance Testing: From Strategy to Tools. O'Reilly Media, Sebastopol (2014)
15. Newman, S.: Monolith to Microservices: Evolutionary Patterns to Transform your Monolith. O'Reilly Media, Sebastopol (2019)
16. Oracle: VirtualBox (2024). https://www.virtualbox.org/. Accessed 15 Aug 2024
17. Park, J., Choi, B., Lee, C., Han, D.: GRAF: a graph neural network based proactive resource allocation framework for SLO-oriented microservices. In: Proceedings of the 17th International Conference on emerging Networking EXperiments and Technologies (CoNEXT), pp. 154–167 (2021)
18. Quéval, P.J., Zdun, U.: Extracting the architecture of microservices: an approach for explainability and traceability. In: Tekinerdogan, B., Trubiani, C., Tibermacine, C., Scandurra, P., Cuesta, C.E. (eds.) ECSA 2023. LNCS, vol. 14212, pp. 346–353. Springer, Cham (2023). https://doi.org/10.1007/978-3-031-42592-9_24
19. Rota, G.C.: The number of partitions of a set. Am. Math. Mon. **71**(5), 498–504 (1964)
20. Saleh Sedghpour, M.R., Klein, C., Tordsson, J.: An empirical study of service mesh traffic management policies for microservices. In: ACM/SPEC International Conference on Performance Engineering (ICPE), pp. 17–27. AMC (2022)
21. Shadija, D., Rezai, M., Hill, R.: Microservices: granularity vs. performance. In: Companion Proceedings of the 10th International Conference on Utility and Cloud Computing (UCC), pp. 215–220 (2017)
22. Singh, S., Werle, D., Koziolek, A.: ARCHI4MOM: using tracing information to extract the architecture of microservice-based systems from message-oriented middleware. In: Gerostathopoulos, I., Lewis, G., Batista, T., Bureš, T. (eds.) ECSA 2022. LNCS, vol. 13444, pp. 189–204. Springer, Cham (2022). https://doi.org/10.1007/978-3-031-16697-6_14
23. Soldani, J., Muntoni, G., Neri, D., Brogi, A.: The μTOSCA toolchain: mining, analyzing, and refactoring microservice-based architectures. Softw. Pract. Exp. **51**(7), 1591–1621 (2021)
24. Spring: Spring Modulith (2024). https://spring.io/projects/spring-modulith. Accessed 15 Aug 2024
25. Su, R., Li, X.: Modular Monolith: Is This the Trend in Software Architecture? arXiv preprint arXiv:2401.11867 (2024)
26. The Kubernetes Authors: Minikube (2024). https://minikube.sigs.k8s.io/docs/. Accessed 15 Aug 2024
27. Vera-Rivera, F.H., Gaona, C., Astudillo, H.: Defining and measuring microservice granularity–a literature overview. PeerJ Comput. Sci. **7**, e695 (2021)

Improving Comprehensibility of Event-Driven Microservice Architectures by Graph-Based Visualizations

Sven Schoop[1,2](✉), Erik Hebisch[2], and Thomas Franz[1]

[1] University of Applied Sciences Duesseldorf, Düsseldorf, Germany
sven.schoop@study.hs-duesseldorf.de, thomas.franz@hs-duesseldorf.de
[2] Ausbildung.de GmbH, Bochum, Germany
erik.hebisch@ausbildung.de

Abstract. This paper presents an industrial case study examining challenges faced by software architects and developers during planning and assessing the correctness of event-driven communication behaviour. We present a novel approach for visualizing the communication paths of an event-driven microservice system. Service-local documentation is used to create a global view of the runtime behaviour. We show that this visualization approach can significantly reduce the time required to understand an architecture, while additionally improving accuracy and confidence.

Keywords: Software Architecture · Microservices · Event-Driven · Visualization

1 Introduction and Motivation

The Microservice pattern is conceptually simple. Each domain of the system is organized into independently maintainable and deployable components. This constitutes a distributed system and needs a way to coordinate work while remaining decoupled. One of the options is to make every service emit and subscribe to events when it needs some work to be done or needs to fetch some information. Events are passed to an event broker which manages the event propagation to every subscribed microservice. Implementing such a system is straightforward. Developers write handlers for each event their microservice subscribes to and emit relevant events back to the system. A microservice's contract consists of the events it subscribes to and will emit in response.

In order to prove that a microservice system exhibits its intended behavior it is however not enough to verify that each service by itself fulfills its contract. As the number of microservice and events increases, it becomes harder to follow all possible interactions and thus message loops and deadlocks in the communication become more likely. It is therefore advised to be able to prove correctness before deployment.

M. Galster et al. (Eds.): ECSA 2024, LNCS 14889, pp. 207–214, 2024.
https://doi.org/10.1007/978-3-031-70797-1_14

This paper presents findings on the real-life challenges of understanding a system composed of microservices during development and maintenance. We present a solution for visualizing the dynamic behavior of an event-driven microservice system. We evaluate the suitability of our solution in a real-world study focussing on improving software architects' and developers' ability to reason about their system's runtime behavior.

2 Problem Observation and Research Hypothesis

We interviewed members of a team responsible for a system with an event-based microservice architecture. We asked them to describe typical tasks, practices and obstacles during maintaining and extending the system. This revealed two major challenges regarding design-time comprehension.

Comprehension is Limited: Every interviewee considered getting an overview of the system as a whole as very challenging. More specifically, they all rated the task of comprehending event chains as the most difficult task: While understanding the execution of a single service and its event emission was described as straightforward, following all resulting event emissions was considered highly complex, even *unmanageable.*

Change is Rare and Tedious, thus Costly: Interviewees reported that changes to events occur infrequently. The rarity of the task, combined with its complexity, made adaptation tedious: Most details of the event emission behavior needs to be re-learned each time. It was believed that everyone would know how to reverse-engineer the event chains from the source code. However, the interviewees were aware of the costs. Particularly, as that would require understanding source code written by others. They also stated that this task was prone to errors.

From these results we infer a need for support in *overviewing* the runtime interaction of services *at design time*, e.g. recognizing communication loops, and *being able to analyze* the event chains triggered from specific events.

Hypothesis: *The efficiency of architects and developers dealing with event-driven microservice architectures can be increased via visualization of the inter-system interaction. Efficiency improvements should be measurable as a reduction of i) the time taken to analyze a specific service interaction, ii) the error rate of the analysis, iii) the duration for decision-making and iv) increased confidence in decisions made.*

3 Visualization as Graphs

To visualize the inter-system interaction, events, services and their connections can be modeled using graph theory [2]. Services can be represented as nodes and events exchanged between services as edges, connecting the nodes.

However, we chose to represent events as nodes, with directed edges representing the *triggers-emission-of*-relationship between the events. This representation allows us to show the complete communication behavior of the system as a collection of event chains. We emphasize the relationship between events in order to make event interaction visible, in contrast to regarding them solely as conveyor of service interaction [3,4,6,13].

We further define two types of events: intra-service and inter-service. *Inter-service* events are consumed to trigger a service behavior. *Intra-service* events are emitted from *within* a service as a result of the consumption of an inter-service event (Fig. 1).

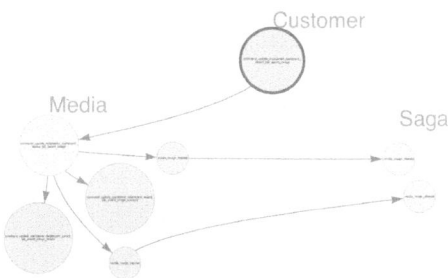

Fig. 1. Example graph showing a single event chain spanning over 3 services and 5 events, 8 nodes in total. Outgoing events $\mathcal{E}$ are shown in red and incoming events $\mathcal{K}$ are green. Events of the same service are drawn closer together. Directional arrows show the flow between events. The starting point of an event chain is highlighted with a red outline. (Color figure online)

Definition 1. *Let $\mathcal{G}$ be the graph to be represented as a set $(\mathcal{E}, \mathcal{V})$. Let $\mathcal{G}$ be a directed multigraph. Let $\mathcal{E}$ be a set of nodes (i.e. the events) $\mathcal{E}(e_1, ..., e_n)$. $\mathcal{V}$ are the edges or connections between the events as ordered pairs of events. $\mathcal{V} \subseteq \mathcal{E} \times \mathcal{E}$ is a subset of the Cartesian product of the events.*

Definition 2. *Services are able to 1.) consume $(\mathcal{K})$ and 2.) to emit events. The outgoing events are divided into two categories: 2.1) Events $\mathcal{E}_k$ that may be emitted after another event has been consumed are modeled as a property of the consumed event. This means that these events may be listed more than once. 2.2) Events $\mathcal{E}_d$ that may be emitted without first consuming another event are listed separately and equal to the incoming events.*

With Definition 2 $\mathcal{K} \subseteq \mathcal{E}$, $\mathcal{E}_d \subseteq \mathcal{E}$ and $\mathcal{E}_k \subseteq \mathcal{E}$ applies. In order to construct a whole-system overview we require only service-local documentation on the consumed events K, the emitted intra-service events E_k and the emitted inter-service events E_d of each separate service. Some (service-internal, intra-service) connections v between events can be derived directly from this documentation: There is an edge between each $\mathcal{K}$ event and its respective $\mathcal{E}_k$ events. To create the inter-service connections, it is necessary to iterate over each service-local documentation and its outgoing events ($\mathcal{E}_k$ and $\mathcal{E}_d$) and match them with the corresponding incoming events ($\mathcal{K}$) from every other service.

4   Evaluation

We conducted a summative evaluation to examine whether the visualization described in Sect. 3 improves the understandability of a microservice system. As our study example we chose one of the systems developed by the team from Sect. 2.

The system in question consists of 17 microservices, of which 14 are consuming or emitting events. The services were defined along domain boundaries, with most services providing and managing access to a top-level domain object.

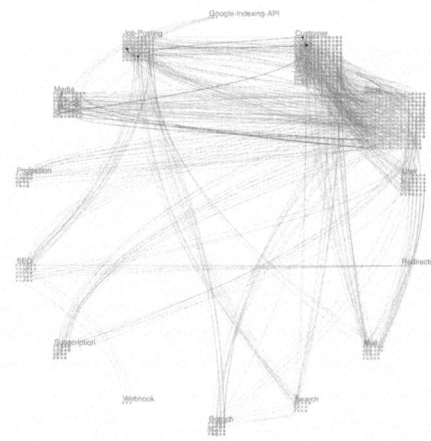

Fig. 2. Overview of the system with all services and events. The events are placed on the circumference of a circle, grouped by service. There are 14 services visible with 316 different events represented as 705 nodes with 702 connecting edges. A potential loop in an event chain is detected and highlighted in black. Another event chain is selected and highlighted in red. Please note that this figure is only intended for illustrative purposes. (Color figure online)

The system in question uses 316 different events. The mean and median number of emitted events is 26 and 12 per service respectively. The mean and median number of consumed events is 24 and 13. The longest event chain comprises 9 services and 33 events.

Figure 2 shows an overview of the system through the visualization with all 14 microservices that consume and/or emit events and their connections.

Evaluation Hypothesis. Our hypothesis H_1 is that using our visualization improves the understandability of the system. Participants should be able to answer relevant questions about the system's communication architecture in less time than in their currently chosen development environment. Thus, our null hypothesis H_0 is that our visualization does not improve the understandability of the system.

Participants. Ten software developers familiar with the system participated in the evaluation. Eight of the ten developers had experience with both back- and frontend development, two were mainly focussed on frontend. Work experience varied between two and ten years of professional software development.

We split the participants into two groups, A and B, each with four full-stack and one frontend developer. Median years of experience in both groups was five years.

Participants of both groups were given an introduction into the background of the evaluation, its goals and challenges. We interviewed each person using their own setup and instructed them to have the latest version of the source code of all services checked out on their own development environment. We let them set up any debugging tools they might need, e.g. for viewing the consumer groups in the event broker. Additionally, all participants were allowed access to any other source code, development environment configurations or project documentation available to them. By letting them prepare this way, we made sure that each participant was as comfortable as possible with his or her setup before starting the test.

Other than personal preference and familiarity in the development setup, the only difference between group A and B was access to the visualization. Only participants in group A were given access to our tool and an overview of the functionality before starting the exercise.

Tasks. As our study assignment we choose two knowledge retrieval tasks. For each, we choose a single event from the 316 total events and ask which services are part of the event chain triggered by the first event. For the first one we choose a "simple" event, the resulting event chain consists of 3 services and 4 messages in total, with 3 of the 4 events being handled by only one service. For the second event we choose a more complex one. Here, 8 services are involved, and 6 events are emitted in total in order to complete a Saga [5].

Task 1. *A redirect has been created in the admin dashboard. For debugging purposes, the business logic that is triggered as a result of the redirect creation should be examined. Please list which services can be addressed and which events can be emitted, based on the Saga event* `command_create_redirect`*.*

Task 2. *The same should now be done for an update of a job posting text, starting with the Saga event* `command_update_job_posting_general_tab`*. Please enumerate which services can be addressed and which events can be emitted.*

5   Results and Discussion

Group A took an average of 43.2 s to complete each task, group B took 7.5 min on average, with Task 2 being aborted after 10 min each time (i.e., a difference of at least one order of magnitude compared to group A). Both tasks were answered correctly every time by group A while group B answered Task 1 correctly 3 out

Table 1. Results for Task 1 and 2.

		Average	Median	σ	σ^2	Correctness
Task 1	Group A	51.2 s	41 s	34 s	1156.7	100
	Group B	294.8 s	247 s	111 s	12405.7	60
Task 2	Group A	35.2 s	47 s	10.3 s	107.2	100
	Group B	600.0 s	600 s	0 s	0	33.3

of 5 times, and in Task 2 only 2 out of 6 services were covered in the time for each subject, resulting in progress of 33.3% after the maximum duration.

We tested the results of both tasks separately using a one-sided t-test with different variances, since H_1 and the calculated means lead to a directional hypothesis, with a significance level of $\alpha = 5\%$. Additionally, the effect size d is examined according to Cohen [1].

According to Welch [12], the p-value of the t-test for Task 1 is approximately $0.0027 < \alpha$ (see values in Table 1), which is well below the significance level. Cohen's d is $d = 2.1 > 0.8$, which is a large effect size.

For Task 2, the p-value of the t-test is about $1.4 \times 10^{-8} < \alpha$, which is well below the significance level (see values in Table 1). Cohen's d is $d = 54.6 > 0.8$, which is a very large effect size.

This indicates that the results of both Task 1 and Task 2 are statistically significant and that the hypothesis H_1 is fulfilled, thus rejecting the null hypothesis H_0.

Construct Validity. In our case study we are using speed and accuracy as proxy metrics for understanding: The faster a participant acquires the required information, and the fewer errors they make, the better our visualization helped them in understanding the system. There are other metrics to measure understanding, e.g. retention over time. However, in the case study's setting we focussed on speed and accuracy, since these were the most prominent performance indicators to the system's architects and developers.

Internal Validity. We verified our findings with a one-sided t-test to get a significance level α of 5% and an effect size d of 2.1 and 54.6. These values suggest that our results are statistically significant. We can thus reject our null hypothesis H_0.

Conclusion Validity. We can reject our null hypothesis and claim that our visualization does indeed support understanding the communication architecture of the system. However, we could only compare our approach to the status quo of the documentation available to the development team. We did not determine if the development team did not know or not use any other, better solution. Thus, we cannot directly compare our approach to others'. We tried to mitigate this by choosing a null hypothesis that allows for any and all other solutions to be used.

External Validity. Although we performed a t-test, the very small sample size ($n = 10$) is not sufficient to obtain a general and meaningful result. Thus, according to Lehr [8], to approximate the required sample size, the sample size should be at least $n = 16\frac{s^2}{\delta^2} = 16\frac{1156.7^2}{243.2^2} \approx 181$ (calculated using the results from Task 1). Therefore, we cannot generalize our result statement to a larger population. Additionally, we only studied the applicability to a specific software system. Since the construction of our visualization is not dependent on any particular feature of the system other than its architecture, the applicability to other systems of the same architecture is implied. In future work we aim to extend our approach of visualizing the runtime behavior for an improved understandability to other software architectures.

6 Related Work

The visualization of software architecture by graphs has long been proposed, e.g. in [2,7], and graph theory has been successfully applied to support use cases ranging from architecture comprehension, over antipattern identification to architecture transformation [3,6,9,13]. The impact of graph-based software visualization with respect to developer-efficiency has been explored in [10]. In contrast to research that addresses comprehensibility of event-based systems at runtime, addressing traceability and monitoring use cases, in this paper, we contribute insights on *design-time* use cases. Design-time support has already been proven successful in the field of robotic systems [11]. We complement such research with the concept of a practical, developer-intuitive support for recurrent architecture decisions, that we evaluated with architects and developers of an industrial software system that realizes an event-driven communication paradigm.

7 Conclusion and Future Work

The trend towards microservice architectures and scalable event-driven architectures impacts architecture work, especially the understanding of the dynamics during runtime. Common architecture diagrams, e.g. from the UML, offer means to explain and describe both static as well as dynamic aspects of a software system. However, due to the amount of events and possible connections between services, architects and developers seem to have trouble documenting the crucial information of the event-driven architectural style with existing diagrams.

We conjecture that there is a need for complementary diagrams or visualizations that support architecture-specific reasoning. In the future, we will further explore the space of runtime architecture visualization. We propose visualizing the dynamic behavior of a system derived from its specific architectural design choices.

Data Availibility Statement. The anonymized protocols from our initial interviews and the raw data from the evaluation are available for download (https://zenodo. org/records/12519817). The service-local documentations of events which is used to evaluate our visualization and the source code of our visualization is proprietary information and not publicly available.

Disclosure of Interests. Authors A and B are employed by the company that develops and maintains the system and team studied.

References

1. Cohen, J.: Statistical power analysis. Curr. Dir. Psychol. Sci. **1**(3), 98–101 (1992)
2. Dąbrowski, R., Stencel, K., Timoszuk, G.: Software is a directed multigraph. In: Crnkovic, I., Gruhn, V., Book, M. (eds.) ECSA 2011. LNCS, vol. 6903, pp. 360–369. Springer, Heidelberg (2011). https://doi.org/10.1007/978-3-642-23798-0_38
3. Farsi, H., Allaki, D., En-Nouaary, A., Dahchour, M.: A graph-based solution to deal with cyclic dependencies in microservices architecture. In: 2022 9th International Conference on Future Internet of Things and Cloud (FiCloud), pp. 254–259 (2022). https://doi.org/10.1109/FiCloud57274.2022.00042
4. Gamage, I.U.P., Perera, I.: Using dependency graph and graph theory concepts to identify anti-patterns in a microservices system: a tool-based approach. In: 2021 Moratuwa Engineering Research Conference (MERCon), pp. 699–704 (2021). https://doi.org/10.1109/MERCon52712.2021.9525743
5. Garcia-Molina, H., Salem, K.: Sagas. ACM Sigmod. Record **16**(3), 249–259 (1987)
6. Guth, J., Leymann, F.: Pattern-based rewrite and refinement of architectures using graph theory. SICS Softw.-Intensive Cyber-Phys. Syst. **35**(1), 115–126 (2020). https://doi.org/10.1007/s00450-019-00416-7
7. Le Metayer, D.: Describing software architecture styles using graph grammars. IEEE Trans. Software Eng. **24**(7), 521–533 (1998). https://doi.org/10.1109/32.708567
8. Lehr, R.: Sixteen s-squared over d-squared: a relation for crude sample size estimates. Stat. Med. **11**(8), 1099–1102 (1992). https://doi.org/10.1002/sim.4780110811
9. Raj, V., Ravichandra, S.: A service graph based extraction of microservices from monolith services of service-oriented architecture. Softw. Pract. Exp. **52**(7), 1661–1678 (2022). https://doi.org/10.1002/spe.3081
10. Shahin, M., Liang, P., Li, Z.: Architectural design decision visualization for architecture design: preliminary results of a controlled experiment. In: Proceedings of the 5th European Conference on Software Architecture: Companion Volume. ECSA 2011. Association for Computing Machinery, New York (2011). https://doi.org/10.1145/2031759.2031762
11. Timperley, C.S., Dürschmid, T., Schmerl, B., Garlan, D., Le Goues, C.: Rosdiscover: statically detecting run-time architecture misconfigurations in robotics systems. In: 2022 IEEE 19th International Conference on Software Architecture (ICSA), pp. 112–123 (2022). https://doi.org/10.1109/ICSA53651.2022.00019
12. Welch, B.L.: The significance of the difference between two means when the population variances are unequal. Biometrika **29**(3/4), 350–362 (1938). http://www.jstor.org/stable/2332010
13. Winchester, G., Parisis, G., Berthouze, L.: On the temporal behaviour of a large-scale microservice architecture. In: NOMS 2023-2023 IEEE/IFIP Network Operations and Management Symposium, pp. 1–6 (2023). https://doi.org/10.1109/NOMS56928.2023.10154427

Sustainability

Energy Consumption of IoT Monitoring Software Architectures in the Edge

Juan Sebastian Ochoa[1]📖, Jennifer Pérez[1]([✉])📖, Javier García[1]📖,
Daniel Guamán[2]📖, Norberto Cañas[1]📖, and Vanessa Rodriguez-Horcajo[1]📖

[1] Universidad Politécnica de Madrid, Madrid, Spain
{js.ochoa,vanessa.rodriguez.horcajo}@alumnos.upm.es,
{jenifer.perez,javier.garciam,norberto.canas}@upm.es
[2] Universidad Técnica Particular de Loja, Loja, Ecuador
daguaman@utpl.edu.ec

Abstract. IoT has led energy consumption to be a critical issue for sustainability due to the huge amount of power connections of IoT devices and the huge amount of generated data that must be transmitted and stored. As a result, the data monitoring process and its processing activities must be performed by the nodes and servers of the Edge in a sustainable way. The IoT monitoring process of data consists of four main activities: listening, filtering, translating and routing. Once the data has been monitored, it can be stored and transmitted to the Edge/Cloud. In order to address sustainable monitoring processes in the Edge, it is necessary to evaluate how the configuration of the components that compose the Edge monitoring architecture may influence its energy consumption. In this paper, we present the experimental results of an exploratory study and its findings, in which the energy consumption of four different software architecture configurations of an indoor environmental monitoring IoT system is measured. From the execution of 24 experiments, the study reveals the importance of balancing the monitoring activities between the Edge nodes and servers, and evidences the energy consumption increment that data storage implies for the Edge.

Keywords: Software Architecture · Edge · IoT · Data Monitoring · Sustainability · Energy Consumption

1 Introduction

Despite the fact that technology has brought significant advances that have facilitated current human life, it is responsible for digital pollution. To deal with this problem, reducing the energy consumption of this technology is mandatory. To that end, significant advances in the design of hardware and communication protocols have been made [21]. However, there is still much work to be done to achieve a significant reduction in the energy consumption of the software running on such hardware [13,31]. It is expected that the 21% of the energy consumption of the planet will come from Information and Communication Technologies

M. Galster et al. (Eds.): ECSA 2024, LNCS 14889, pp. 217–232, 2024.
https://doi.org/10.1007/978-3-031-70797-1_15

(ICT) in 2030 and, in the worst case, this percentage may reach the 50% [8]. Although the industry is addressing the design of sustainable hardware, a lot of work is still pending in software [18]. This is specially critical in Internet of Things (IoT) systems, since they are exponentially increasing, and it is estimated that the number of licensed cellular IoT connections will increase from 3.5 bn in 2023, to 5.8 bn in 2030 [24]. Therefore, IoT software architectures must be sustainable-aware.

Sustainability refers to "the 'capacity' of a system 'to endure'" [7] and "meeting the needs of the present without compromising the ability of future generations to meet their own needs" [18]. Becker et al. define a sustainability manifesto [10] and five dimensions: social, individual, environmental, economic, and technical. The technical dimension refers to "longevity of information, systems, infrastructure, and their adequate evolution with changing surrounding conditions. It includes maintenance, innovation, obsolescence, data integrity, etc." This dimension promotes practices, criteria and metrics to support green-aware activities during software construction and evolution [10]. As a result, it is important to use green practices for designing software architectures [26]. This research work is focused on finding green design practices for IoT Software Architectures. IoT software architectures are constructed by applying Edge [33,34], Fog [11] and Cloud [9] computing.

Edge computing emerged to deal with the specific IoT needs [16]. IoT is a computing paradigm for monitoring data sources as near to them as possible. The monitoring process consists in retrieving data from the IoT devices and processing, storing and transmitting it. The processing may consist in computing data for real-time systems, or filtering or transform data in order to complete data or withdraw replications in order to transmit or store the correct data. The Edge is implemented by the architectural layer between the physical and Cloud layers (see Fig. 1) [14,33]. However, there are many configurations regarding where to locate the nodes and/or servers at the Edge layer, as well as where to deploy the software components that implement the monitoring process of the Edge. In this work, we focus on the study of the energy consumption of the Edge layer when it is used for data transmission or storage [33], and we evaluate how different architectural configurations may influence energy consumption.

This paper presents the experimental results of an exploratory study and its findings to assist a green-aware Edge-architecture design. The energy consumption measurements of 4 different software architecture configurations of an indoor environmental monitoring IoT system are presented and compared among them. In particular, the 4 architectures have been measured with 24 experiments and their results evidence the importance of balancing the monitoring activities between the Edge nodes and servers and the energy consumption increments that data storage implies for the Edge.

The paper has been structured as follows: Sect. 2 describes the related work about studies on the energy consumption of IoT monitoring Edge Architectures, Sect. 3 describes the exploratory study design, Sect. 4 presents the results of the 4 evaluated architectures and their 24 measurement experiments, Sect. 5 presents

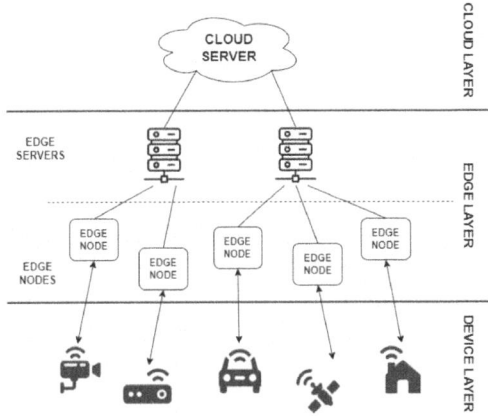

Fig. 1. 3-tier IoT architecture

the threats to validity, and finally, conclusions and future work are presented in Sect. 6.

2   Related Work

There is an incremental concern about the power consumption of IoT monitoring systems [25]. The survey of Jalali et al. [19] sets out the relevance of the location of resources at the architectural level and to study the design of the energy consumption of IoT architectures at the different layers (device, Edge, Fog, Cloud, etc.). In addition, it points out the need of performing architectural studies of energy consumption of IoT Systems. There are works that evaluate the energy-consumption of IoT systems comparing the Edge Architectures with the Cloud, Fog or hybrid architectural proposals [4,20,23]. Also, there are works that construct sustainable solutions applied to specific domains such as buildings [29], health [12], agriculture [6], etc. Other works present studies about models, schemes or algorithms to improve the energy consumption of specific IoT systems by considering different variables and alternatives for their construction [5,15,28,30]. However, these works do not focus on studying how the architectural alternatives influence in the energy consumption of the Edge. The work of Jinaporn et al. focuses on studying the impact of the energy consumption for data processing and the gateway placement on the maximum lifetime routing (MLR) network problem. But they do not focus on the energy consumption of the complete Edge architecture and the monitoring process and its software components. As a result, this work deals with the need of comparing the Edge energy consumption of the different architectural solutions to design the IoT monitoring process.

3 Exploratory Study Design

This research work analyzes how the configuration of the data monitoring components of IoT architectures influences the energy consumption of the Edge layer avoiding software architect bias. To that end, an exploratory study was conducted and reported following the guidelines of the literature [27,32]. In addition, since this study requires a rigorous process for measuring the energy consumption of the IoT Monitoring Architectures and its subsequent analysis, the study design has also taken into account the measuring energy consumption guidelines of Mancebo et al. [22] for analysing the energy efficiency of software. In the Edge, the IoT data monitoring process consists of several activities and may have different purposes. Taking into account these activities and purposes, this exploratory study was decomposed into the study of the energy consumption of four different Edge software architectures. This section describes the Context and Scope of the study, the research objective and questions, the construction of the four architectures, and the definition of the experiments to execute the study.

3.1 Context and Scope

How the IoT monitoring activities are decomposed into software components and where these components are located in the Edge nodes and servers, generate different software architecture configurations (see Fig. 1). In the Edge, the IoT data monitoring process consists of four main activities: listening (collect data from IoT Devices), filtering (filter data), translating (transform data into the required formats) and routing (send data). Then, in addition to processing and computing the recovered data, the monitoring process may have two different purposes: to store the data in the Edge or to transmit the data to the Cloud. Depending on the purpose, the architecture is also designed in a different way. This exploratory study is focused on measuring how these variations in the architecture influence the energy consumption. Therefore, this exploratory study was decomposed into the study of the energy consumption of four different Edge software architectures in order to evaluate the energy consumption depending on both variables: the location of monitoring components and the purpose of the Edge [33]. These architectures are designed considering each activity of the monitoring process as a different software component in order to avoid tangled concerns and facilitate its reuse, maintenance and change location in the architecture; i.e. there are four different components to implement the listening, filtering, translating and routing activities, respectively. As a result, the scope of this research work is to study the Edge layer's energy consumption of the four following architectural variants:

- Architecture 1: The complete IoT monitoring process is performed by the Edge nodes and its purpose is to store the data in the Edge.
- Architecture 2: The listening and routing activities are performed by the Edge nodes, the filtering and translating activities are performed by the Edge server, and its purpose is to store the data in the Edge.

- Architecture 3: The complete IoT monitoring process is performed by the Edge nodes and its purpose is to transmit the data to the Cloud.
- Architecture 4: The listening and routing activities are performed by the Edge nodes, the filtering and translating activities are performed by the Edge server, and its purpose is to transmit the data to the Cloud.

Edge Computing is applied to many IoT domains, such as smart homes and buildings, smart cities, smart grids, smart-transportation, e-Health, crowd-sensing, etc. [14,33]. In this paper, we have focused on smart buildings, specifically on an indoor environmental monitoring IoT system.

3.2 Objectives and Research Questions

The main goal of this study is to identify sustainable-aware design decisions for IoT monitoring systems in the Edge and to evidence that the architectural design of the IoT system's Edge layer influences the energy consumption. To that end, this study addresses the following research questions:

RQ1. Does the Edge energy consumption vary depending on whether the monitoring components are deployed on the Egde nodes or on the Edge servers?

RQ2. Does the Edge energy consumption vary depending on whether data is stored in the Edge or transmitted to the Cloud?

The architectures 1 and 3 deploy the four data monitoring components on the Edge nodes; whereas architecture 2 and 4 deploy two of the four components on the Edge nodes and the other 2 on the Edge server. Therefore, comparing architectures 1 and 3 with architectures 2 and 4 enables us to address RQ1. Additionally, architectures 1 and 2 share the purpose of storing data in the Edge, while architectures 3 and 4 aim to transmit the data to the Cloud. Thus, comparing architectures 1 and 2 with architectures 3 and 4 allows us to address RQ2.

3.3 Construction of the Software Architectures

Architectural Description of the IoT System. This exploratory study is based on an indoor environmental monitoring IoT system, whose purpose is to monitor the environmental quality of a building. The physical layer is composed of a wireless sensor network that measures environmental metrics with different kinds of sensors. In particular, these sensors are slave sensors that collect the following measures: temperature, relative humidity, light, CO_2, movement and noise. These slave sensors communicate with a master sensor using the Zigbee communication protocol [17]. The master sensor, usually an IoT device, connects the Device layer and Edge layer. It is in charge of collecting all the data transmitted by the slave sensors and sending it through a serial bus to the Edge nodes (see Fig. 2). These measures are data that must be encapsulated into frames for its correct communication. To that end, a communication frame was designed for this exploratory study. It consists of the following three elements (see Fig. 3):

- HEADER: It provides information about the frame structure, and it is composed of 5 bytes. The byte 0 is to determine if it is a control frame or a measure frame. The bytes 1–3 contain the master sensor identifier, and the byte 4 indicates the number of slave sensors that send data in the frame.
- DATA: It contains the data obtained from sensors. Its size varies according to the number of slave sensors that communicate with the master. A single slave sensor needs 5 bytes for its data, i.e. in the case of a master sensor that is communicated with 2 slave sensors, the DATA element size is 10 bytes. The first byte of each slave sensor identifies the sensor type, and the last 4 bytes contain the measurement's value.
- END: It identifies the end of the frame, and it has 2 bytes.

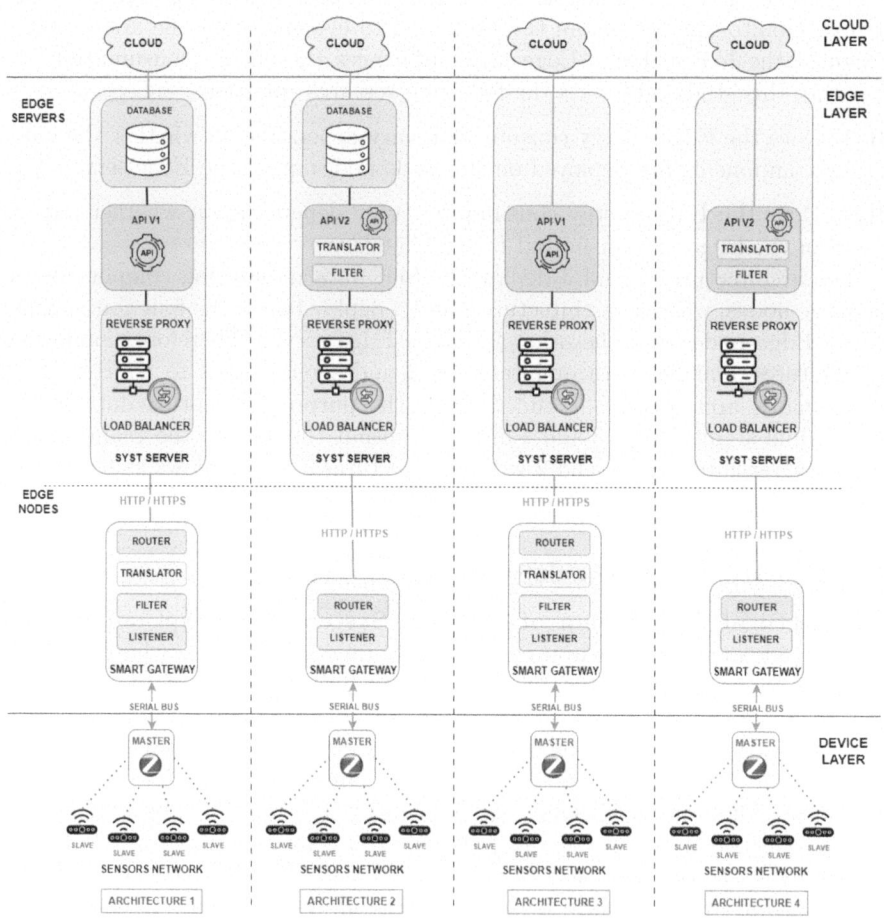

Fig. 2. The four IoT architectures under evaluation

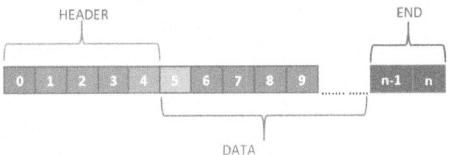

Fig. 3. Design of the communication frame

The Edge layer is composed of Edge nodes and Edge servers. The Edge nodes in this case are smart gateways connected to the master sensor of the device layer. A smart gateway is a device with a greater computing capacity than the end IoT device [3], since they usually preprocess the data collected from IoT devices before forward them to the Edge/Cloud servers. Smart gateways are also called IoT gateway in IoT systems [33]. The Edge servers are in charge of communicating the IoT devices with the Cloud, and they can be used for storage purposes. In this case study, the two purposes are going to be evaluated (see Fig. 2).

With regard to the data monitoring process, as it has been mentioned in the Sect. 3.1, the four activities of data monitoring have been designed as different software components which perform the following functionalities:

– Listener: This component listens to and receives the raw data sent by the master sensors.
– Filter: This component verifies the integrity of the data received from the Listener.
– Translator: This component translates the raw data into the appropriate format for being correctly processed for its next transmission and storage. In this case, data is translated into a JSON format.
– Router: This component sends the JSON formatted data to the server using http/https requests.

Configuration of Software Architectures. The exploratory study consists in evaluating the Edge layer's power consumption of the four different architectures. Figure 2 illustrates the architectural differences and similarities among them. The four architectures share the physical and Cloud layers, since the exploratory study is focused on the Edge layer. At this layer, all architectures have a proxy component that balances the load of data received from the smart gateway for being transmitted to the Cloud or stored in the database through an API. With regard to the Edge server design, the architectures 1 and 2 share the purpose of storing the data in the Edge. Therefore, the Edge servers of both architectures have a database component to store the collected data from the sensors. On the other hand, the architectures 3 and 4 are designed to transmit data to the Cloud. As a result, they do not have the database component. However, thanks to the different implementation of APIs, the architectures 1 and 2 are different between them, and also the architectures 3 and 4. The architectures 1 and 3 implement the API v1, and Architectures 2 and 4 implement the API v2. In particular, this API v2 is different due to the fact that the Edge server implements the components Filter and Translator.

Concerning the Edge node design, i.e. the smart gateway, we have designed two different variants. One smart gateway contains the components Listener, Filter, Translator and Router, which is implemented in architectures 1 and 3. The other smart gateway contains the components Listener and the Router, which is implemented in architectures 2 and 4. This variability in the Edge node will allow us to study if it is more sustainable to process data in the gateway or in the server.

3.4 Description of the Energy Measurement Environment

In order to measure the Edge energy consumption, the Edge server and the Smart Gateway have to be monitored. Specifically, to monitor the energy consumption of the Smart Gateway, it is used an INA219 sensor, which allows to measure the current and voltage of the device that it is connected to. To that end, the INA219 sensor was connected to the smart gateway, communicating one to each other through the I2C protocol.

On the other hand, to monitor the server we are going to use a software energy measurement tool called jRAPL (java Running Average Power Limit) [2]. This tool is able to calculate the energy consumption of Java applications executed on Linux in computers with an Intel processor with the architecture Sandy Brigde. This kind of tools apply calculation models from the computer/device's processor to obtain the power consumption, and despite the fact that it is not an energy measurement from the current as the INA219 performs, they provide the facility of calculating the energy of a specific process ID. However, in the case of INA219, the power consumption provided corresponds to the consumption of the device and not the specific consumption of a process. Therefore, to extract the specific energy consumption of the architectural components of the IoT Data Monitoring, instead of the energy consumption of all the processes executed on the Smart Gateway, it is also necessary to know the energy consumption of the Smart Gateway when it is idle.

3.5 Definition of the Energy Measurement Experiments

To evaluate the energy consumption of the different architectures, 24 measurement experiments of energy consumption were executed. Firstly, we defined three kinds of experiments for each architecture, sending 1000 frames in each one and varying the time elapsed between each transmission to avoid the bias that a constant transmission data could derive. In addition to these 12 measurement experiments, we made a replication to avoid bias in the conclusions. This experimental planning is detailed in Table 1, which detail the frame transmission frequency of each experiment and architecture (every 100, 150 or 200 ms).

In addition to the experimental design for the energy measurements of each architecture, it is necessary to perform measurements to determine the basal

Table 1. Definition of the energy measurement experiments

Experiment	Architecture 1	Architecture 2	Architecture 3	Architecture 4
Experiment 1	100 ms	100 ms	100 ms	100 ms
Experiment 2	150 ms	150 ms	150 ms	150 ms
Experiment 3	200 ms	200 ms	200 ms	200 ms
Experiment 4 (replication exp. 1)	100 ms	100 ms	100 ms	100 ms
Experiment 5 (replication exp. 2)	150 ms	150 ms	150 ms	150 ms
Experiment 6 (replication exp. 3)	200 ms	200 ms	200 ms	200 ms

consumption of the Smart Gateway. In this case, we defined to collect the basal consumption during 6 experiment to monitoring 6 days, 4 of them consecutive in date and during all the day, and 2 of them not consecutive in date and without monitoring the complete day. This variability avoids specific problems of the power network, that could appear in certain periods of time and could generate bias in the analysis of results. Specifically, the basal consumption of the Smart Gateway is monitored 24 h during the days 1–4, and 16 h during the days 5 and 6. Experiments 1–5 were consecutively executed at the same week, whereas the experiment 6 was executed the next month.

4   Results Analysis and Findings

All the raw data of the 24 measurement experiments and the analysis and synthesis of them are provided in the repository that supports this research work [1] (see the Data Availability section). In addition, the basal consumption study of the Smart Gateway, i.e. the raw data of the energy measurements of the 6 days and the basal consumption calculation, are also provided in the folder "Dummy" of this repository. From this study, we observed that there is no important differences from the measurements of the 6 days, and we calculated the basal energy consumption of the Smart Gateway from the average of the 411635 energy measurements collected, i.e. 2,541895493 Ws.

The launch of the architectures that share the same purpose, storing or transmitting data, were executed simultaneously. The experiments of architecture 1 and 2 were executed at the same time and the same day, and the experiments of the architectures 3 and 4 were executed the next day at the same hour. In this section, we present the results of power consumption registered in the 24 experiments. From the Smart Gateway collected data, i.e. the time elapsed, the current consumption and the voltage, the energy consumption is calculated for analyzing its energy consumption; whereas the energy consumption of the server is calculated from the time elapsed and the CPU power consumption. Moreover, the INA219 and jRAPL provide additional data from the Smart Gateway and the server measurements as the CPU usage, the resident memory size (RSS), the virtual memory size (VSZ) and the DRAM power consumption. They are provided in the folder "Additional Metrics" of the repository [1] (see the Data Availability section).

The total energy consumption for the four architectures is detailed in Table 2. For each of the six above mentioned experiments (see Table 1), Table 2 shows the energy consumption measurements obtained in each of the four architectures. Specifically, it indicates the energy consumption in Ws that has been measured in each of the two components of the Edge layer (Smart Gateway and Edge server) during the transmission of the 1000 frames. From these collected data is possible to answer the research questions defined for this study. In addition, these data evidence that the experiment 3 and its replication (experiment 6) are the most consuming ones, whereas experiment 1 and its replication (experiment 4) are the less consuming ones. From these data, it is possible to extract the first finding F1.

> **F1.** *Data transmission frequency influences the energy consumption. A smaller data transmission frequency helps to reduce the energy consumption.*

Table 2. Power consumption at the Edge of the four architectures

EXPERIMENT	ARCHITECTURE 1		ARCHITECTURE 2		ARCHITECTURE 3		ARCHITECTURE 4	
	SMART GATEWAY - POWER CONSUMPTION (Ws)	EDGE SERVER - POWER CONSUMPTION (Ws)	SMART GATEWAY - POWER CONSUMPTION (Ws)	EDGE SERVER - POWER CONSUMPTION (Ws)	SMART GATEWAY - POWER CONSUMPTION (Ws)	EDGE SERVER - POWER CONSUMPTION (Ws)	SMART GATEWAY - POWER CONSUMPTION (Ws)	EDGE SERVER - POWER CONSUMPTION (Ws)
Exp. 1	36,34048209	59,50068700	32,70540138	58,57431100	37,18898367	12,46125500	35,22781752	12,08009500
Exp. 2	40,94209779	62,53107100	39,22505890	63,84908800	41,88284644	14,85491900	39,96978700	14,84774400
Exp. 3	49,96156479	66,53659400	48,63696401	66,47980600	49,32720644	19,34011000	48,29217887	18,89854100
Exp. 4 (Replication Exp. 1)	32,47185197	60,14417000	33,17763271	59,94928800	35,35421527	12,35153800	32,00710656	11,66530000
Exp. 5 (Replication Exp. 2)	36,69879056	62,38997700	36,75387125	64,42251200	37,72393507	15,20404000	36,58862261	14,87124300
Exp. 6 (Replication Exp. 3)	42,54305609	66,67256800	41,85754982	67,77891600	44,00361766	19,32323500	41,00548767	19,10857900

RQ1. *Does the Edge energy consumption vary depending on whether the monitoring components are deployed on the Egde nodes or on the Edge servers?*

Architectures 1 and 3 design the monitoring process including the four components in the Smart Gateway; whereas in architectures 2 and 4 the components Filter and Translator are deployed in the Edge server, and the Listener and the Router are located in the Smart Gateway. To answer this question, we are going to compare the data of these different designs. So, firstly the energy consumption of the smart gateway and the server are analyzed individually, and then the total energy consumption of both architectural elements is studied.

Figure 4 presents the current consumption of the Smart Gateway. In first place, the power consumption analysis of the smart gateway isolating the purpose of the architecture is performed. Therefore, we compare the architectures that share the same purpose and different monitoring design, i.e. the architectures 1 and 2, and the architecture 3 and 4. Architecture 1 consumes more energy than Architecture 2 sharing the purpose of storing the data in the Edge layer, except in the replications of the experiments 4 and 5, that it is higher and equal, respectively. There is a minimum energy saving of 0,7 Ws and a maximum energy saving of 3,6 Ws. On the other hand, Architecture 3 consumes more energy than

Architecture 4 sharing the purpose of transmitting the data in the Edge layer. There is a minimum energy saving of 1 Ws and a maximum energy saving of 3,4 Ws. From this data analysis, the finding F2 can be derived.

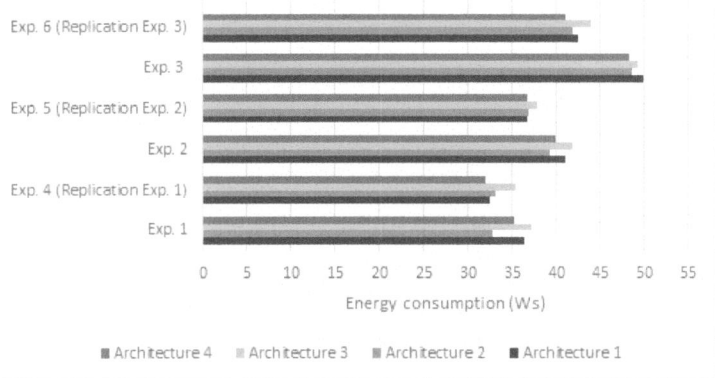

Fig. 4. Power consumption analysis of the Smart Gateway

F2. *The architectural element where the IoT monitoring process components are deployed influences the power consumption of the Edge layer.*

If we analyzed together all the architectures, it is evidenced that the smart gateways of architectures 2 and 4 are less power consuming than architectures 1 and 3. However, it is important to also analyze the power consumption of the Edge servers. This is necessary to determine if the energy saving of the smart gateways of architectures 2 and 4 not imply a high increase of its corresponding Edge servers due to the movement of the Filter and Translator components to them. This fact is evidenced in Table 2 and Fig. 5, which illustrate that it does not imply an increase in the power consumption of the server, since the architectures with the Filter and Translator in the Edge server have approximately the same energy consumption than those that have not include them in the rest of the element servers and are designed in the same way, i.e. the server's energy consumption of architectures 1 and 2 are mostly the same, and energy consumption of architectures 3 and 4 are almost equal. As a result, we can answer the RQ1 and conclude the findings F3 and F4.

F3. *The deployment of the Filter and Translator components in the Edge server does not imply an increment in the energy consumption of the Edge server.*

F4. *The less power consuming architectural design of the IoT monitoring process consist in deploying the Filter and Translator components in the Edge server, and the Listener and the Router in the Edge Smart Gateway.*

RQ2. *Does the Edge energy consumption vary depending on whether data is stored in the Edge or transmitted to the Cloud?*

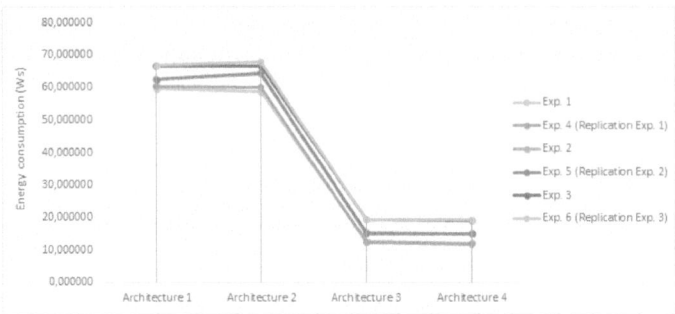

Fig. 5. Energy consumption of the Edge server

The transmitting and storing purposes of the Edge layer are reflected in the Edge server design, since it implies the inclusion or absence of a database and their corresponding APIs, i.e., architectures 1 and 2, and architecture 3 and 4. Figure 5 evidences the affirmative response of RQ2 and the findings F5 and F6. F6 is also an important economical and green information for the company of the IoT system, because it implies an energy and economical saving from the Edge layer that it is included in the Cloud payment per its use.

F5. *The purpose of storing data in the Edge or transmitting data to the Cloud influences the power consumption.*

F6. *Storing the data in the Edge server is more energy consuming for the Edge layer than transmitting it to the Cloud*

Once both research questions have been answered affirmatively, it is necessary to determine which is the most sustainable design of the Edge in order to have this architectural guidance. To that end, Table 3 presents the total power consumption of the Edge layer of each experiment in each architecture; i.e. the sum of the current consumption of the Edge server and the Edge Smart Gateway. In addition, Fig. 6 illustrates the average grouped by architecture to facilitate the power consumption comparison among the different experiments and the architectures in Fig. 7. These data reveal that Architecture 1 is the most consuming for the Edge layer and Architecture 4 is the less consuming one (see finding F7).

F7. *The greener design architecture of an IoT monitoring process at the Edge layer consist in: (i) transmitting the data for being stored in the Cloud, and (ii) deploying the Filter and Translator components at the Edge server, and the Listener and Router components at the Smart Gateway*

5   Threats to Validity

To improve the internal validity of the presented results, we design the IoT system for automatically store in log files the energy measurements collected by

Table 3. Total power consumption at the Edge of the four architectures

Experiment	Architecture	Smart Gateway	Server	Total
Exp.1	Architecture 1	36,34048209	59,501	95,8411691
	Architecture 2	32,70540138	58,574	91,2797124
	Architecture 3	37,18898367	12,461	49,6502387
	Architecture 4	35,22781752	12,080	47,3079125
Exp.2	Architecture 1	40,94209779	62,531	103,473169
	Architecture 2	39,2250589	63,849	103,074147
	Architecture 3	41,88284644	14,855	56,7377654
	Architecture 4	39,969787	14,848	54,817531
Exp.3	Architecture 1	49,96156479	66,537	116,498159
	Architecture 2	48,63696401	66,537	115,173558
	Architecture 3	49,32720644	19,340	68,6673164
	Architecture 4	48,29217887	18,899	67,1907199
Exp.4	Architecture 1	32,47185197	60,144	92,616022
	Architecture 2	33,17763271	59,949	93,1269207
	Architecture 3	35,35421527	12,352	47,7057533
	Architecture 4	32,00710656	11,665	43,6724066
Exp.5	Architecture 1	36,69879056	62,390	99,0887676
	Architecture 2	36,75387125	64,423	101,176383
	Architecture 3	37,72393507	15,204	52,9279751
	Architecture 4	36,58862261	14,871	51,4598656
Exp.6	Architecture 1	42,54305609	66,673	109,215624
	Architecture 2	41,85754982	67,779	109,636466
	Architecture 3	44,00361766	19,323	63,3268527
	Architecture 4	41,00548767	19,109	60,1140667

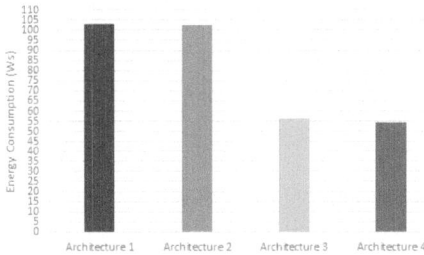

Fig. 6. Edge layer consumption of the four architectures

the sensor INA219 and the tool jRAPL. As a result, this automatic information management avoided personal bias. The replicated executions have been performed in the same infrastructures and the experiments with architectures with the same purpose have been executed at the same time. In addition, 6 experiments with different variations considering the date and the time of measurements have been performed to calculate the basal consumption in order to provide data as accurate as possible and to avoid bias in the data.

Construct validity was dealt with using tools that allow us to calculate mathematical operations such as averages, subtractions and additions of the different variables, as well as graphics to represent them. Since the data are quantitative and the synthesized data was calculated by tools, author bias have been avoided.

Finally, to address external validity, the model was constructed using 24 experiments. Four different architectural designs have been performed and 3 variations of execution have been implemented varying the data transmission frequency. These twelve experiments have been measured two times to avoid

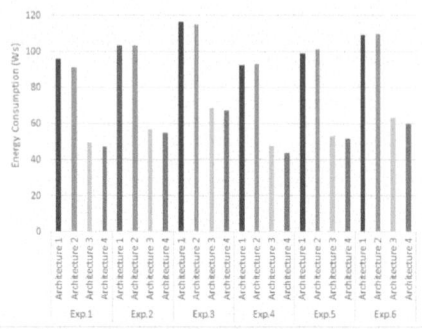

Fig. 7. Total power consumption by architecture and experiment

bias. However, it will be interesting to make more replications in future works in order to guarantee the power consumption of the smart gateway, since the sensor INA219 measurements of the smart gateway can be affected by other power consumption processes. In addition, to replicate the experiment with other hardware setup and other tools different to jRAPL, it would be interesting to generalize the results. In order to facilitate the replication and reuse of data, all of them are provided in the repository associated to this work [1].

6   Conclusions

This paper presents the result of an exploratory study to determine how the IoT monitoring process design influences the power consumption. We have constructed an indoor environmental monitoring IoT system with 4 different designs, i.e. 4 architectures, in which 24 experimental measurements have been performed. From these experiments, it has been possible to evidence the importance of balancing the monitoring activities between the Edge nodes and servers, and the energy consumption increment that data storage implies for the Edge. This study analysis has been synthesized into 7 findings that can be used for software architects in order to apply a green design decision making process. In addition, an study of the basal consumption of the Smart Gateway has been executed.

Finally, we will work on replicating this study in other domains, circumstances of the network, communication protocols or technologies, in order to reinforce the results and to determine if these factors must be taken into account during the IoT monitoring architectural design. In addition, we plan to use other energy measurement tools for measuring the energy of the server in order to determine the influence of them in the results.

Acknowledgments. This work is supported by the Spanish Ministry of Science and Innovation (MICINN) through "SIoTCom: Sustainability-Aware IoT Systems Driven by Social Communities" (PID2020-118969RB-I00).

Data Availibility Statement. In order to promote future works from this study, all the raw and synthesized data are provided in a dataset repository published in the ECSA Zenodo Community https://doi.org/10.5281/zenodo.12516350 and GitHub [1].

References

1. Energy consumption of IoT monitoring software architectures in the edge (2024). https://github.com/jenniferperezbenedi/EnergyConsumption_ IoTMonitoringEdge. Architectures. https://doi.org/10.5281/zenodo.12516350
2. jRAPL: A framework for profiling energy consumption of java programs (2024). https://kliu20.github.io/jRAPL/
3. Aazam, M., Huh, E.N.: Fog computing and smart gateway based communication for cloud of things. In: 2014 International Conference on Future Internet of Things and Cloud, pp. 464–470 (2014). https://doi.org/10.1109/FiCloud.2014.83
4. Ahvar, E., Orgerie, A.C., Lebre, A.: Estimating energy consumption of cloud, fog, and edge computing infrastructures. IEEE Trans. Sustain. Comput. **7**(2), 277–288 (2022)
5. Al Aidaros, O., Kardjadja, Y., Bouida, Z., Ibnkahla, M.: Energy and time-effective computation offloading for edge computing-enabled IoT networks. In: 2023 IEEE Sensors Applications Symposium (SAS), pp. 1–6 (2023)
6. Alharbi, H.A., Aldossary, M.: Energy-efficient edge-fog-cloud architecture for IoT-based smart agriculture environment. IEEE Access **9**, 110480–110492 (2021)
7. Amsel, N., Ibrahim, Z., Malik, A., Tomlinson, B.: Toward sustainable software engineering: NIER track. In: 2011 33rd International Conference on Software Engineering (ICSE), pp. 976–979 (2011)
8. Andrae, A.S.G., Edler, T.: On global electricity usage of communication technology: trends to 2030. Challenges **6**(1), 117–157 (2015)
9. Armbrust, M., et al.: A view of cloud computing. Commun. ACM **53**(4), 50–58 (2010)
10. Becker, C., et al.: Sustainability design and software: the karlskrona manifesto. In: 2015 IEEE/ACM 37th IEEE International Conference on Software Engineering, vol. 2, pp. 467–476 (2015)
11. Bonomi, F., Milito, R., Zhu, J., Addepalli, S.: Fog computing and its role in the internet of things. In: Proceedings of the First Edition of the MCC Workshop on Mobile Cloud Computing, MCC 2012, pp. 13–16. Association for Computing Machinery, New York (2012)
12. Borujeni, A.M., Fathy, M., Mozayani, N.: Developing and evaluating a real time and energy efficient architecture for an internet of health things. In: 2020 4th International Conference on Smart City, Internet of Things and Applications (SCIOT), pp. 106–111 (2020)
13. Capra, E., Formenti, G., Francalanci, C., Gallazzi, S.: The impact of MIS software on it energy consumption (2010)
14. Chen, S., Li, Q., Zhang, H., Zhu, F., Xiong, G., Tang, Y.: An IoT edge computing system architecture and its application. In: 2020 IEEE International Conference on Networking, Sensing and Control (ICNSC), pp. 1–7 (2020)
15. Cui, L., et al.: Joint optimization of energy consumption and latency in mobile edge computing for internet of things. IEEE Internet Things J. **6**(3), 4791–4803 (2019)

16. El-Sayed, H., et al.: Edge of things: The big picture on the integration of edge, IoT and the cloud in a distributed computing environment. IEEE Access **6**, 1706–1717 (2018)
17. Farahani, S.: ZigBee Wireless Networks and Transceivers. Newnes, USA (2008)
18. The Climate Group: Smart 2020: enabling the low carbon economy in the information age (2008)
19. Jalali, F., Khodadustan, S., Gray, C., Hinton, K., Suits, F.: Greening IoT with fog: a survey. In: 2017 IEEE International Conference on Edge Computing (EDGE), pp. 25–31 (2017)
20. LE, H.: Optimizing energy in fog computing architecture based on offloading mechanism for IoT devices. In: 2023 Asia Meeting on Environment and Electrical Engineering (EEE-AM), pp. 1–6 (2023)
21. Madhura, S.: A review on low power VLSI design models in various circuits. J. Electron. **4**, 74–81 (2022)
22. Mancebo, J., García, F., Calero, C.: A process for analysing the energy efficiency of software. Inf. Softw. Technol. **134**, 106560 (2021)
23. Mebrek, A., Merghem-Boulahia, L., Esseghir, M.: Efficient green solution for a balanced energy consumption and delay in the IoT-fog-cloud computing. In: 2017 IEEE 16th International Symposium on Network Computing and Applications (NCA), pp. 1–4 (2017)
24. Global System for Mobile Communications Association: The mobile economy 2024. GSMA (2024). https://www.gsma.com/solutions-and-impact/connectivity-for-good/mobile-economy/wp-content/uploads/2024/02/260224-The-Mobile-Economy-2024.pdf
25. Muniswamaiah, M., Agerwala, T., Tappert, C.C.: Green computing for internet of things. In: 2020 7th IEEE International Conference on Cyber Security and Cloud Computing (CSCloud)/2020 6th IEEE International Conference on Edge Computing and Scalable Cloud (EdgeCom), pp. 182–185 (2020)
26. Noureddine, A., Rouvoy, R., Seinturier, L.: A review of energy measurement approaches. SIGOPS Oper. Syst. Rev. **47**(3), 42–49 (2013)
27. Runeson, P., Host, M., Rainer, A., Regnell, B.: Case Study Research in Software Engineering: Guidelines and Examples, 1st edn. Wiley Publishing, Hoboken (2012)
28. Thanh, N.H., Trung Kien, N., Hoa, N.V., Huong, T.T., Wamser, F., Hossfeld, T.: Energy-aware service function chain embedding in edge-cloud environments for IoT applications. IEEE Internet Things J. **8**(17), 13465–13486 (2021)
29. Verde Romero, D.A., Villalvazo Laureano, E., Jiménez Betancourt, R.O., Navarro Álvarez, E.: An open source IoT edge-computing system for monitoring energy consumption in buildings. Results Eng. **21**, 101875 (2024)
30. Wang, C., Zhai, D., Zhang, R., Kaddoum, G., Singh, S.: Energy consumption minimization in dynamic UAV-assisted mobile edge computing networks. In: ICC 2023 - IEEE International Conference on Communications, pp. 4671–4676 (2023)
31. Webb, M., et al.: Smart 2020: Enabling the low carbon economy in the information age. The Climate Group, London, vol. 1, no. 1, p. 1 (2008)
32. Wohlin, C., Runeson, P., Hst, M., Ohlsson, M.C., Regnell, B., Wessln, A.: Experimentation in Software Engineering. Springer, Heidelberg (2012). https://doi.org/10.1007/978-3-642-29044-2
33. Yu, W., Liang, F., He, X., Hatcher, W.G., Lu, C., Lin, J., Yang, X.: A survey on the edge computing for the internet of things. IEEE Access **6**, 6900–6919 (2018)
34. Zhang, J., Chen, B., Zhao, Y., Cheng, X., Hu, F.: Data security and privacy-preserving in edge computing paradigm: survey and open issues. IEEE Access **6**, 18209–18237 (2018)

Software Architecture Assessment for Sustainability: A Case Study

Iffat Fatima$^{(\boxtimes)}$ and Patricia Lago

Vrije Universiteit Amsterdam, Amsterdam, The Netherlands
{i.fatima,p.lago}@vu.nl

Abstract. Software Architecture (SA) assessment provides an analysis of the quality of a high-level view of software-intensive systems, serving as a quality assurance mechanism. Sustainability is a crucial quality for digital ecosystems and as such, it presents assessment challenges due to the multi-dimensional nature of sustainability. This study addresses the challenge of sustainability assessment in SA. Due to a lack of guided sustainability assessment methods, we use an SA evaluation blueprint which we tailored for sustainability assessment. We use a blend of experience-based and quantitative assessment techniques for the assessment of design decision options. The SA assessment is performed on a case study integrating a SaaS-based solution, a learning management system called Canvas, within an educational institute. Our assessment provides an overview of trade-offs between design decision options. We use elements from an open-source toolkit (SAF Toolkit) and a Sustainability Impact Score (SIS) to identify the possible trade-offs and sustainability impacts across different sustainability dimensions. The assessment results identify the trade-offs between QAs and sustainability dimensions (mainly environmental) per design option. This information can help architects make informed decisions about sustainable design choices. Our evaluation method is designed to allow for the assessment of other SAs.

Keywords: software architecture · design decision · sustainability · software architecture assessment · software architecture evaluation · sustainability score · trade-off

1 Introduction

Software Architecture (SA) assessment is crucial for preventive quality assurance in the software development life cycle (SDLC). It is essential to ensure that the chosen architecture design decisions contribute to the quality of software. SA assessment serves as a means to assess quality at an early stage of SDLC, thereby reinforcing the significance of careful architecture planning in software development processes.

Lago *et al.* [17] put forward a case for treating sustainability as a software quality, allowing for comparisons with other quality attributes (QAs) for potential trade-offs. While maintainability is the most widely evaluated QA in SA assessment [9], sustainability can be viewed through the achievement of several QAs in combination, to support multiple sustainability dimensions (economic, environmental, social, and technical) [7].

ⓒ The Author(s), under exclusive license to Springer Nature Switzerland AG 2024
M. Galster et al. (Eds.): ECSA 2024, LNCS 14889, pp. 233–249, 2024.
https://doi.org/10.1007/978-3-031-70797-1_16

While several case studies exist on the SA assessment for different QAs such as maintainability, performance, availability [4, 8], the sustainability assessment case studies are only carried out in the context of maintainability and evolvability QAs [14, 18, 19].

In our review of SA assessment methods, we observed a lack of SA assessment methods or their application for sustainability [9]. We identified a gap in trade-off analysis across 4D-sustainability. Using the findings and data from this work, we developed an SA evaluation blueprint tailored for sustainability assessment [10]. In this study, we instantiate this blueprint and apply it to assess the design decisions for sustainability. We aim to answer the research question, *"How to guide software architects in assessing SA for sustainability?"* We carry out the SA assessment on the integration of Canvas, a learning management system (LMS) at an educational institute. We leverage the expertise of the architects at the educational institute to perform experience-based SA assessment by analyzing the relevant design decisions for their sustainability.

The results of this study provide, (i) a step-by-step guide with specific techniques for SA assessment for sustainability, (ii) Sustainability-Quality Model (SQ Model) [5] representing dependency of sustainability dimension on QAs, (iii) Dependency Matrix (DMatrix) [6] identifying the relationship between QAs and their trade-offs, and (iv) Sustainability Impact Score (SIS) for sustainability support per design decision option. We further elaborate on the implications of our results, challenges, and future directions.

2   Background

In this section, we define the background knowledge of the research area. We define software sustainability in general and then contextualize it for SA. We also provide contextual information about our case study subject.

Software Sustainability. Sustainable development is defined as meeting 'the needs of the present without compromising the ability of future generations to meet their own needs' [3]. In the context of software, Venters *et al.* [21] link this 'need' to the time dimension of sustainability, *i.e.*, evolution of software due to changing requirements of stakeholders over time causing direct, enabling, and systemic impacts [12]. Lago *et al.* [17] represent sustainability across four dimensions; economic, environmental, social, and technical. Further, the definition is elaborated as "the preservation of the beneficial use of digital solutions, in a context that continuously changes" [16]. Hence, the notion of change makes sustainability assessment of digital solutions, imperative.

Software Architecture and Sustainability. Architecture sustainability is defined as the ability of the architecture to endure changes throughout the SDLC [1]. These changes can stem from shifts in requirements and business goals, operating environment, and technologies. In software architecture, sustainability refers to both (i) Architecture Knowledge Sustainability, *i.e.*, preservation of architecture knowledge with continuous change *e.g.*, consistency of architecture views, rationalization completeness, requirement fulfillment, change scenario robustness, and decision traceability [19], and (ii) Sustainability through Architecture Practices, *i.e.*, the adoption of SA practices that enable the development of a sustainable software product, *e.g.*, tactics from Vos *et al.* [22] like edge computing, using spot instances, auto-scaling, etc.. In this study, we focus on the latter.

Case Study Subject - *Canvas*. Canvas is a learning management system (LMS) developed by Instructure[1]. It supports online and blended learning environments to create, manage, and deliver academic content. Canvas is an off-the-shelf software-as-a-service (SaaS) solution integrated with an educational institute's internal authentication and information system. Third-party learning and teaching tools are integrated with Canvas through a Learning Tools Interoperability (LTI) interface. The details of the SA components can be found in our replication package [11] (under /docs). We use this case of Canvas integration as a subject for SA assessment to support the educational institute for its new sustainability reporting requirements (See Sect. 5-II).

3   Related Work

Extensive work has been done in SA assessment with methods like ATAM [13], and SARA [20]. This section outlines the existing work in SA assessment specifically for sustainability and identify the focus areas and gaps.

Koziolek *et al.* [14] present MORPHOSIS, a metric-based SA evaluation method for sustainability assessment at the source code level. They use Architecture Level Modifiability Analysis (ALMA) [2] as an underlying evaluation technique. Their approach tracks architecture-level metrics for evolution scenarios and focuses on modifiability, reusability, modularity, and testability metrics. Sehestedt *et al.* [19] evaluate SA sustainability using architecture-level quantitative metrics, defining sustainability as maintainability and evolvability. Their 4C criteria (completeness, consistency, correctness, clarity) assess architecture models, decisions, and requirements, focusing on decomposition quality, best practice adherence, view consistency, rationalization completeness, requirement fulfillment, change scenario robustness, and decision traceability. The evaluation focuses directly on *Architecture Knowledge Sustainability*. Ojameruaye *et al.* [18] present an SA evaluation based on Cost Benefit Analysis Method (CBAM), Portfolio theory, and Technical Debt, linking architecture assessment to system goals. They provide a desirability rating for each architecture strategy for realistic cost-benefit analysis, helping identify risk and debt-based trade-offs across five sustainability dimensions.

The current SA assessments for sustainability focus on either, the sustainability of architecture at the source code level or the sustainability of the documented architecture knowledge, *i.e.*, models, views, and requirements. Quantification-based techniques like architecture metrics or ratings are used. None of these methods analyze the high-level architecture (e.g. at the level of architecture views or design decisions) and the implications of the chosen decisions on sustainability, *i.e.*, having positive or negative effects on different sustainability dimensions. Our study evaluates design decisions based on prioritized evaluation criteria, *i.e.*, QAs for sustainability. We use a mix of experience-based and quantitative approaches.

4   Study Design

We describe the goal of our study based on the template, as shown in Table 1. We use the SA evaluation blueprint [10] as a guide to carry out the SA assessment. The SA assessment requires, as major input (i) Architecture Documentation, (ii) Tacit Knowledge of

---

[1] https://www.instructure.com/canvas.

Table 1. Goal of the study—Wohlin *et al.* [23]

Analyze	Software Architecture of an LMS
For the purpose of	Assessment
With Respect to	Sustainability
From the point of view of	Software Architects
In the context of	An Educational Institute

Architects, and (iii) QAs as evaluation criteria. The expected output of this evaluation is information about inter-QA tradeoffs and inter-sustainability-dimension trade-offs for design concerns. As a side product of this study, we also get an instantiation (step-by-step method) of the SA evaluation blueprint for sustainability in the form of concrete steps, which can be used for other SA assessments. Two experts were chosen for this case study, both working as solution architects at the educational institute for 6 years with an approximate total of 17 years of experience working in IT. Experts were chosen based on their experience, knowledge of the system, and availability. One of the experts has worked for the educational institute for 17 years and proved to be a rich resource for tacit knowledge. We asked both experts questions regarding the input elements to drive each phase of the SA assessment. The authors of this study facilitated the evaluation process and performed the quantitative analysis which was again discussed with the experts. The quantitative analysis was performed by defining a Sustainability Impact Score (see Eq. 1) supported by the elements of the SAF Toolkit (via D-Matrix).

5   Study Execution and Results

The SA evaluation blueprint [10] provides a set of eleven general steps to support the sustainability assessment process. It is independent of the techniques used in each step, *e.g.*, the techniques for identification of approaches, prioritization, generation, and evaluation of data. In our study, we elaborate on each step of the blueprint and provide step-by-step details of the execution of each step. We use a mix of experience-based and quantitative techniques. Figure 1 shows the SA assessment steps from the blueprint in the middle together with the inputs and outputs (I/O) of our case study. The steps are tagged with specific techniques used (experience-based and quantitative), and the sustainability perspective is injected in each step.

I. Preparation. In the preparation phase, the authors met with the management of the project to discuss the proposed SA assessment and their expectations. Together we selected the relevant experts from the project for their expertise. We prepared a slide set for the explanation of the study execution and presented it to all involved stakeholders.

II. Requirement Identification. In this phase, we identified the requirements of the SA assessment, *i.e.*, the evaluation criteria. We identified the QAs for SA of Canvas integration from the technical requirements document. These QAs were checked for their

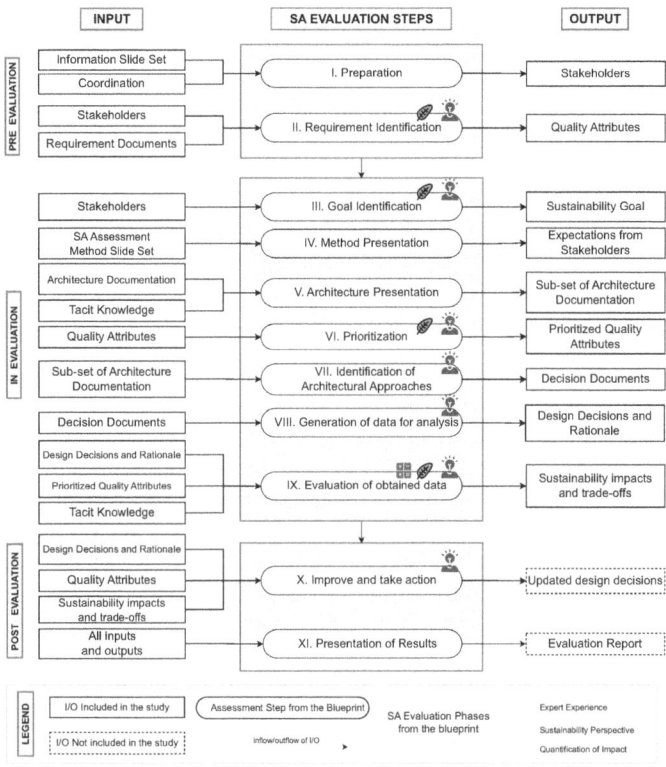

Fig. 1. Instance of the SA Evaluation Blueprint in [10]

relevance with input from the experts. The management team at the educational institute requested a sustainability requirement to be added for the procurement of Canvas as an LMS, to comply with the future EU regulations for reporting carbon emissions[2]

Hence, the educational institute aims to identify the sustainability-related improvement points on which they require support internally, or from Canvas. We suggested resource utilization as an additional QA for this research. The final list of QAs was identified in consultation with the experts and management team at the educational institute.

III. Goal Identification. The goal of this SA assessment was a sustainability assessment of the Canvas integration with the information system at the educational institute. We treated Canvas itself as a single black box component. We did not analyze the SA of the sub-components inside Canvas.

IV. Method Presentation. We presented the SA assessment method to the experts to explain all the steps and information needed from them during a one-hour meeting ses-

---

[2] https://consilium.europa.eu/en/press/press-releases/2024/02/05/environmental-social-and-gov ernance-esg-ratings-council-and-parliament-reach-agreement/.

sion. In this session, we exchanged information and communicated the expectations from both ends about the SA assessment.

V. Architecture Presentation. The experts presented the SA as a component diagram (see /docs [11]) and explained all the components, their interconnections, and the functionality they support. The experts also shared internal documentation. As the original documentation was in Dutch, we translated the documents using Google Translate. The experts being native Dutch speakers helped resolve any discrepancies in translation for the correctness of technical and contextual information.

VI. Prioritization. The SA evaluation blueprint [10] presents two types of prioritization. At first, we performed *prioritization of evaluation criteria*. It involved an iterative process wherein two experts (*i.e.,* architects) were consulted. These evaluation criteria were defined by the QAs identified in Step II. Initially, we presented the experts with the list of QAs with their respective definitions sourced from the documentation. The experts ranked the QAs, placing the most important ones at the top and the least important at the bottom (see Table 2). Through mutual discussion concerning the role of QAs for the SA, adjustments to the prioritization of QAs were made. Once the QAs were prioritized, a sustainability-specific QA 'resource utilization', was introduced. The experts determine the placement of this new QA considering its possible trade-offs with high-priority QAs, at rank 4 (1 = lowest). During this prioritization process, some resource-intensive scenarios were also identified where the high-priority QAs are vital.

Table 2. Prioritized Evaluation Criteria

Quality Attribute	Priority Rank (13 = highest)
Availability	13
Scalability	12
Time Behavior	11
Security	10
Interoperability	9
Portability	8
Modifiability	7
Reliability	6
Testability	5
Resource Utilization	4
Analysability	3
Modularity	2
Reusability	1

QA definitions are available in Online Material (/data/sqmodel.xlsx) [11]

VII. Identification of Architecture Approaches. We utilized the tacit knowledge of the experts to explore the documentation, choose an architecture view to start our research and identify the documents useful for our goal. We used their expertise to extract architecture design decisions from the documents and their experience to validate the extracted data and its relevance to the architecture. The existing documentation included (i) Component diagrams (ii) Decision Documents (not necessarily architecture) (iii) Other documents about requirements, scenarios, procurement needs, etc.. We read these documents and use the tacit knowledge of the experts to understand the available artifacts. The *Decision Documents* are system-wide decision documents with decision information about the functional, non-functional, UI-specific, and architecture aspects. Further, the scenarios identified during the prioritization steps were used to list design concerns and their respective design options for those scenarios.

VIII. Generation of Data for Analysis. In this step, we extracted design decisions from the identified architecture approaches which acted as data for evaluation.

Creating a Design Decision Template. To document the design decisions consistently, we created a template based on, (i) information available in documents and (ii) information needed for evaluation, *e.g.*, QAs, IDs (see data/DD_Template.xlsx [11]).

Data Extraction. We extracted information mainly from the decision documents of the system into the Design Decision Template. For information extraction, we only used decision documents flagged for architecture advice in the internal dashboard documents. We used other documents for finding contextual information about the design decisions (*e.g.*, about a specific functional requirement or goals, requirements behind a certain decision, relevant scenarios, etc..).

Populating the Template. We populated the information from these decision documents into our template as nine design decisions. The extracted elements included (i) design concern, (ii) design options, (iii) the rationale per design option, (iv) issues (if any), (v) supported QAs, and (vi) chosen decision. As the documents under analysis were not in a single template or format, our strategy ensured a systematic recovery of design decisions for effective SA assessment. We used the knowledge and expertise of the experts to identify the QAs (from Table 2) supporting each design option.

Validation of Data. The nine extracted decisions were validated by the experts as being correct and having architectural implications. We gathered feedback from the experts for each step of data generation and analysis going forward.

IX. Evaluation of Obtained Data. The extracted design decisions were used as input for this step. We used a combination of two techniques, (i) Experience-based technique – by using feedback from the experts for analyzing alternatives and trade-offs, and (ii) Quantification-based technique – by using elements of open source SAF Toolkit [15] and Sustainability Impact Score (SIS) to identify sustainability impact.

SQ Model. We use the QAs from Table 2 to build a Sustainability Quality (SQ) Model [7] as shown in Table 3. The SQ model helps map the QAs to 4D-sustainability; Economic

(Ec), Environmental (En), Social (S), and Technical (T). This mapping helps us understand how each QA can have an impact across one or more sustainability dimensions. This is essential to observe inter-dimension sustainability trade-offs. It is important to note that the mapping can vary based on the context of the software under evaluation. The detailed SQ model with definitions of the QAs can be found in our Replication Package [11] under docs/sqmodel.xlsx.

Table 3. SQ Model

Quality Attribute	Sustainability Dimensions			
	Economic	Environmental	Social	Technical
Availability			x	x
Scalability	x			x
Time Behavior		x		x
Security	x		x	x
Interoperability	x		x	x
Portability	x		x	x
Modifiability	x			x
Reliability	x		x	x
Testability				x
Resource Utilization	x	x		x
Energy Efficiency		x		x
Analysability				x
Modularity				x
Reusability	x	x		x

DMatrix. We create a dependency matrix (DMatrix) [6] for each design option to see the inter-dimension impact of QAs. We take the QAs representing each sustainability dimension from the SQ Model. Using DMatrix, we map the QAs of technical dimension with QAs of economic (see Table 5), environmental (see Table 6) and social (see Table 7) sustainability. The impact values in DMatrix are identified based on the scenario(s) represented by the design concern. The information about these scenarios comes from the documentation as well as the architects. The DMatrix is a representation of the inter-QA and inter-dimension dependencies. The prioritized QAs act as the evaluation criteria. We create 3 DMatrices (T-Ec, T-En, T-S) per design option. For each design option, we compare the DMatrices for the same dimension pair. For example, DMatrix (T-Ec) for design option O-5.1 is only compared to DMatrix (T-Ec) for design option O-5.2, and so on. The QAs to analyze are chosen based on the set of QAs supported by all design options. Cross-dimension scores (*e.g.*, between T-Ec and T-En) can only be compared if the same list of QAs is being used in all DMatrices.

Sustainability Impact Score (SIS). Next, we prioritize one design option among available alternatives based on our evaluation criteria, *i.e.*, list of QAs. We develop a Sustainability Impact Score (see Eq. 1) to represent what is the impact of QAs within the technical dimension T with respect to QAs within the other sustainability dimensions dim (Ec, En, S). We use the QA priority values as weights for the QA impact. We calculate the sum of the product of QA priority with the impact value for all QAs. We do not normalize the value of the score to retain the representation of the number of QAs contributing to a dimension, which becomes apparent from the relatively high magnitude of the score and vice versa (see QA number and index values in Table 5 vs Table 6). For the analysis, we compare the value of the SIS across the same dimension. The score with a relatively higher value represents higher sustainability support. Currently, we do not compare cross-dimension scores.

$$\text{Sustainability Impact Score}_{T,dim} = \sum_{\substack{i=1 \\ j=1}}^{n,m} (\text{Priority}_{T\_i} + \text{Priority}_{dim\_j}) \times \text{Impact}_{ij} \quad (1)$$

where n=total QAs per dimension T, m=total QAs per dimension dim, dim $\in \{$Ec, En, S$\}$, impact = $\{+1, -1, 0\}$

The SIS can be used to prioritize between design options by making a trade-off between the chosen evaluation criteria, *i.e.*, QAs. Hence, SIS can help prioritize design decisions *based on evaluation criteria*. Here, we present the evaluation of two design concerns. See the replication package [11] for all design concerns and their evaluation results.

Example 1. Here we present the detailed evaluation of design concern DC-5 (see Table 4), the design options considered and their SISs. Here, a design decision needs to be made for choosing whether independent or shared production instances need to be implemented for different institutions that collaborate and have shared data.

Table 4. Extracted Design Decisions for DC-5

Design Concern ID: DC-5				
Design Concern: Choice of the number of production instances				
Design Option ID	Design Option Description	Rationale	Supported QA(s)	Choice
O-5.1	Implement two separate Canvas production instances, allowing each institution to have autonomy over implementation settings and branding.	Both institutions seek autonomy in decision-making and distinct branding. However, certain crucial Canvas settings can only be adjusted at the account level, requiring separate instances to accommodate differing preferences.	Availability, Interoperability, Portability, Modifiability	Yes
O-5.2	Utilize a single Canvas instance shared by both institutions, compromising on implementation autonomy and branding flexibility	Requires fewer resources and reduces the overhead of policy syncronization	Resource Utilization	No

Economic Sustainability. Table 5 shows the DMatrix for the two design options of the design concern DC-5. It shows the impacts of QA support across the technical dimension (T) on QAs across the economic dimension (Ec). For design option O-5.1, resource utilization (T) has a neutral impact on modifiability (Ec) and vice versa. For O-5.2, resource utilization(T) has a negative impact on modifiability (Ec) and vice versa. The negative impact is also clear from the lower $SIS_{\text{T-Ec}}$ for O-5.2. Other QAs have the same impacts across both design options. It means that O-5.2 is less favorable for economic sustainability as it will be difficult to change the system per the different requirements of both institutes. As the QAs for O-5.1 hold a higher priority, the choice of this design decision remains unchanged.

Table 5. DMatrices Technical-Economic (T-Ec)

(a) O-5.1

Technical	Priority	Economic			
		Interoperability	Portability	Modifiability	Resource Utilization
		9	8	7	4
Availability	13	o	o	-	o
Interoperability	9		+	+	+
Portability	8	+		+	o
Modifiability	7	+	+		o
Resource Utilization	4	o	o	o	

$SIS_{\text{T-Ec}} = 89$

(b) O-5.2

Technical	Priority	Economic			
		Interoperability	Portability	Modifiability	Resource Utilization
		9	8	7	4
Availability	13	o	o	-	o
Interoperability	9		+	+	+
Portability	8	+		+	o
Modifiability	7	+	+		-
Resource Utilization	4	o	o	-	

$SIS_{\text{T-Ec}} = 67$

Environmental Sustainability. Table 6 shows the DMatrix for the two design options of the design concern DC-5 for the impact of QA support across technical dimension (T) on QAs across environmental dimension (En). For design option O-5.1, modifiability (T) has a negative impact on resource utilization (En). For design option O-5.2, modifiability (T) has a neutral impact on resource utilization (En). This is also clear from the lower $SIS_{\text{T-En}}$ for O-5.1. Other QAs have the same impacts across both design options. It means that O-5.1 is less favorable for environmental sustainability as it uses more resources. However, the decision still remains in favor of O-5.1 due to the higher priority of QAs supporting the economic dimension.

Table 6. DMatrices Technical-Environmental (T-En)

(a) O-5.1

Technical	Priority	Environmental: Resource Utilization
		4
Availability	13	o
Interoperability	9	+
Portability	8	o
Modifiability	7	-
Resource Utilization	4	

$SIS_{\text{T-En}} = 2$

(b) O-5.2

Technical	Priority	Environmental: Resource Utilization
		4
Availability	13	o
Interoperability	9	+
Portability	8	o
Modifiability	7	o
Resource Utilization	4	

$SIS_{\text{T-En}} = 13$

Social Sustainability. Table 7 shows the DMatrix for the two design options of the DC-5 for the impact of QA support across technical dimension (T) on QAs across social dimension (S). The impacts are the same for both design decisions as well as their SIS_{T-S}. Hence, both of these design options have no impact on social sustainability.

Table 7. DMatrices Technical-Social (T-S)

(a) O-5.1

Technical	Social	Availability	Interoperability	Portability
Priority	13	9	8
Availability	13		o	o
Interoperability	9	+		+
Portability	8	o	+	
Modifiability	7	o	+	+
Resource Utilization	4	U	o	o

$SIS_{T-S} = 87$

(b) O-5.2

Technical	Social	Availability	Interoperability	Portability
Priority	13	9	8
Availability	13		o	o
Interoperability	9	+		+
Portability	8	o	+	
Modifiability	7	o	+	+
Resource Utilization	4	U	o	o

$SIS_{T-S} = 87$

Trade-Off. The results of our assessment for DC-5 shows that economic sustainability is favored over environmental sustainability by the choice of O-5.1. However, given the rationale for the decisions, this choice is still favorable. This does not mean that the chosen design decision is not sustainable, but it helps architects think other sustainability dimensions (environmental in this case) which are compromised for the chosen trade-off (in favor of economic sustainability).

Example 2. Here we present the detailed evaluation of design concern DC-6 (see Table 8), the design options considered and their SISs. This design concern refers to decision-making regarding the choice of data access and retention period for inactive users, *e.g.,* students who have graduated.

Table 8. Extracted Design Decisions for DC-6

Design Concern ID: DC-6				
Design Concern: Lifecycle and retention of course data				
Design Option ID	Design Option Description	Rationale	Supported QA(s)	Choice
O-6.1	User access period should be aligned with data retention period (7 years) to keep course history available, allowing users to retain access to Canvas even after they are no longer active within the institution.	To allow availability of data to users	Availability	Yes
O-6.2	Do not retain course history in the LMS for an extended period. Only provide access to active users.	Users may not have the option to extract materials from Canvas after they are no longer active, hence the data is using extra resources. Deleting the data may impact accreditation sources.	Security, Resource Utilization	No

Economic and Environmental Sustainability. Tables 9 and 10 show the DMatrix for the two design options of the design concern DC-6 for the impact of QA support across technical dimension (T) on QAs across economic dimension (Ec) and environmental dimension (En), respectively. For design option O-6.1, availability (T) has a negative impact on resource utilization (Ec, En) and vice versa. For O-6.2, availability (T) has a neutral impact on resource utilization (Ec, En). The negative impact is also clear from the lower $SIS_{T\text{-}Ec}$ for O-6.1. The same impacts are also seen across the environmental dimension with the same $SIS_{T\text{-}En}$. Other QAs have neutral impacts across both design options. It means that O-6.1 is less favorable for economic and environmental sustainability because of additional resource utilization due to long-term data storage.

Table 9. DMatrices Technical-Economic (T-Ec)

Technical		Economic	Security	Resource Utilization		Technical		Economic	Security	Resource Utilization
	Priority	10		4			Priority	10		4
Availability	13	o		-		Availability	13	o		o
Security	10			o		Security	10			o
Resource Utilization	4	o				Resource Utilization	4	o		
$SIS_{T\text{-}Ec = -17}$						$SIS_{T\text{-}Ec = 0}$				

(a) O-6.1 (left) / (a) O-6.2 (right)

Social Sustainability. Availability (T) and Security (T) have a positive impact on each other (S) across the social dimension (see Table 11) while Resource Utilization (T) has a negative impact on availability (S). Relatively, O-6.1 has a higher $SIS_{T\text{-}S}$ showing support for social dimension.

Trade-Off. The results of our assessment for DC-6 show that social sustainability is favored over economic and environmental sustainability by the choice of O-6.1. In the currently chosen option, social sustainability is favored as the availability of data is given more importance based on the institute's policy decisions (also reflected by the high priority of Availability QA).

Table 10. DMatrices Technical-Environmental (T-En)

Technical		Environmental	Resource Utilization		Technical		Environmental	Resource Utilization
	Priority		4			Priority		4
Availability	13		-		Availability	13		o
Security	10		o		Security	10		o
Resource Utilization	4				Resource Utilization	4		
$SIS_{T\text{-}En = -17}$					$SIS_{T\text{-}En = 0}$			

(a) O-6.1 (left) / (b) O-6.2 (right)

Table 11. DMatrices Technical-Social (T-S)

(a) O-6.1

Technical		Social	Availability	Security
	Priority		13	10
Availability	13			+
Security	10		+	
Resource Utilization	4		o	o
$SIS_{T-S} = 46$				

(b) O-6.1

Technical		Social	Availability	Security
	Priority		13	10
Availability	13			o
Security	10		o	
Resource Utilization	4		o	o
$SIS_{T-S} = -17$				

X. Improve and Take Action. Table 12 displays the SIS for each design option and highlights the inter-dimension sustainability trade-offs made for each design concern. Among all design concern options, O-1.1 stands out for its lack of sustainability support, prompting a recommendation to change the decision to O-1.2.

While O-5.1 prioritizes the economic dimension, it makes a trade-off across the social dimension. We do not recommend any changes here as the economic dimension is represented by high-priority QAs.

Although O-9.1 offers strong sustainability support across two dimensions (Ec and S), O-9.2 provides comparatively balanced sustainability support. The choice of O-9.1 is better as it supports maximum sustainability dimensions however, it could be argued that O-9.2 could provide a balanced sustainability support as all dimensions are relatively equally supported. In such a case, the use of expert knowledge becomes imperative for decision-making.

Table 12. Chosen design options and sustainability trade-offs

Design Concern ID	Design Option ID	Choice	SIS_{T-Ec}	SIS_{T-En}	SIS_{T-S}	Is sustainability addressed?
DC-1	O-1.1	Yes	0	0	0	No
	O-1.2	No	41	78	49	
DC-2	O-2.1	Yes	11	0	0	Yes
	O-2.2	Yes	0	0	0	
DC-3	O-3.1	Yes	43	0	43	Yes
	O-3.2	Yes	0	0	0	
DC-4	O-4.1	Yes	34	0	55	Yes
	O-4.2	Yes	31	0	51	
DC-5	O-5.1	Yes	89	2	87	Partially
	O-5.2	Yes	67	13	87	
DC-6	O-6.1	Yes	-17	-17	46	Partially
	O-6.2	Yes	0	0	-17	
DC-7	O-7.1	Yes	66	0	81	Yes
	O-7.2	Yes	0	0	0	
DC-8	O-8.1	Yes	0	-15	0	Yes
	O-8.2	Yes	0	-15	0	
DC-9	O-9.1	Yes	68	0	51	Partially
	O-9.2	Yes	32	38	32	

□ = higher SIS, ■ = lower SIS

We conducted a reflective session with an architect to discuss our results, the implications of SIS, and to gather feedback. We suggested revisiting the design decisions in cases with a negative impact on sustainability. The architect found the SIS useful for objective decision-making. They highlighted that different stakeholders may prioritize QAs in a different order and suggested that the prioritization should give more weight to the choice of those stakeholders most affected by a design concern. Our analysis helped the experts view the design decisions in the context of their sustainability support through different QAs.

XI. Presentation of Results. Detailed results are presented as part of the replication package [11]. The architects confirmed that the provided templates are useful and they plan to use them in their future decision-making process. We plan to provide a formal report to present to the technical team of the project for taking appropriate action.

6 Discussion

In this section, we present insights from our results, the challenges faced in the SA assessment, and the future direction of our research.

Insights. Our assessment helped the experts see two types of trade-offs: (i) Inter-QA trade-offs (*i.e.*, trade-offs between two different QAs, *e.g.*, maintainability and resource utilization) through the D-Matrix in terms of inter-QA dependencies, and (ii) Inter-dimension trade-offs (*i.e.*, trade-offs between two sustainability dimensions *e.g.*, economic and environmental) through the SIS. Our study helps the evaluators visualize the interdependencies for such trade-offs for informed decision-making. The evaluators agreed that the SQ Model helped them visualize the sustainability dimensions per QA which they were not so sure about before the assessment.

The SIS aids decision-making by quantifying the sustainability support across dimensions, hence enabling evaluators to consider alternative design options for better trade-offs. Experts find this approach clear and easy to use, with SIS quantification enhancing understanding of decision impacts and enabling informed decision-making.

Looking at the design decisions and assessing them for sustainability based on experience-based technique becomes harder when the number of QAs and decisions increase [9]. Our approach helps address this complexity issue thanks to the SIS, which streamlines the assessment process by providing a consolidated evaluation of sustainability dimensions and highlighting trade-offs that may not be immediately evident. By using a blend of experience-based and quantitative technique, we support evaluators in the decision-making process.

Challenges. Specifically for our case study, the assessment posed certain challenges. Navigating the case study's SA and documentation was difficult. With steps VII and VIII of our method, we uncovered missing information in the architecture documentation. This encompassed both documented and undocumented design decisions, including design options, rationales, and relevant QAs. This recovered information is crucial for mapping the right QAs with design options and facilitating informed decision-making.

The experts noted that some decisions are made without having a choice, *i.e.,* there are no alternatives available. For such cases, the design decisions are typically not documented. However, it is possible that such decisions could inadvertently impact a QA not considered initially. Hence, missing the inter-QA or inter-dimension trade-offs.

In our SA assessment, we do not prioritize the sustainability dimensions. However, there might be an implicit priority associated with them based on the sustainability mapping in the SQ Model. Table 3 shows that most of the chosen QAs inherently support economic and technical sustainability while a few support environmental and social sustainability. Hence, sustainability dimensions mapped to low-priority QAs or less number of QAs might get overlooked. For informed decision-making, a re-prioritization might be appropriate after making the SQ Model. Such re-prioritization is supported inherently in the blueprint due to the oscillating nature of Step VI–Prioritization. We observe that a higher value of the SIS is due to either the high priority of the chosen QAs or a relatively higher number of the total QAs for design option(s).

Future Directions. We only use 3 values of impacts $(+1, -1, 0)$ in the DMatrix. However, the positive impact of one QA may be stronger than the other, making the representation of this relative strength apparent. It would be valuable to quantify a comparable impact score across different dimensions (say S and Ec) to see by what margin a dimension is favored over another during the trade-off. We aim to calculate SIS across other combinations of dimensions, *i.e.,* Ec-En, Ec-S etc.. and refine it to improve its support for a more objective SA assessment, and evaluate the effectiveness of this approach through systematic feedback from the evaluators, *e.g.,* through questionnaire survey.

7   Threats to Validity

We use the classification from Wohlin *et al.* [23] to identify threats to validity and present our mitigation actions.

Construct Validity. The possibility of internal bias of the experts was mitigated by cross-examining their decisions and rationale and questioning them. Further, the study design and result of each evaluation phase were cross-checked by the second author. We do not normalize the SIS values to avoid the loss of meaningful variation due to the number of QAs considered and their respective priority.

Internal Validity. The choice of design concerns for analysis was based on the availability of data hence the choice is not biased by any selection criteria. The design decisions were evaluated after being translated from Dutch to English, hence creating a risk of loss of contextual information or correctness. We mitigated this issue by the review of translated documents by the architects who are native Dutch speakers and have knowledge of the SA. The results of the evaluation show that the chosen design options and QA priorities match, hence confirming the evaluation results. The quantification was done by the research team and then we went back to confirm the results from the experts, who agreed. We also provide a replication package for the study.

External Validity. The case study focuses on a specific software system in one organization. While the SA assessment method is reusable and not tied to a particular SA, the evaluation results are specific to the studied SA.

8   Conclusion and Future Work

In this study, we present, (i) an instantiation of an SA assessment method for sustainability with a blend of experience-based and quantitative techniques, (ii) a Sustainability Impact Score for informed decision-making, and (iii) application of our SA assessment method on a real case. Our results provide insights to the architects on possible sustainability trade-offs that they must consider to make design decisions. The SIS score shows whether a design choice is made based on prioritizing a certain dimension over another or balancing the trade-offs across all dimensions. The quantitative assessment simplifies the process of complex decision-making for the architects. The assessment results help architects in observing the inter-QA and inter-dimension trade-offs of their choices.

Acknowledgements. This publication is part of the project SustainableCloud (OCENW. M20.243) of the research programme Open Competition which is (partly) financed by the Dutch Research Council (NWO). This work is partially funded by Enabling Energy Efficient Community through Context-Aware IoT framework and Sustainable Software (E-CAISS) Project (SPARC/2019-2020/P2323/SL) supported under the SPARC scheme. We are grateful to our colleagues and the solution architects at Vrije Universiteit Amsterdam, for donating their time, expertise, and support towards this project.

Data Availability Statement. The data for the study has been made available as a replication package online [11].

References

1. Avgeriou, P., Stal, M., Hilliard, R.: Architecture sustainability [guest editors' introduction]. IEEE Software **30**(6) (2013). https://doi.org/10.1109/MS.2013.120
2. Bengtsson, P., Lassing, N., Bosch, J., van Vliet, H.: Architecture-level modifiability analysis (ALMA). J. Syst. Softw. **69**(1–2), 129–147 (2004)
3. Brundtland, G.H.: Our common future-call for action. Environ. Conserv. **14**(4), 291–294 (1987)
4. Christensen, H.B., Hansen, K.M., Lindstrøm, B.: Lightweight and continuous architectural software quality assurance using the aSQA technique. In: Babar, M.A., Gorton, I. (eds.) ECSA 2010. LNCS, vol. 6285, pp. 118–132. Springer, Heidelberg (2010). https://doi.org/10.1007/978-3-642-15114-9_11
5. Condori-Fernandez, N., Lago, P.: Characterizing the contribution of quality requirements to software sustainability. J. Syst. Softw. **137** (2018). https://doi.org/10.1016/j.jss.2017.12.005
6. Condori-Fernandez, N., Lago, P., Catala, A., Luaces, M.R.: Defining Interdimensional Dependencies of the Sustainability-Quality Model. Technical report, VU Amsterdam (2024)
7. Condori-Fernandez, N., Lago, P., Luaces, M.R., Places, A.S.: An action research for improving the sustainability assessment framework instruments. Sustainability **12**(4) (2020). https://doi.org/10.3390/su12041682
8. Dayanandan, U., Kalimuthu, V.: A fuzzy analytical hierarchy process (FAHP) based software quality assessment model: maintainability analysis. Int. J. Intell. Eng. Syst. (2018). https://doi.org/10.22266/ijies2018.0831.09
9. Fatima, I., Lago, P.: A review of software architecture evaluation methods for sustainability assessment. In: 2023 IEEE 20th International Conference on Software Architecture Companion (ICSA-C). IEEE (2023)

10. Fatima, I., Lago, P.: Towards a sustainability-aware software architecture evaluation for cloud-based software services. In: Tekinerdoğan, B., Spalazzese, R., Sözer, H., Bonfanti, S., Weyns, D. (eds.) ECSA 2023. LNCS, vol. 14590. Springer, Cham (2023). https://doi.org/10.1007/978-3-031-66326-0_13

11. Fatima, I., Lago, P.: Software Architecture Assessment for Sustainability: A Case Study (2024). https://doi.org/10.5281/zenodo.11655904

12. Hilty, L.M., Aebischer, B.: ICT for sustainability: an emerging research field. In: Hilty, L.M., Aebischer, B. (eds.) ICT Innovations for Sustainability. AISC, vol. 310, pp. 3–36. Springer, Cham (2015). https://doi.org/10.1007/978-3-319-09228-7_1

13. Kazman, R., Barbacci, M., Klein, M., Carrière, S.J., Woods, S.G.: Experience with performing architecture tradeoff analysis. In: International Conference on Software Engineering. IEEE/ACM (1999). https://doi.org/10.1145/302405.302452

14. Koziolek, H., Domis, D., Goldschmidt, T., Vorst, P., Weiss, R.J.: MORPHOSIS: a lightweight method facilitating sustainable software architectures. In: 2012 Joint Working IEEE/IFIP Conf. on Software Architecture and European Conference on Software Architecture (2012). https://doi.org/10.1109/WICSA-ECSA.212.40

15. Lago, P.: SAF Toolkit (2024). https://github.com/S2-group/SAF-Toolkit

16. Lago, P.: The digital society is already here – pity it is 'unsustainable'. In: Connected World - Insights from 100 Academics on How to Build Better Connections. VU University Press (2023). https://vuuniversitypress.com/product/connected-world

17. Lago, P., Koçak, S.A., Crnkovic, I., Penzenstadler, B.: Framing sustainability as a property of software quality. Commun. ACM 58(10) (2015). https://doi.org/10.1145/2714560

18. Ojameruaye, B., Bahsoon, R., Duboc, L.: Sustainability debt: a portfolio-based approach for evaluating sustainability requirements in architectures. In: 38th International Conference on Software Engineering Companion (ICSE-C). IEEE/ACM (2016)

19. Sehestedt, S., Cheng, C.H., Bouwers, E.: Towards quantitative metrics for architecture models. In: Proceedings of the WICSA 2014 Companion Volume. ACM (2014). https://doi.org/10.1145/2578128.2578226

20. Tekinerdogan, B., Sozer, H., Aksit, M.: Software architecture reliability analysis using failure scenarios. J. Syst. Softw. (2008). https://doi.org/10.1016/j.jss.2007.10.029

21. Venters, C.C., et al.: Software sustainability: research and practice from a software architecture viewpoint. J. Syst. Softw. 138 (2018). https://doi.org/10.1016/j.jss.2017.12.026

22. Vos, S., Lago, P., Verdecchia, R., Heitlager, I.: Architectural tactics to optimize software for energy efficiency in the public cloud. In: 2022 International Conference on ICT for Sustainability (ICT4S) (2022). https://doi.org/10.1109/ICT4S55073.2022.00019

23. Wohlin, C., Runeson, P., Höst, M., Ohlsson, M.C., Regnell, B., Wesslén, A.: Experimentation in Software Engineering. Springer, Heidelberg (2012). https://doi.org/10.1007/978-3-642-29044-2

Trustworthiness

Modeling and Analyzing Zero Trust Architectures Regarding Performance and Security

Nicolas Boltz$^{(\boxtimes)}$, Larissa Schmid, Bahareh Taghavi, Christopher Gerking, and Robert Heinrich

Karlsruhe Institute for Technology (KIT), Karlsruhe, Germany
{nicolas.boltz,larissa.schmid,bahareh.taghavi,christopher.gerking, robert.heinrich}@kit.edu

Abstract. Zero Trust is considered a powerful strategy for securing systems by emphasizing distrust of all resource access requests. There are different approaches to integrating ZTAs into a system, differing in their components, assembly, and allocation. Early evaluation and selection of the right approach can reduce the costs of resources. In this paper, we propose a novel zero trust architecture (ZTA) metamodel based on literature and industry applications. We introduce our proposed metamodel elements and provide a model instance using the Palladio Component Model (PCM). We describe the requirements for enabling two existing approaches to performance simulation and security data flow analysis on the architectural level and outline how we realize them in our PCM-based implementation. Our evaluation demonstrates the applicability of our ZTA metamodel. It can represent real-world ZTA approaches in various domains, enabling the simulation of performance impact and analysis of the correct implementation of zero trust principles at the architectural level.

1 Introduction

The advancement of the Internet of Things, 5G networks, and cloud technologies has allowed systems and critical infrastructures to shift to more distributed architectures. This shift allows for increased efficiency, for example, by introducing virtual power plants [5] that integrate multiple distributed power-generating components, energy-storing units, and management systems. In the business sector, work from home has become a widely accepted practice, implying that business resources must be accessed off-site and through various devices. Both these trends lead to a larger attack surface of the systems and consequently to an increased number of cyber attacks [20].

A widely adopted approach to coping with such security problems is to separate the system using different access interfaces, firewalls, and intrusion detection

---

N. Boltz, L. Schmid and B. Taghavi—The main authors contributed equally.

M. Galster et al. (Eds.): ECSA 2024, LNCS 14889, pp. 253–269, 2024.
https://doi.org/10.1007/978-3-031-70797-1_17

systems. However, once an adversary manages to infiltrate the internal domain of the system, they can escalate their privileges and gain access to critical system components and resources. To prevent internal attacks and malicious access, the zero trust paradigm requires per-request access control for non-public resources, limiting access for unauthorized entities and enforcing granular access control. Contextual factors like the identity of the requesting subject and the environmental context determine access. Factors can be provided by several security mechanisms, like identity and event management systems. Zero trust also upholds the Principle of Least Privilege (PoLP), ensuring subjects are granted only the necessary permissions for resource access, preventing unauthorized propagation within the system. Zero Trust Architectures (ZTAs) incorporate principles of the zero trust cybersecurity paradigm. Existing research shows, that ZTAs can be applied to enhance security in virtual power plants [1], smart healthcare [7], smart manufacturing industries [26], and other more general business domains [7,16,38]. While standards regarding ZTA exist [11,24,30,40], an approach that unifies the ZTA structure is missing.

Designing a new system incorporating ZTA or migrating an existing system to use ZTA is not straightforward [36]. The architecture introduces new components and requires additional authentication and authorization checks. How ZTA should be integrated and to what extent is highly dependent on the system under consideration [30]. Moreover, integrating ZTA into the system may lead to overhead due to additional user authentication and authorization, potentially causing performance and availability issues. However, as ZTA covers the entire system, the design and testing of alternatives are complex and may not be viable for systems in active use. Software architecture simulators enable the simulation of a system's quality attributes, such as performance and access control violations, at design-time [6,28]. However, it is unclear if architecture simulators are applicable to ZTAs and if their predictions regarding quality attributes are accurate.

In this paper, we present our approach to modeling and analyzing ZTAs at the architectural level. We utilize existing technologies to check for performance and security issues. Our main contributions are:

C1 We propose a ZTA metamodel derived from literature and real-world applications.
C2 We provide reusable modeling templates for ZTA architectures.
C3 We enable analysis of multiple quality attributes for our ZTA models.

Our ZTA metamodel (**C1**) aligns different abstract approaches with the terminology provided by the zero trust architecture standard of the National Institute of Standards and Technology (NIST) [30]. The model templates of **C2** are instances of our ZTA metamodel and are created using the Palladio Component Model (PCM) [28]. Compared to other approaches we can simulate the performance impact of integrating the ZTA components into a system, while at the same time analyzing the system for security violations that indicate an improper or insufficient implementation of ZTA.

2   Background

In this section we provide an overview of the background on which our approach is based. It focuses on approaches to modeling and performing analyses, as well as the meaning of ZTA.

2.1   Software Architecture Quality Predictions

Palladio is a tool-supported software architecture simulation approach, that is used to predict a modeled software's Quality of Service (QoS) properties. The Palladio Component Model (PCM) is a metamodel of component-based software architectures [28]. The software architect defines components and their interfaces as reusable system building blocks in the *repository model*. Additionally, the architect describes how instances of the components from the repository model are connected to form a system in the *assembly model*. A software developer uses so-called *Service Effect Specifications* (SEFFs) to provide a coarse-grained execution logic of the services offered by the components in the repository model. The deployment expert adds information about available hardware in the *resource environment model* and describes the deployment of the component instances of the assembly model in the *allocation model*. Different usage profiles and workloads for types of users are defined by a domain expert in the *usage model*. Using this information, Palladio allows for the simulation of a modeled system's behavior under load [4]. Using a variety of measuring points, performance metrics such as response time and hardware utilization can be simulated. Based on these metrics, performance bottlenecks, scalability issues, and reliability threats can be identified and counteracted.

Data Flow Diagrams (DFDs) [12] are unidirectional graphs representing the data flow and processing in systems. Graph nodes are Actors, Processes, or Stores, connected by data flows. DFDs are widely used to analyze various types of security aspects [2,32,34,37] and represent an established way of modeling software architecture. To enable a more general analysis of information security, an extended notation [6,32] integrates node behavior and labels as first-class entities. Labels represent security-relevant properties, that can be assigned to nodes, and are grouped into label types according to semantic commonalities. Node behavior defines how labels are propagated along the data flow. Security analyses use the node behavior to propagate labels and iterate the resulting DFD to check constraints, e.g., if security-relevant labels can reach certain nodes, or if different labels flow together in a node. In addition, Seifermann *et al.* [32] define model annotations for the PCM, enabling label and behavior definition for various PCM elements, allowing for the extraction of data flows and subsequent security analysis of PCM instances. Building upon these concepts, Boltz *et al.* [6] provide an extensible data flow analysis framework for information security. While these analyses have been successfully applied to check information security properties, like confidentiality and privacy [6,32], they have not yet been applied to check alignment to zero trust principles.

256   N. Boltz et al.

2.2 Zero Trust Architecture

The National Institute of Standards and Technology (NIST) defines zero trust Architectures (ZTAs) as architectures that apply the concepts of zero trust (see Sect. 1) in their structure and functionality [30]. To control access to resources, ZTAs contain two core elements: The *Policy Enforcement Point* (PEP) and the *Policy Decision Point* (PDP), as shown in Fig. 1. The PEP is responsible for intercepting the access request and forwarding the requests to the PDP. It also monitors the access to resources and, if access is not granted, terminates the connection. The PDP evaluates static policies and the context of the request and creates required access credentials.

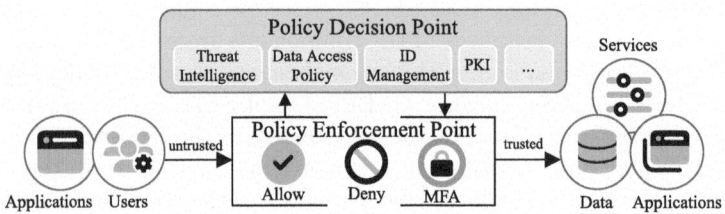

Fig. 1. Overview of ZTA elements [30].

Besides the NIST standard, several other approaches to ZTA exist: Microsoft outlines *zero trust strategies*, which include similar principles regarding zero trust [23]. They also describe logical components of a ZTA, that can be mapped to the PEP, PDP and context providers of the NIST standard. BeyondCorp is a real-world zero trust solution of Google aimed at enterprise applications [16,38]. Similar to the NIST standard, the BeyondCorp ZTA includes *Data Sources* which provide contextual information, *Access Intelligence components* which handle the access control and policies, *Gateways* such as network switches and web proxies, and the protected resources [25]. The Software Defined Perimeter (SDP) of the Cloud Security Alliance [39,40] is an approach to granular access control and follows zero trust principles. SDP provides a way to segment the network using software-defined perimeters and hide resources, by the use of policy-based, identity-centric access control to prevent unauthorized access to segments.

3 ZTA Metamodel

We propose a metamodel of ZTA based on literature and real-world approaches [24,30,38,40]. As it provides the most abstract description of ZTA, we align our terminology and overall workflow with the terminology and workflow used in the NIST standard (see Subsect. 2.2). Some elements in the NIST standard, such as requests, policies, and contexts, are less formally specified. To be able to capture all relevant information on the architectural level, however, we include them as first-class entities in our model. Figure 2 shows our proposed metamodel.

An access *Request* is sent by a subject that wants to access a *Resource* (1). The *PolicyEnforcementPoint* (PEP) intercepts requests for the resources that the PEP is responsible for (2) and forwards the request to a *PolicyEngine* (PE) of its *PolicyAdministrator* (PA). A PE is responsible for assessing the access request (3). To do so, the PE checks the context of the request based on *Contexts* provided by *ContextProviders* (4). ContextProviders are systems for which different solutions exist, e.g. identity management systems and device databases. These systems are also used outside of a ZTA but are integrated into a ZTA to provide context information that covers zero trust principles. The NIST standard also defines so-called trust algorithms as special ContextProviders, that provide trust rating to the PE, based on other contexts. *ContextEvaluator* is a specialized ContextProvider that evaluates contexts from other providers before providing a context resulting from the evaluation. An example can be the prior aggregation of multiple contexts. Using the context information, the PE evaluates static *Policies* provided by its available *PolicyProviders* (5) and decides whether the request should be granted. If granted, the PE provides instructions to the PA, including the exact authorization level. The PA is responsible for managing and creating corresponding access credentials for the request, such as access tokens (6). If access is denied, the PA is also responsible for signaling to the PEP to terminate the connection. An extended discussion of the metamodel is provided in the master's thesis [9].

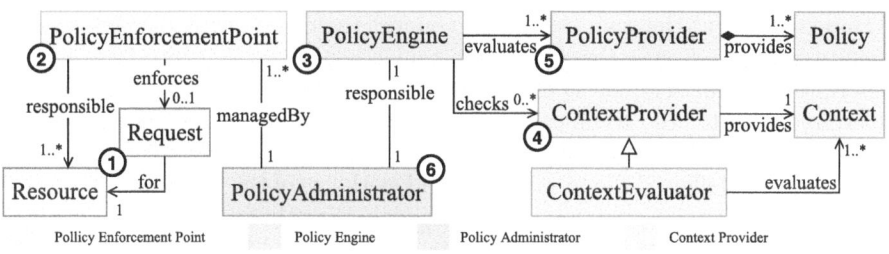

Fig. 2. Structure of our proposed ZTA metamodel.

4   Modeling Templates

To aid software architects in designing or migrating systems with zero trust in mind, we provide instances of our proposed metamodel (see Sect. 3) that follow three general ZTA approaches. These instances represent reusable modeling templates that include components, interfaces, and some assembly information. We provide templates for three approaches: the NIST standard regarding ZTA [30], the BeyondCorp approach of Google [16,38], and the Software-Defined Perimeter (SDP) [40] (see Subsect. 2.2). We use the PCM (see Subsect. 2.1) as a component-based architectural design language for creating the instances. Several elements

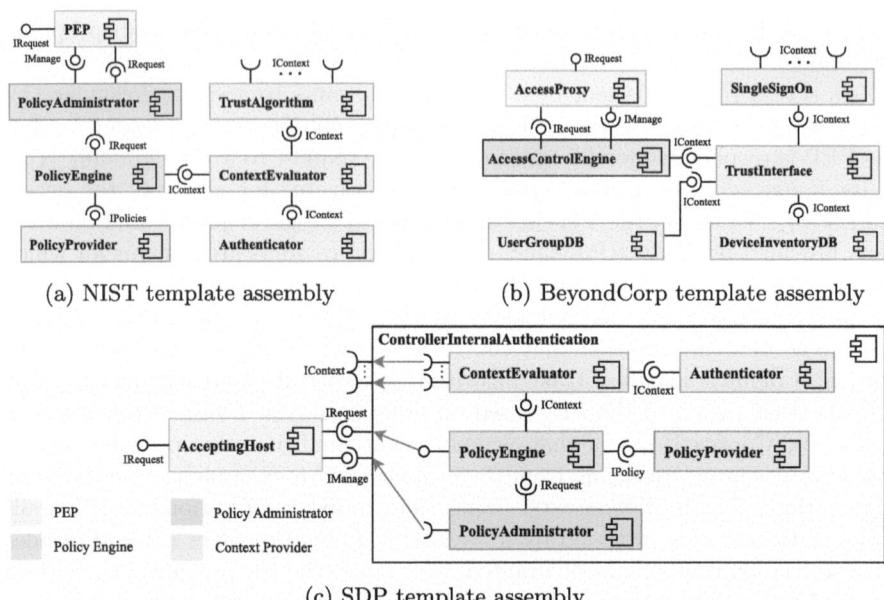

(a) NIST template assembly   (b) BeyondCorp template assembly

(c) SDP template assembly

Fig. 3. Overview of assembly of modeling template components.

from our metamodel can be directly mapped to components, while other elements like *Request, Context,* and *Policy* are encapsulated in interfaces (e.g. IRequest, IContext, IPolicy) that can be used to connect the components. For the sake of brevity, we only discuss the most important components and interfaces in this description of our templates. Components, interfaces, and data types that are solely required for the technical realization have been omitted. Please refer to our dataset (see Sect. 9) for a full overview of the PCM instances.

As we have followed the terminology and workflow of the NIST ZTA standard [30] when creating our metamodel, the template directly follows the structure of our metamodel. As shown in Fig. 3a, the *PEP, PolicyAdministrator,* and *PolicyEngine* are connected by requiring and providing the *IRequest* interface. This interface defines services that, for each component, continue to process an incoming request. The *PolicyProvider* component provides the *IPolicy* interface that the *PolicyEngine* requires. In our NIST standard template, there is always a *ContextEvaluator* component that aggregates context information of authentication and a trust algorithm for the *PolicyEngine* component. The *Authenticator* component represents a subject authentication using some form of credentials. The *TrustAlgorithm* component is also an instance of ContextEvaluator from our metamodel. It aggregates several other contexts from the system. While the kind of context providers and trust calculation logic must be fitted to the system under consideration, we provide some generic components for context providers mentioned in the NIST standard, e.g., device authentication, a security information and event management system, an ID management system, and an activity log system.

The BeyondCorp ZTA defines an access proxy that intercepts requests for resources and an access control engine that manages the requests and access policy checking. The *AccessProxy* is an instance of the PEP of our metamodel and mirrors the *PEP* component in our NIST template. The *AccessControlEngine* component combines the behavior of the *PolicyAdministrator, PolicyEngine*, and *PolicyProvider*. In our template, we model this by creating a composite component and encapsulating the corresponding components from our NIST template. In their concrete implementation, the BeyondCorp architecture of Google includes two databases, one for device registration and one for user and role management. In our template, we have added components for both databases as instances of ContextProvider that are connected to the *TrustInterface*. Beyond-Corp also uses a single-sign-on system to authenticate subjects using multi-factor authentication. We include a *SingleSignOn* component that is an instance of ContextEvaluator of our metamodel. By reusing two *Authenticator* components from our NIST template, one for each of its required interfaces, we can model two-factor authentication. The components can be further customized to the specific type of authentication required for the system under consideration. Similar to our NIST templates, contexts are always evaluated by an instance of ContextEvaluator before providing the result to the *PolicyEngine*. In the BeyondCorp ZTA, this component is called *TrustInterface*. It calculates a trust context based on the contexts provided by the connected ContextProviders, similar to the *TrustAlgorithm* in the NIST standard. However, it also takes into account the authentication context of the *SingleSignOn* component.

SDP defines *Accepting Host* (AH), *Initiating Host* (IH), and *Controller* elements. AHs evaluate requests for resources based on policies provided by the *Controller* prior to the request, while IHs represent the subjects making a request. The *Controller* authenticates each requesting IH and deploys access instructions to the AHs of the requested resources. While not following the terminology of the NIST standard, the components and functionality that make up SDP can be mapped to our metamodel and components from the NIST template shown in Fig. 3a. Figure 3c shows the allocation of our SDP template. AHs represent the PEPs in the NIST ZTA due to their similar functionality. Since the controller authenticates the IHs and decides which resources to allow access to, the *Controller* represents a combination of PolicyAdministrator, PolicyEngine, and PolicyProvider, similar to the BeyondCorp template. However, the SDP specification includes the authentication as part of the *Controller*. Figure 3c shows the resulting composite component that makes up the *Controller*. To evaluate and authorize the IH, the controller may gather information from other services, such as identity and device management systems, geolocation services, or host validation services. Like for the NIST and BeyondCorp templates, these services represent ContextProviders in our metamodel. As they match, we can simply reuse the context provider components we described for the NIST template. While the SDP template might look more straightforward than the NIST or BeyondCorp templates, one central aspect of SDP is that many AHs are used to define very small perimeters and that each AH is either deployed with the resources it is responsible for or on the direct path to them.

5   Analyzing Quality of Zero-Trust Architectures

Using various analysis tools within a modeling environment involves defining the interaction (tool orchestration) between them. There are different types of orchestration strategies [18], but none of them exactly fits our purpose. Therefore, based on those strategies, we developed a new strategy called Separate Multiple Quality Analysis. This strategy allows tools to run completely independently within a modeling environment to analyze two or more specific quality attributes. In this strategy, the modeling environment provides input to a first analysis tool. By applying a transformation and adding additional requirements to the input, it is then transformed into the inputs of specific tools included in the same workbench. The outputs of these analyses demonstrate the results of different quality attribute assessments. For performing quality analyses on the software architecture models created with the PCM (see Sect. 2.1), the system architect needs to assemble the modeled components into an application. Besides this assembly model, the performance simulation and security analysis have additional requirements. The analyses themselves are independent of each other but work on the same model artifacts. In this section, we describe these requirements and how we extend our ZTA metamodel or modeling templates to meet the requirements for the performance analysis (see Sect. 5.1) and the security analysis (see Sect. 5.2).

5.1   Enabling Performance Analysis

While our ZTA model templates contain components needed to model ZTA along with their service effect specification (SEFF), component developers need to add resource demands to the specifications to instantiate components for a concrete system and implementation [4]. The SEFFs describe how the provided services of a component relate to its required services, modeling their implementation on an abstract level. The component developers specify the resource usage of internal actions, possibly depending on the parameters the service is called with. How often an internal action is executed can depend on loop iterations and branch transitions. As all of our ZTA components are regular PCM components, the performance simulation fully supports their execution logic.

Additionally to the assembly model, the performance simulation requires a resource environment model, an allocation model, and a usage model. While the resource environment model is not specific to the application, it specifies the hardware used and its processing power, whereas the allocation model specifies where components are deployed. The simulation of the models starts from usage models that contain different usage scenarios, describing how many users are expected to interact with the system in what way, i.e., which services they call.

5.2   Enabling Data Flow Security Analysis

We want to ensure that the security analysis can check that the modeled system aligns with the zero trust principles. As described in Subsect. 2.1, the data flow analysis is based on the propagation and annotation of security labels. To align with the principles, we need to define label types and labels that represent if a request is *authorized* and whether a subject of a request is *authenticated* and *trusted.* For authentication, we enable checking of multi-factor authentication by defining the *subject authentication* label type, which contains labels that represent different authentication factors. As it is a common requirement for ZTA [24, 30], we define two labels (*first factor* and *second factor*) to represent 2-factor authentication. When applying our metamodel and modeling templates to other systems with higher authentication requirements, the *subject authentication* label type can be extended or adapted to the needs of the system under consideration. To represent different kinds of authorization, we define label types for the *subject authentication level, device authentication,* and *trust.* For the device authentication and trust label types, we each define the label *authenticated* that represents either the successful authentication of a device or signals a successful trust calculation. We define four abstract authorization levels *Level 0, Level 1, Level 2, Level 3,* each representing a higher authorization level in ascending order. Similarly to the authentication factors, these labels can be adapted to the particular system under consideration, for example, to map to organizational roles for role-based access control [14]. The corresponding labels are used for the label types that represent the rights of the subject (*SubjectAuthorization*), the authorization needed to access a resource (*ResourceRequiredAuthorization*), and the authorization that has been assigned to the request of a subject (*RequestAuthorization*). Using the same labels allows the security analysis to check if a label is set and to compare two labels of different types with each other.

Understanding the behavior of our proposed ZTA components in handling our defined labels is crucial for effective security analysis. The behavior defines whether and how a component alters the labels and, in turn, how they are propagated along the data flow. For the sake of simplicity, we only describe the behavior in natural language. A complete definition can be found in the PCM instance in our dataset (see Sect. 9). For our ZTA components introduced in Sect. 3, we define the following behavior: The *DeviceAuthenticator* and *TrustAlgorithm* each alter their respective label type. In our implementation, both set the label *authenticated* if they execute successfully, e.g., when data flows through them. The *Authenticator* components alter the *SubjectAuthentiaction* label type, depending on the used authenticator and the number of authentication factors. The *PolicyEngine* alters the labels of the *SubjectAuthorization* label type. To align with the PoLP, it considers the rights of the subject and dependencies of the requested resource. Most other components do not change labels; they simply forward all labels that flow into them. Our other implementations of ZTA approaches like BeyondCorp [38], and SDP [40] follow the same principles, labels, and label types, based on the mapping described in Subsect. 2.2.

6   Evaluation

In this section, we evaluate our approach regarding applicability and ability to simulate performance and security impact of integrating zero trust components into the software architecture. We investigate the following Research Questions:

RQ1 How applicable is our ZTA metamodel when modeling systems based around the principles of zero trust?

RQ2 Can we simulate the performance impact induced by ZTA elements and different execution flows?

RQ3 Can we identify security violations based on principles of zero trust?

We answer RQ1 with a discussion on the applicability of our proposed ZTA metamodel to represent different real-world approaches from related work. For RQ2 and RQ3 we use a Media Store case study from literature. We answer RQ2 by comparing response times of simulations of different scenarios with varying integration of ZTA elements. We answer RQ3 by calculating precision and recall values for different scenarios with manually added violations of zero trust principles.

6.1   Case Study: Media Store

We use the Media Store system [35] that has been widely used for software architecture research [4,17,19] to illustrate and evaluate our approach. The Media Store models a file-hosting system to which users can upload and download audio files. It provides basic media management, as well as file encoding and watermarking. The Media Store model does not include any security measures as-is.

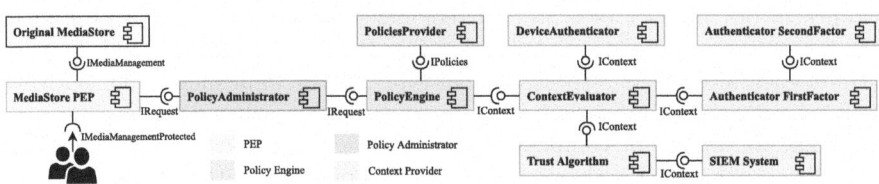

Fig. 4. Overview of the integration of ZTA components into the MediaStore.

We integrate ZTA into the Media Store following the steps presented in the Zero Trust Guide provided by the United Kingdom National Cyber Security Center (NCSC) [24]. Figure 4 shows an overview of our ZTA integration. First, we identify the upload and download functionality and the file access as assets to protect, all provided by the *IMediaManagement* interface of the original *MediaStore*. To protect them, we have to intercept requests to them. We therefore place the *MediaStorePEP* in front of the original *MediaStore* component. To consider the behavior of users when making decisions, we instantiate the *TrustAlgorithm* component that queries data about previous requests by the user from the *Security Information and Event Management (SIEM)* system for decision-making. We instantiate both components from our NIST model template. Next, we integrate policies and policy evaluation into the system. While a *PolicyEngine* is added to evaluate policies, a *PolicyAdministrator* creates the configurations based on the decisions and manages the PEP. We use the *PoliciesProvider* component as policy source for the *PolicyEngine*. Further, we connect a *ContextEvaluator* component to the *PolicyEngine* and supply the device and user authentication as well as the trust algorithm to its required *IContext* roles. As the NCSC guide requires multi-factor authentication for users, we instantiate our *Authenticator* component twice and connect them to each other. Also, we instantiate a *DeviceAuthenticator* component, for device authentication.

We extend the already existing resource environment, consisting of two containers for the *ApplicationServer* and *DatabaseServer*, with a third container for the *ClientDevice*. We allocate all databases to the *DatabaseServer*, the *CredentialsProvider* to the *ClientDevice*, and all other components to the *ApplicationServer*.

6.2   RQ1: Applicability

To evaluate applicability, we focus on two different aspects: First, the ability of our proposed ZTA metamodel to represent different real-world approaches from related work. Second, we discuss the effort needed to integrate ZTA principles in an existing software architecture model using our ZTA model templates. We also discuss the quality of ZTA after integration using the CISA maturity model [11].

Representation of ZTA Approaches Using Our Metamodel: BeyondCorp [38] and SDP [40] presented in Subsect. 2.2 represent the most prominent real-world approaches. Especially BeyondCorp seems to have found widespread use, as it is offered as part of the Google Cloud for enterprises [16]. We already show that our metamodel can represent systems based on both approaches with the reusable modeling templates we provide in Sect. 4. Several other approaches exist in the literature beyond the typical enterprise scenarios for zero trust. Paul *et al.* [26] propose a zero trust model for industrial IoT ecosystems. While they focus on the production domain and, for example, include cyber-physical systems on the architecture level, the approach is largely comparable to BeyondCorp's. Therefore, our metamodel can be easily applied. Ramezanpour *et al.* [27] propose a ZTA for 5G/6G networks that integrates machine learning in some of their proposed components. While this offers new capabilities during runtime, on the architectural level, the introduced 'intelligent' components only represent two new *context providers* when mapped to our ZTA metamodel. Chen *et al.* [8] also propose a software-defined ZTA for 6G networks. Their approach focuses more on the elastic scalability of the architecture while allowing adaptive collaborations among control domains. Similar to the approach of Ramezanpour *et al.* [27], their introduced components for trust calculation, like a Vulnerability Database (VDB), Cybersecurity Event Ledger (CEL) and Anomalous Behaviour Detector

(ABD) represent different *context providers* in our ZTA model. Figure 5 shows an excerpt of the resulting architecture using our ZTA metamodel and already defined components of our modeling templates. Lee *et al.* [22] propose situational awareness-based risk adaptable access control for enterprise networks, which follows the general zero trust principles. Their proposed elements can all be mapped to components of our modeling templates, e.g., the *ContextHandler* from the SDP template. Thus our metamodel can be applied. We provide an in-depth description of how we modeled these systems in our dataset (see Sect. 9).

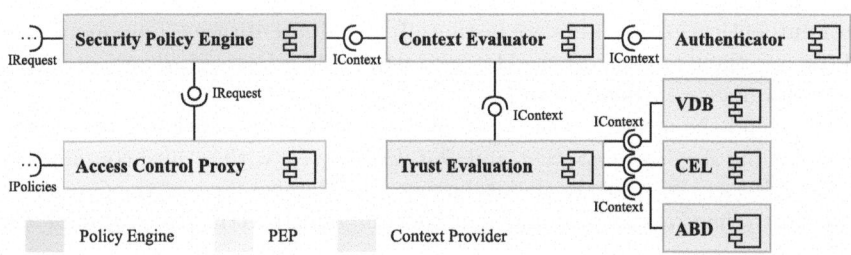

Fig. 5. Excerpt of architecture of Chen *et al.* [8] mapped to our ZTA components.

Integration of ZTA into Existing Software Architecture: We specified a rather broad defense perimeter for our case study (see Subsect. 6.1). Compared to the original media store case study, only minimal changes to the system were required to make it compatible with our ZTA components. When trying to define finer defense perimeters, more changes would be required. However, when integrating a specific overarching concept, like zero trust, into a system that was originally designed without this concept in mind, more extensive changes are to be expected. When assessing the maturity of our implemented ZTA integration based on the CISA Zero Trust Maturity Model [11], a maturity model that defines requirements to achieve increasingly demanding levels of maturity of ZTA-relevant aspects, we conclude that most aspects of our integration have an *initial* or *advanced* rating. A thorough description of our maturity rating and a discussion on how certain aspects can be improved is included as part of the dataset (see Sect. 9).

6.3   RQ2: Simulating Performance Impact of ZTA Components

We evaluate our model's ability to simulate the ZTA components' performance impact and the different execution flows of the request evaluation process. To get deterministic results, we define the usage of the system so that every user requests the download of four files at a time, each with a size of 40,568,000. We use the Media Store system without ZTA as baseline (S0) and evaluate different scenarios with ZTA. In Scenario S1, all requests are policy-authorized but not context-authorized. This implies that the complete path is executed, with each part of the context being evaluated for every request. Each request should pass the evaluation and be forwarded to the initial Media Store components. In Scenario S2, we turn off usage of the trust algorithm, assuming all subjects accessing the system have the same level of trust.

To obtain accurate performance predictions, component developers need to add resource demands to the model according to the component implementation (see Sect. 5.1). As we do not add new types of components to the model, we can assume that their simulation behavior is accurate [3]. Therefore, our evaluation focuses on relative differences introduced by the ZTA components rather than the correctness of the absolute response time values. We expect *S1* to show the slowest response times as the configuration of ZTA requires the execution of the complete path for every request, thus adding the most overhead. *S2* should show better performance than *S1* as it does not use the trust algorithm, skipping some parts of the execution process, which results in less overhead and faster response time. Lastly, we expect the baseline *S0* to have the fastest response times as there is no ZTA integrated, and requests are directly passed to the *MediaManagement* component of the MediaStore.

Figure 6 shows the response times of the different evaluation scenarios after simulating their execution for 15.000 time units each. The results confirm our expectations: *S0* is fastest with a median response time of 27.25 s. *S1* adds the most overhead, resulting in a median response time of 28.57 s. While *S2* still adds overhead to *S0*, it is faster than *S1* with a median response time of 28.38 s due to the disabling of the trust algorithm.

6.4   RQ3: Analyzing Security Violations

We evaluate our model's ability to be analyzed for security violations. The data flow analysis of Boltz *et al.* [6] is already able to analyze sufficiently well-specified models for violations of access control policies. However, zero trust principles go beyond traditional access control, e.g. by enforcing the PoLP. For our evaluation, we focus on violations of zero trust principles that indicate faulty or insufficient implementation of ZTA. Using the labels defined in Subsect. 5.2, we define four

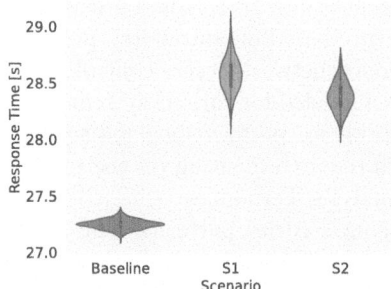

Fig. 6. Performance results of different evaluation scenarios.

independent reusable constraints. The constraints cover correct authorization, adherence to the PoLP, subject authentication, and trust calculation.

We use these constraints and compare if the data flow analysis can correctly identify violations. The evaluation is based on a manually created gold standard based on our media store case study system with ZTA. We define a baseline *S0* without violations and four violation scenarios, each containing a violation of a core zero trust principle (see Sect. 1). As violations, we introduce a flow with unauthorized access *(S1)*, a flow violating the PoLP *(S2)*, a flow skipping subject authentication *(S3)*, and two flows with a lack of trust calculation based on the context *(S4)*. Using our constraints, the analysis correctly identified the introduced violations for all scenarios without any false positives. This results in a precision, recall, and F_1 score of 1.0.

6.5   Threats to Validity

We discuss the validity and reliability of our evaluation as characterized by Runeson *et al.* [31]. Our main threat to external validity is the limited generalizability due to the case study-based evaluation. We mitigate this threat by using a well-known case study from literature. Our main threat to internal validity lies in the design decisions about the elements included in our metamodel. We mitigate this threat by basing most of our metamodel on the descriptions of the NIST standard [30]. Regarding construct validity, our evaluation does not comprehensively cover all aspects of zero trust. We cannot fully mitigate this threat but have chosen the examined aspects based on the commonalities and general understanding of the principles described in related work.

7   Related Work

ZTA models have been proposed for various systems, such as for smart industry systems [26], next generation networks [27], and 6G networks based on communities of user equipment [8]. Jung *et al.* [21] propose a ZTA for blocking malicious access to enterprise resources. Three components form the architecture – a PDP, PEP, and an Authentication Server Function. Lee *et al.* [22] incorporate security situational awareness into Risk Adaptable Access Control. Ghate *et al.* [15] describe an architecture based on automated policy generation definition to achieve low-cost fine-grained network access control. These works primarily discuss how a ZTA could be modeled for various systems and implemented, without considering a model suitable for performance and security analysis.

Furthermore, there is research focusing on the modeling of systems for performance and security analysis. Fernandez *et al.* [13] analyze ZTAs and evaluate security architectures using security patterns. The authors propose a Security Reference Architecture of a ZTA and extract elements from different security patterns to form a concept model of a ZTA. They evaluate some proposed ZTAs by answering questions about its performance and security. The evaluation, however, depends only on experiences since there are no quantitative measures of ZTA systems, and the proposed model is simply a concept. Rodigari et al. [29] study the performance of communication between microservices in terms of latency and physical resources with enabling Zero Trust in a multicloud environment. Sharma et al. [33] employ Discrete Time Markov Chains to model software systems and predict performance, security, and reliability based on its software architecture. Cortellessa et al. [10] introduce a framework for modeling the performance and security aspects of software architecture. They create a UML library, which models the basic security mechanisms of encryption, decryption, signature generation, and verification. This is followed by a modeling of composite security mechanisms. Despite some attempts to model ZTA for quality requirements analysis, there is a lack of an approach to modeling a reusable ZTA model while evaluating the impact of integrating the proposed ZTA model into other systems.

8 Conclusion

In this paper, we present our approach to modeling and analyzing ZTAs regarding their performance impact and security. We discuss several principles of zero trust and provide an overview of general approaches to ZTA. Based on these approaches, we devise a ZTA metamodel. By creating instances of our metamodel with the Palladio Component Model (PCM), we provide reusable modeling templates based on the NIST [30], BeyondCorp [38], and SDP [40] architectures. We outline the requirements for enabling PCM performance simulation as well as data flow-based security analysis, and describe how we extend our ZTA model instances to meet these requirements. Based on existing literature, we show that our ZTA metamodel can represent several real-world approaches to ZTA from different domains. Moreover, we show that by using different quality checks, we can simulate the performance impact of integrating ZTA components and identify security problems resulting from an incorrect implementation of ZTA in a system.

Acknowledgements. This publication is partially based on the research project SofDCar (19S21002), which is funded by the German Federal Ministry for Economic Affairs and Climate Action. This work was also supported by funding from the pilot program Core Informatics at KIT (KiKIT) and the topic Engineering Secure Systems of the Helmholtz Association (HGF), KASTEL Security Research Labs, and the German Research Foundation (DFG) under project number 499241390 (FeCoMASS).

Data Availability. We provide a data set (https://doi.org/10.5281/zenodo.11580654) containing all code artifacts, PCM instances of our ZTA modeling templates, and the used case study model instances.

References

1. Alagappan, A., Venkatachary, S.K., Andrews, L.J.B.: Augmenting zero trust network architecture to enhance security in virtual power plants. Energy Rep. **8**, 1309–1320 (2022)
2. Alshareef, H., et al.: Precise analysis of purpose limitation in data flow diagrams. In: ARES (2022)
3. Becker, M., Becker, S., Meyer, J.: SimuLizar: design-time modeling and performance analysis of self-adaptive systems (2013)
4. Becker, S., Koziolek, H., Reussner, R.: Model-based performance prediction with the palladio component model. In: WOSP, pp. 54–65 (2007)
5. Bhuiyan, E.A., et al.: Towards next generation virtual power plant: technology review and frameworks. Renew. Sustain. Energy Rev. **150** (2021)
6. Boltz, N., et al.: An extensible framework for architecture-based data flow analysis for information security. In: Tekinerdoğan, B., Spalazzese, R., Sözer, H., Bonfanti, S., Weyns, D. (eds.) ECSA 2023. LNCS, vol. 14590, pp. 342–358. Springer, Cham (2024). https://doi.org/10.1007/978-3-031-66326-0_21
7. Chen, B., et al.: A security awareness and protection system for 5G smart healthcare based on zero-trust architecture. IEEE IoT J. **8**(13), 10248–10263 (2020)
8. Chen, X., et al.: Zero trust architecture for 6G security. IEEE Netw. (2023)
9. Cholakov, E.: Modelling and analysing zero-trust-architectures regarding performance and security. Master's thesis (2024). https://doi.org/10.5445/IR/1000171583
10. Cortellessa, V., Trubiani, C., Mostarda, L., Dulay, N.: An architectural framework for analyzing tradeoffs between software security and performance. In: Giese, H. (ed.) ISARCS 2010. LNCS, vol. 6150, pp. 1–18. Springer, Heidelberg (2010). https://doi.org/10.1007/978-3-642-13556-9_1
11. Cybersecurity and Infrastructure Security Agency (CISA), CISA Zero Trust Maturity Model (2023). https://www.cisa.gov/sites/default/files/2023-04/zero_trust_maturity_model_v2_508.pdf. Accessed 23 Feb 2024
12. DeMarco, T.: Structure analysis and system specification. In: Tekinerdoğan, B., Spalazzese, R., Sözer, H., Bonfanti, S., Weyns, D. (eds.) ECSA 2023. LNCS, vol. 14590, pp. 255–288. Springer, Cham (1979). https://doi.org/10.1007/978-3-031-66326-0_21
13. Fernandez, E.B., Brazhuk, A.: A critical analysis of zero trust architecture (ZTA). Comput. Stand. Interfaces **89**, 103832 (2024)
14. Ferraiolo, D.F., et al.: Proposed NIST standard for role-based access control. TISSEC **4**(3), 224–274 (2001)
15. Ghate, N., et al.: Advanced zero trust architecture for automating fine-grained access control with generalized attribute relation extraction. IEICE Proc. Ser. **68**(C1-5) (2021)
16. Google Cloud: BeyondCorp (2024). http://cloud.google.com/beyondcorp
17. Gorsler, F., Brosig, F., Kounev, S.: Controlling the Palladio Bench using the Descartes Query Language. In: KPDAYS, pp. 109–118 (2013)
18. Heinrich, R., et al.: Composing Model-Based Analysis Tools. Springer, Heidelberg (2021). https://doi.org/10.1007/978-3-030-81915-6
19. Heinrich, R. et al.: The palladio-bench for modeling and simulating software architectures. In: ICSE-C, pp. 37–40 (2018)
20. IoT - Market data analysis and forecasts. https://de.statista.com/statistik/studie/id/109209/dokument/internet-der-dinge-market-outlook-report/

21. Jung, B.G. et al.: ZTA-based federated policy control paradigm for enterprise wireless network infrastructure. In: APCC, pp. 1–5 (2022)
22. Lee, B. et al.: Situational awareness based risk-adapatable access control in enterprise networks. arXiv preprint arXiv:1710.09696 (2017)
23. Microsoft Corporation, Evolving Zero Trust (2021). https://query.prod.cms.rt. microsoft.com/cms/api/am/binary/RWJJdT. Accessed 23 Feb 2024
24. National Cyber Security Centre UK, ZTA design principles. https://www.ncsc.gov. uk/collection/zero-trust-architecture. Accessed 23 Feb 2024
25. Osborn, B. et al.: BeyondCorp: design to deployment at google. USENIX Association: login: Magazine (2016)
26. Paul, B., Rao, M.: Zero-trust model for smart manufacturing industry. Appl. Sci. **13**(1), 221 (2022)
27. Ramezanpour, K., Jagannath, J.: Intelligent ZTA for 5G/6G networks: principles, challenges, and the role of machine learning in the context of O-RAN. Comput. Netw. **217**, 109358 (2022)
28. Reussner, R.H., et al.: Modeling and Simulating Software Architectures: The Palladio Approach. MIT Press, Cambridge (2016)
29. Rodigari, S., et al.: Performance analysis of zero-trust multi-cloud. In: 2021 IEEE 14th International Conference on Cloud Computing (CLOUD), pp. 730–732 (2021)
30. Rose, S., et al.: Zero Trust Architecture. NIST Special Publication (2020). https:// doi.org/10.6028/NIST.SP.800-207
31. Runeson, P., et al.: Case Study Research in Software Engineering: Guidelines and Examples. Wiley, Hoboken (2012)
32. Seifermann, S., et al.: Detecting violations of access control and information flow policies in data flow diagrams. J. Syst. Softw. **184**, 111138 (2022)
33. Sharma, V.S., Trivedi, K.S.: Quantifying software performance, reliability and security: an architecture-based approach. J. Syst. Softw. (2007)
34. Sion, L. et al.: Solution-aware data flow diagrams for security threat modeling. In: SAC, pp. 1425–1432 (2018)
35. Strittmatter, M., Kechaou, A.: The media store 3 case study system. KIT (2016)
36. Teerakanok, S., Uehara, T., Inomata, A.: Migrating to zero trust architecture: reviews and challenges. Secur. Commun. Netw. (2021)
37. Tuma, K., Scandariato, R., Balliu, M.: Flaws in flows: unveiling design flaws via information flow analysis. In: ICSA, pp. 191–200 (2019)
38. Ward, R., Beyer, B.: BeyondCorp: a new approach to enterprise security. USENIX Association: login: Magazine (2014)
39. WG: SDP and Zero Trust, Integrating SDP and DNS Enhanced Zero Trust Policy Enforcement. CSA (2022). https://cloudsecurityalliance.org/artifacts/integrating-sdp-and-dns-enhanced-zero-trust-policy-enforcement/
40. WG: SDP and Zero Trust, SDP Specification v2.0. CSA (2022). https://cloudsecurityalliance.org/artifacts/software-defined-perimeter-zero-trustspecification-v2/

Towards Secure Management of Edge-Cloud IoT Microservices Using Policy as Code

Samodha Pallewatta[1,2]([⊠]) and Muhammad Ali Babar[1,2,3]

[1] CREST - The Centre for Research on Engineering Software Technologies,
Adelaide, Australia
[2] The University of Adelaide, Adelaide, Australia
{samodha.pallewatta,ali.babar}@adelaide.edu.au
[3] Cyber Security Cooperative Research Centre (CSCRC), Joondalup, Australia

Abstract. IoT application providers increasingly use MicroService Architecture (MSA) to develop applications that convert IoT data into valuable information. The independently deployable and scalable nature of microservices enables dynamic utilization of edge and cloud resources provided by various service providers, thus improving performance. However, IoT data security should be ensured during multi-domain data processing and transmission among distributed and dynamically composed microservices. The ability to implement granular security controls at the microservices level has the potential to solve this. To this end, edge-cloud environments require intricate and scalable security frameworks that operate across multi-domain environments to enforce various security policies during the management of microservices (i.e., initial placement, scaling, migration, and dynamic composition), considering the sensitivity of the IoT data. To address the lack of such a framework, we propose an architectural framework that uses Policy-as-Code to ensure secure microservice management within multi-domain edge-cloud environments. The proposed framework contains a "control plane" to intelligently and dynamically utilise and configure cloud-native (i.e., container orchestrators and service mesh) technologies to enforce security policies. We implement a prototype of the proposed framework using open-source cloud-native technologies such as Docker, Kubernetes, Istio, and Open Policy Agent to validate the framework. Evaluations verify our proposed framework's ability to enforce security policies for distributed microservices management, thus harvesting the MSA characteristics to ensure IoT application security needs.

Keywords: Microservice Architecture · Internet of Things · Policy as Code · Edge Computing

1 Introduction

Internet of Things (IoT) solutions are rapidly expanding across a large range of domains (i.e., healthcare, industrial, commercial, and agriculture) to extract

M. Galster et al. (Eds.): ECSA 2024, LNCS 14889, pp. 270–287, 2024.
https://doi.org/10.1007/978-3-031-70797-1_18

valuable information from the data. International Data Corporation estimates a total generation of nearly 80 Billion zettabytes of data in 2025 from 55.7 billion connected devices [13]. The vast amount and variety of data processed by IoT applications make the security of the IoT ecosystem one of the critical aspects [1]. IoT applications handle data with different sensitivity levels, such as highly sensitive, personal, and mission-critical data (e.g., health records and security camera footage) and low-sensitive data like weather data. Hence, the processing of IoT data needs to be done within trust boundaries depending on data sensitivity levels [17,22,24]. Owing to the highly resource-constrained nature of the IoT devices, computationally expensive data processing tasks are often offloaded to distributed computing resources with more resource availability (i.e., edge data centres, private data centres or cloud) [1]. To optimally utilise these heterogeneous resources, the state-of-the-art research explores the federation of geo-distributed edge and cloud resources managed by multiple infrastructure providers (i.e., multi-domain edge-cloud) along with the distributed deployment of IoT applications across them [6,8].

Thus, the current IoT landscape sees a rise in MicroService Architecture (MSA) as the preferred software architecture for IoT applications accompanied by federated edge and cloud computing for application deployment [18]. Edge computing brings cloud-like processing, networking and storage capabilities closer to the network edge. With this, Edge computing not only improves QoS parameters like latency and throughput of IoT services but also enhances the security and privacy of IoT data. Processing data using edge resources closer to the data sources reduces the data exposure during transmission and provides higher flexibility to perform processing within trusted boundaries [2]. Meanwhile, MSA provides data segmentation based on sensitivity and well-defined boundaries among independently deployable and scalable microservices [16]. Such characteristics of MSA support the implementation of granular security controls at the microservices level with secured boundaries among microservices, which result in isolation among data types with varying sensitivity levels [10]. However, the deployment and composition of distributed microservices within multi-domain edge-cloud computing resources pose some novel security challenges due to geographical and administrative divisions among resources. We highlight these challenges using a motivating scenario in the following section.

1.1 Motivating Scenario

We consider a UAV Path Finding Application developed using MSA (see Fig. 1). The application is decomposed into microservices to achieve data segmentation, such that the data with varying sensitivity levels remains isolated for processing by distinct microservices. The IoT application is deployed across the geo-distributed edge and cloud resources managed by multiple resource providers, creating a *multi-domain edge-cloud environment* characterised by geographical and administrative divisions, resulting in security scenarios that affect microservice placement, scaling, migration and composition-related decisions:

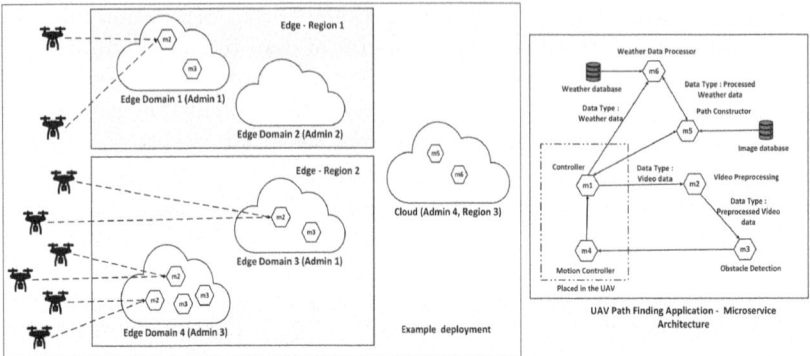

Fig. 1. Motivating Scenario - Deployment Scenario of UAV Path Finding Application

Domain-Aware Microservice Placement Restrictions - Due to the high sensitivity of the live video data (i.e., footage containing private properties, critical infrastructure and individuals), placement of m2 and m3 are restricted to trusted Edge domains. In the example environment, this restricts the dynamic placement of m2 and m3 to domains managed by Admin 1 and Admin 3.

Locality Constraints Between the Data Source and Ingress Microservice - To reduce the exposure of sensitive data during transmission, the real-time video data sent from UAVs need to be processed by m2 microservice instances placed within the same domain the UAV connects to. Thus, Edge Domain 4 should contain more microservice instances to satisfy the request volume generated from a comparatively higher number of UAVs connected to Edge Domain 4 to meet performance requirements without forwarding requests to m2 instances in other domains.

Locality Constraints on Microservice Composition - m2 pre-processes privacy-sensitive video data and forwards it to m3 for obstacle detection. Pre-processing actions like aggregation, anonymizing and transformation make the data output of m2 less privacy-sensitive than its input. However, as it still can contain some privacy-sensitive contextual data, it should be processed without leaving the region to reduce the exposure during data transmission between collaborating microservices (m2 and m3 in this case). In this example, resource-constrained Edge Domain 4 has insufficient resources to host m3 instances to support all its directly connected UAVs. Thus, more m3 instances are placed within Edge Domain 3 within the same region as Edge Domain 4 to satisfy the throughput requirements without violating the security policy.

The above scenarios demonstrate that microservices can have granular security constraints depending on the sensitivity of the data processed by/transmitted among microservices. Security constraints govern two types of microservice management activities: where the microservices are deployed during their initial placement, scaling and migration, and request routing among horizontally scaled microservices placed across domains. The following research questions must be addressed to ensure the security constraints are satisfied during above activities:

1. How to define and integrate such granular security constraints into the microservice deployment decision process in the context of multi-domain deployments? With the popularity of MSA as a cloud-native software architecture, DevSecOps practices promote the use of Policy-As-Code (PaC) to decouple security policies from the application code [3] to improve dynamic policy enforcement. However, existing research in multi-domain Edge-Cloud environments [4,7,18] does not explore the use of PaC to define microservice deployment and composition-related security constraints and integrate PaC to control planes for microservice deployment decision making.
2. How to enforce the defined security constraints during dynamic composition and request routing among multi-domain microservices? Microservices utilize the cloud-native technology stack encompassing container technologies (e.g., Docker), container orchestration (e.g., Kubernetes) and service mesh (e.g., Istio) to facilitate their dynamic deployment and composition within distributed computing environments. Thus, the method of security policy enforcement is intertwined with these underlying technologies and needs to be dynamically handled by the intelligent control plane logic. Existing architectural frameworks for edge and cloud integrated environments [4,5,18] mainly focus on proposing control planes to ensure QoS requirements (e.g., latency, throughput, etc.) and lack focus on dynamic security policy enforcement through intelligent configuration of underlying technologies.

1.2   Contributions

To overcome these limitations, we propose an architectural framework that facilitates the integration and enforcement of security policies during the deployment and composition of microservices across multi-domain edge and cloud resources.

Main contributions of our work are as follows:

1. Derive the requirements of a secure microservice management framework for multi-domain Edge-Cloud environments by analyzing a microservices-based IoT application scenario to resolve the above research questions.
2. Design an architectural framework that integrates PaC into an intelligent control plane to satisfy the identified requirements.
3. Implement a prototype system based on the proposed framework using open-source cloud-native technologies.
4. Integrate a PaC-integrated heuristic placement algorithm to evaluate the proposed architectural framework for secure management of microservices.

Our proposed framework can be used by IoT application developers to integrate PaC to define and enforce security requirements, thus extending DevSecOps practices to the edge.

Edge-cloud infrastructure providers and researchers can use the architectural framework to implement platforms for secure management of microservices-based IoT applications.

2   Related Work

This section compares state-of-the-art edge application management frameworks and platforms with our proposed framework.

Early edge-cloud frameworks like FogBus [21] focused on application execution directly on physical resources or virtual machines. However, with the emergence of MSA and lightweight deployment technologies like Docker containers, Edge computing frameworks were developed for efficient deployment and orchestration of containers. FogPlan [25], Con-Pi [14], and FogBus2 [4] proposed frameworks for dynamic provisioning of containerised applications. Works such as Wang et al. [23], FogAltas [9], and Foggy [20] support the management of containers at a scale by integrating Kubernetes orchestrator. Sophos [15] and MicroFog [19] frameworks integrated Istio service mesh with Kubernetes to provide service discovery and load balancing for the dynamic composition of microservices. All of these frameworks consider the placement of microservices across integrated edge and cloud environments. However, only MicroFog supports distributed deployment and compositions of microservices across multiple Kubernetes clusters forming a multi-domain edge-cloud environment.

All above frameworks focus on deploying applications in a performance-aware manner. Frameworks such as FogBus2 [4], FogAltas [9] and MicroFog [19] allow users to implement placement algorithms to meet QoS requirements such as latency and throughput. Out of these frameworks, Fogbus [21] utilises Blockchain technology to add security by ensuring data integrity during data transmission among edge computing nodes. As a data privacy and security measure, FogAtlas [9] allows users to select which data to keep on-premise and which data to send to the cloud. Use of Istio service mesh in MicroFog [19] and Sophos [15] enables mutual TLS authentication by default, which encrypts the communication among microservices. However, none of the above frameworks focuses on integrating security constraints related to the dynamic deployment of microservices or restrictions on their dynamic composition, especially considering multi-domain edge-cloud environments.

Thus, in this paper, we propose a framework designed for the secure management of microservices within multi-domain edge-cloud environments using PaC. It is important to note that our proposed framework does not replace the data security measures achieved through Blockchain technology or data encryption discussed in existing frameworks. Instead, this framework complements these measures by introducing an intelligent control plane that conducts microservice management (including dynamic deployment and composition) in a security policy-aware manner within multi-domain edge-cloud environments.

3   Secure Microservice Management Framework

This section details our *"Secure Microservice Management Framework"*, containing requirement elicitation of the frameworks, system model representing the business domain, architectural framework depicted using Modular Architecture Diagram and Interaction Diagram and a prototype based on the architectural framework.

3.1 Requirement Elicitation

We follow the *"Scenario-based requirement elicitation method"* proposed by Haumer et al. [12] to construct generalized requirements of the framework utilizing the motivating scenario from Sect. 1.1. Requirements aim to resolve the two research questions related to secure microservice management by utilizing PaC. Elicited framework requirements are as follows:

1. Provide an approach to define microservice management-related security constraints and enable their integration into deployment and composition decision process: Establish a method to define different types of security constraints related to MSA such as *Domain-aware microservice placement restrictions, Locality constraints between IoT device and ingress microservice*, and *Locality constraints on microservice composition* using PaC. Policy data and policy evaluation functions included in the definitions should be accessible through APIs to ensure the integration of security policies into edge-cloud control planes.
2. Ensure the utilization of defined policies during microservice deployment within multi-domain resources: Policies should be considered during the initial placement of microservices upon the arrival of new application placement requests and their migration or scaling based on runtime system monitoring metrics (e.g., resource utilization, request rates, etc.). Edge-cloud control planes should have access to policy definitions through APIs to retrieve eligible domains for each microservice and provide them as input to placement, scaling or migration algorithms to determine where to deploy microservices.
3. Ensure that the request routing among microservices complies with the security policies: When deploying microservice instances based on the output of the placement, scaling or migration algorithms, the control plane should configure the underlying service mesh to ensure that traffic routing adheres to locality-related security policies such as *Locality constraints between IoT device and ingress microservice* and *Locality constraints on microservice composition*.
4. Ensure scalable management of policies such that the distributed control plane can efficiently retrieve the policy evaluations or policy-related data across the multi-domain edge and cloud clusters.
5. The framework must be designed for extensibility, enabling the extension of existing policies (e.g., new locality levels, new domain definitions, etc.) and integration of new security policy types.

3.2 Domain Model

Based on the requirements, we construct a domain model (see Fig. 2) to capture the requirements and depict the domain logic managed by our framework.

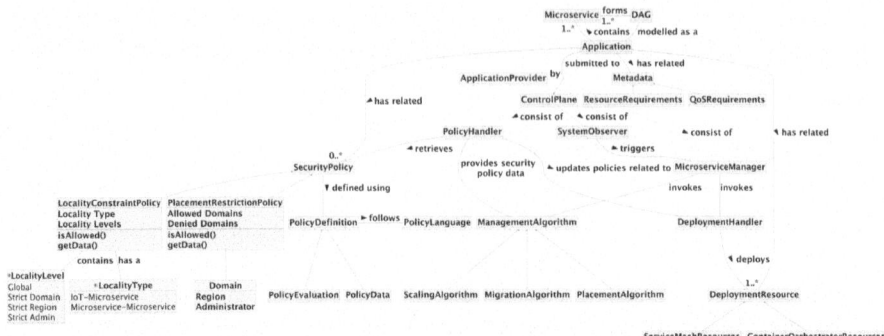

Fig. 2. Domain Model for secure microservice management framework

IoT Application Providers submit requests to the *Control Plane* to conduct the initial placement of MSA-based *Applications*. According to existing literature, MSA applications can be modelled as Directed Acyclic Graphs (DAGS) [7,11], thus enabling the control plane to utilize *DAG* application model, *Deployment Resources* and *Security Policies* along with other *Metadata* such as *QoS Requirements* and *Resource Requirements* to manages microservices across distributed edge-cloud clusters to meet *Security Policies* and *QoS Requirements*.

Control Plane contains three main components: *Microservice Manager, System Observer and Policy Handler*. *Microservice Manager* is responsible for invoking *Management Algorithms* including: 1) *Placement Algorithms* that are invoked during initial placement of the *Application*, 2) *Scaling Algorithms* and *Migration Algorithm* that are invoked based on triggers received from the *System Observer* to optimize the deployment of already placed *Applications*. *System Observer* communicates with components that monitor performance metrics across edge-cloud clusters (i.e., Prometheus, Kubernetes Metric Server, etc.) and trigger *Scaling Algorithms* or *Migration Algorithms* accordingly. *Policy Handler* is responsible for integrating *Security Policy* related *Policy Data* and *Policy Evaluations* when executing *Management Algorithms* to ensure that *Security Policies* are enforced throughout the management cycle of microservices.

Policy Handler acts as the interfacing component that retrieves the data and evaluations from Security Policies defined using PaC. The Policy Definitions are created following a Policy Language and define Policy Data and Policy Evaluation functions. Domain model depicts three secure microservice management-related policy types derived from the motivation scenario as follows:

1. Placement Restriction Policy: Defines the *Domain-aware Microservice Placement Restrictions*, with Policy Data containing allowed/ denied domains for the microservices (exposed through getData function endpoint) and Policy Evaluation (exposed through isAllowed function endpoint) to evaluate and return the eligibility of the microservice to domain mappings.

2. Locality Constraint Policy (Locality Type: IoT-Microservice): Defines the *Locality constraints between IoT Devices and Ingress Microservices*, where Policy Data contains the *Locality Level*. This policy also has two endpoints exposing Policy Data (i.e., locality level for each ingress microservice) and Policy Evaluation calculating if the placement of an ingress microservice is eligible given the domain of the IoT device.

3. Locality Constraint Policy (Locality Type: Microservice-Microservice): Defines the *Locality constraints between communicating microservices*. This contains Policy Data denoting *Locality Levels* between consumer and consumed microservices. Policy Evaluation calculates if the placement of a consumed microservice is allowed given the domain of the consumer microservice.

Microservice Manager uses the policy data and evaluations retrieved through the Policy Handler during two main steps:

1. During the execution of the Management Algorithms: Microservice Manager uses Security Policy data/evaluations queried through the Policy Handler as input to Placement Algorithms, Scaling Algorithms and Migration Algorithms to determine where to deploy each microservice in a security policy-aware manner.

2. During microservice deployment using Deployment Resources: After receiving microservice to device mapping from the Management Algorithm, the Microservice Manager uses Security Policy data/evaluations queried through the Policy Handler to update Deployment Resources to enforce security policies (e.g., update routing rules among microservices by updating Virtual Services, Destination Rules of Istio service mesh). Afterwards, the Microservice Manager invokes the Deployment Handler to execute the updated Deployment Resources.

3.3   Framework Design

Based on the domain model, this section designs the architectural framework for the secure management of microservices within multi-domain edge-cloud environments using two views: 1) a component diagram depicting the modular architecture and 2) an interaction diagram depicting the dynamic behaviour.

Component Diagram: The architectural framework consists of three main layers: Infrastructure Layer, Supporting Services Layer and Control Plane (see Fig. 3).

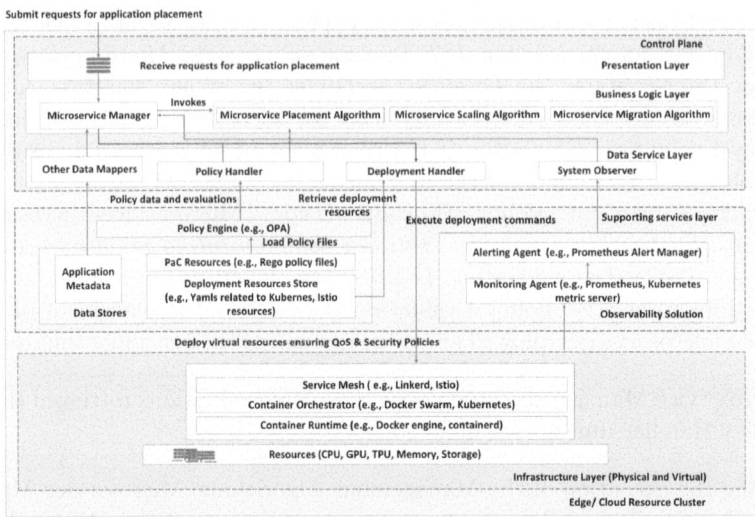

Fig. 3. Component Diagram visualizing the highlevel architecture of the framework

Infrastructure Layer provides the infrastructure-level support for running microservices as containers at a scale within distributed environments. This layer contains physical resources and virtual machines (e.g., CPU, GPU, TPU, Memory and Storage) as the underlying foundation with a Container Runtime (e.g., Docker engine[1], containerd[2], etc.) for running containers. Following the cloud-native practices, a Container Orchestrator (e.g., Docker Swarm[3], Kubernetes[4], etc.) manages the containers at a scale and a Service Mesh (e.g., Istio[5], Linkerd[6]) enables dynamic composition of containerised microservices across domains.

Supporting Service Layer consists of containerised *Observability Solution* and *Data Stores*, providing services to the Control Plane. *Data Stores* maintain data required by the Control Plane to make intelligent microservice management decisions. Data Stores include *Application Metadata* describing the application, *Deployment Resources*, which are the resource definitions used by Container Orchestrator and Service Mesh to deploy containerised microservices and *PaC Resources*, which are the security policy definition files related to each application. *Observability Solution* monitors the infrastructure layer through *Monitoring Agents* and generates alerts using an *Alerting Agent*.

---

[1] https://www.docker.com/.

[2] https://containerd.io/.

[3] https://docs.docker.com/get-started/swarm-deploy/.

[4] https://kubernetes.io/.

[5] https://istio.io/latest/.

[6] https://linkerd.io/.

Control Plane consists of three main layers as follows:

1. *Presentation Layer* provides the users access to the Control Plane through a REST API or a GUI to submit the application.
2. *Business Logic Layer* contains intelligence related to secure microservice management encapsulated in microservice placement, scaling and migration algorithms. Microservice Manager invokes these algorithms on two occasions: 1) When a request for an application placement is received through the presentation layer, 2) When the System Observer communicates the need for microservice scaling or migration. The Microservice Manager also implements the logic to update Deployment Resources according to the algorithm output and communicates with the Deployment Handler to retrieve and deploy the Deployment Resources. To enable the secure management of microservices, Microservice Manager and Management Algorithms obtain security policy-related data using the Policy Handler component in the Data Service Layer.
3. *Data Service Layer* is responsible for communicating with Supporting Services layer and Infrastructure layer. It acts as the abstract layer between the Business Logic Layer and lower layers, and consists of 4 main components: 1) Policy Handler responsible for loading policy files to the Policy Agent and querying the Policy Agent for security policies (i.e., policy data and evaluation results), 2) Deployment Handler responsible for retrieving deployment resource definitions from the Supporting Services Layer, updating them based on the input from the Microservice Manager and executing them on the Infrastructure Layer, 3) System Observer responsible for receiving alerts from Observability Solution and directing the relevant alerts to the Microservice Manager, 4) Other Data Mappers responsible for receiving Metadata used as input for management algorithms.

Interaction Diagram. Figure 4 depicts the ordered interactions among framework components, demonstrating the incorporation of security policies into the decision process.

3.4   Prototype Implementation

To demonstrate and evaluate the viability of the framework, we create a prototype by extending MicroFog [19]: an open-source platform containing a base architecture for distributed placement of microservices across multiple edge-cloud domains. MicroFog contains a distributed controller that communicates with Kubernetes (as Container Orchestrator) and Istio (as Service Mesh) to automatically deploy microservices as Docker containers (as Container Runtime). However, MicroFog focuses on satisfying QoS parameters during the initial placement and does not consider application security requirements. We utilised and extended Microfog components to implement the Supporting Service Layer and Control Plane of our architectural framework to enable secure management of microservices as follows:

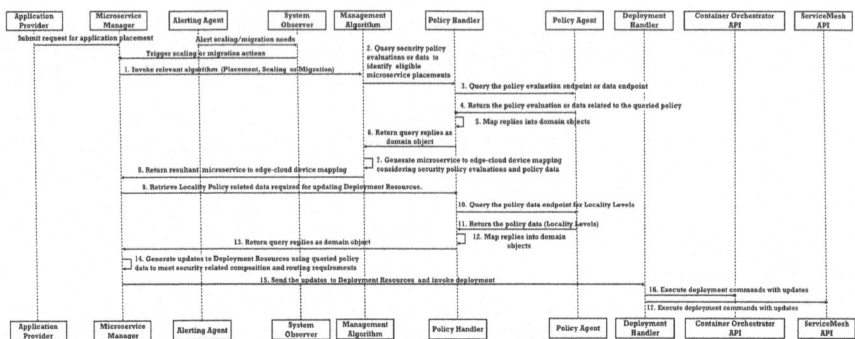

Fig. 4. Interaction Diagram

Supporting Services Layer Implementation -

Data Stores: MicroFog uses Redis[7] to store Application Metadata and MinIO[8] to store Yaml files of the Deployment Resources. To integrate PaC, we extend MinIO to store the PaC Resources (i.e., Policy-as-Code files). The deployment architecture of MinIO in MicroFog contains distributed deployment, replication and fault-tolerance, which matches the requirement of scalable management of policies in distributed environments. We select Rego[9], a declarative language supporting fine-grained policy declaration to define the policies.

Policy Agent: We integrate Open Policy Agent (OPA)[10], a cloud-native policy management tool which reads Rego policies and provides HTTP API endpoints for accessing policy data and evaluations. We deploy OPA as Kubernetes pods and expose it to Control Plane as Kubernetes and Istio services.

Observability Solution: Kubernetes Metric Server as the Monitoring Agent and a custom microservice as the Alerting Agent to retrieve metrics, generate alerts and communicate them to the System Observer.

Control Plane Implementation - We extend the *Control Engine* component of MicroFog to support the architecture of our proposed *Control Plane*. Micro-Fog *Control Engine* contains a data mapper Java object to access Application Metadata and a Deployment Handler object to communicate with the MinIO bucket to load, update and deploy Kubernetes and Istio resources. We utilize these and extend *Control Engine* by implementing Policy Handler and System Observer components to complete our proposed Data Service Layer.

Policy Handler: We implement the *Policy Handler* as a collection of Java objects that provides methods for querying the policy data and evaluations. Policy Handler uses a Reactive REST Client to send *asynchronous requests* to OPA as

---

[7] https://redis.io/.

[8] https://min.io/.

[9] https://www.openpolicyagent.org/docs/latest/.

[10] https://www.openpolicyagent.org/.

parallel queries to improve the performance during policy querying. Moreover, Policy Handler maps the replies received from the OPA into domain objects easily accessible to *Business Logic Layer*.

System Observer: System Observer contains a REST API endpoint to receive alerts from Alerting Agent. Upon receiving the alerts, System Observer directs them to the Microservice Manager to invoke scaling and migration actions.

Microservice Manager and Management Algorithms: We extend MicroFog to consider security policies during deployment algorithm execution and deployment resource updates. Microservice Manager instantiates *Policy Handler* component and loads Rego policies into the OPA for use during business logic execution. We implement the "PaC Integrated Microservice Placement Algorithm" (see Algorithm 1): a heuristic Placement Algorithm invoked by the Microservice Manager. The algorithm queries the loaded security policies through *Policy Handler* to identify eligible placement for microservices. Furthermore, we extend the Microservice Manager to utilize locality-related security policies to update Istio Virtual Services and Destination Rules to ensure that request routing among microservices adheres to locality levels among them.

Algorithm 1. PaC Integrated Microservice Placement

---

procedure PLACEAPPLICATION
    application ← received by Control Plane for placement
    # **Microservices that are placed within IoT devices**
    placedMicroservices ← placed on IoT devices
    # **Microservices that directly receives data/ requests from IoT devices**
    ingressMicroservices ← get ingress microservices of application
    # **Start from ingress microservices and traverse through the application graph iteratively**
    microservicesToPlace ← get eligible microservices for placement based on *placedMicroservices* and *ingressMicroservices*
    # **Sort based on the strictness of the locality constraints to ensure microservices with stringent locality constraints are prioritised**
    sortedMicroservicesToPlace ← sort microservicesToPlace by locality constraints
    while *sortedMicroservicesToPlace* **do**
        microservice ← first in *sortedMicroservicesToPlace*
        eligibleDomains ← get eligible domains based on **Placement Restriction Policy**
        if microservice is an ingress microservice **then**
            eligibleDomains ← refine *eligibleDomains* based on **Locality Constraint Policy (IoT-Microservice)**
        else
            eligibleDomains ← refine *eligibleDomains* based on **Locality Constraint Policy (Microservice-Microservice)**
        # **Device to microservice mapping used for deploying microservices to meet throughput expectation of the application**
        devicesToPlace ← select devices with sufficient resources from *eligibleDomains*
        Update *placementMapping*
        Update *placedMicroservices*
        microservicesToPlace ← update eligible microservices for placement based on *placedMicroservices* and *ingressMicroservices*
        sortedMicroservicesToPlace ← sort microservicesToPlace by locality constraints
    return *placementMapping*

Presentation Layer: MicroFog provides a REST API to submit application placement requests. We extend this API, to support multi-domain application placement by enabling the user to include details about expected request distribution across domains (i.e., throughput distribution).

4 Evaluation

We demonstrate the prototype's ability to meet the requirements identified in Sect. 3.1 which answers the two research questions explored in this work.

4.1   Experiment Setup

Multi-domain Edge-Cloud Infrastructure - Following the motivation scenario, we consider two edge computing clusters (representing the Region 2 in Fig. 1 with Edge Domain 3 and Edge Domain 4) and one cloud cluster (representing the Region 3 in Fig. 1). They are created as KinD[11] Kubernetes (containerised k8s) clusters communicating using Istio service mesh in multi-primary mode.

Deployed Application - We create the UAV Path finding application from Sect. 1.1 using the Workflow Generator from [19]. The security policy definitions are derived from the 3 example scenarios (Sect. 1.1) and mapped onto the 3 policy types defined in Domain Model (see Fig. 5)

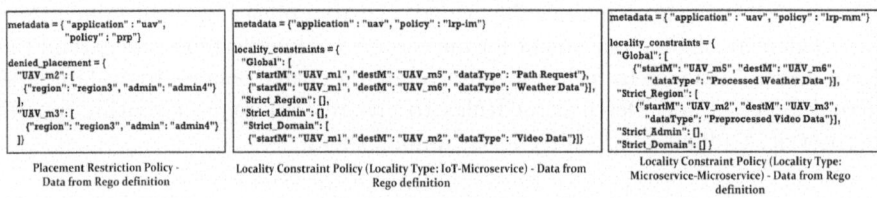

Fig. 5. UAV Path Finding Application - Security Policies

Microservice Management Scenario - Application providers of the UAV Path Finding Application submit a request to deploy the application across *Multi-domain Edge-Cloud Infrastructure*. The request is submitted with the expected throughput per edge domain. We consider a scenario similar to Sect. 1.1 where Edge Domain 4 expects twice the throughput as Edge Domain 3. Afterwards, Microservice Manager in the *Control Plane* guides the deployment of the microservices with updated YAML resources to ensure microservice placement and dynamic request routing satisfy the security policies.

4.2   Analysis of Results

After the control plane completes the application deployment, we evaluate the framework based on the two main management actions performed by the Microservice Manager to resolve our highlighted research questions: 1) microservice placement should satisfy the security policies while meeting the throughput requirements of the application. 2) dynamic request routing among distributed microservices across domains should satisfy the related security policies.

Figure 6(a) contains the Kubernetes API output listing the pods deployed within the UAV application namespace in each edge/cloud domain. Deployed locations of each microservice confirm that the *placementMapping* generated from the *Control Plane* satisfies the security policies defined in Fig. 5. Resultant

---

[11] https://kind.sigs.k8s.io/.

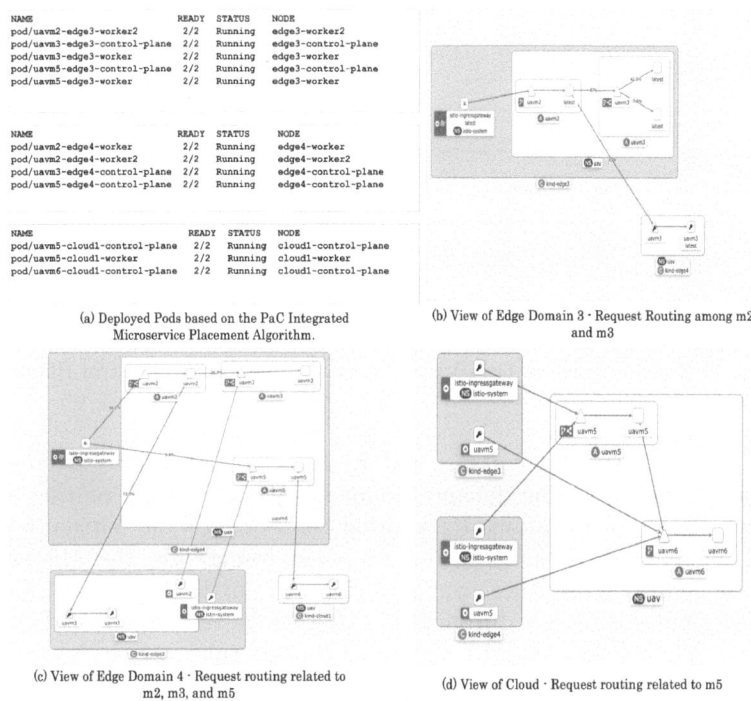

(a) Deployed Pods based on the PaC Integrated Microservice Placement Algorithm.

(b) View of Edge Domain 3 · Request Routing among m2 and m3

(c) View of Edge Domain 4 · Request routing related to m2, m3, and m5

(d) View of Cloud · Request routing related to m5

Fig. 6. Deployed pods and their composition across domains

placement satisfies the Placement Restriction Policy by deploying microservices *m2* and *m3* within Edge Domain 3, and Edge Domain 4. Edge Domain 4 has twice the amount of *m2* instances as Edge Domain 3 to meet the throughput requirement without violating the Locality Constraint Policy (Locality Type: IoT-Microservice). According to Locality Constraint Policy (Locality Type: Microservice-Microservice), *m3* is not limited strictly to the domain of *m2* and can be placed within the region. Thus, Edge Domain 3 and Edge Domain 4 host *m3* to collectively satisfy requests generated from *m2* instances (see Fig. 6 (b)). Hence, the pod deployment demonstrates the ability of the Control Plane to integrate PaC and intelligently utilize them to make secure microservice deployment decisions while satisfying QoS parameters like throughput.

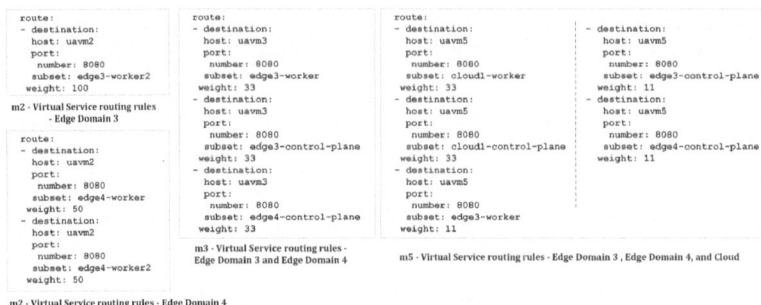

Fig. 7. Routing rules in Istio Virtual Services updated by Control Plane

Istio Virtual Service definitions and Kiali views of the inter-microservice traffic flow demonstrate the dynamic composition and request routing among microservices. Figure 7 contains the routing rules of the Istio Virtual Services deployed within each domain. In our proposed framework, the *Microservice Manager* generates these rules dynamically based on the Placement Mapping and Security policies retrieved from PaC. Afterwards, the Deployment Handler updates the Virtual Services with generated routing rules and deploys them in relevant domains. Figure 7 shows that the routing rules for *m2* differ based on the domain to ensure that the requests received for *m2* do not leave the domain and are served by *m2* instances in the same domain (i.e., ensuring Strict Domain Locality level). Compared to this, *m3* routing rules are the same for both Edge Domains 3 and 4, enabling load balancing among *m3* within the region (i.e., ensuring Strict Region Locality level). Due to the Global locality level, *m5* routing rules allow request routing among regions, covering both edge and cloud clusters.

Kiali views related to traffic routing further demonstrate the impact and validity of the updated routing rules. As Kiali cannot capture all traffic routing in multi-cluster scenarios with a single view, we present multiple views covering all 3 domains (see Fig. 6 - b,c,d). Views of Edge Domain 3 and Edge Domain 4 confirm that inter-domain request routing among edge clusters of the same region occurs for requests from *m2* to *m3*. Meanwhile, requests for *m2*, which are received through the Ingress Gateway of the domain, do not cross domain boundaries. View of the Cloud demonstrates that *m5* receives requests routed across regions which comply with the Global locality level.

Above results demonstrates framework's ability to meet requirements 1–3 of Sect. 3.1.

4.3 Discussion on Extensibility and Scalability of Policy Handling (requirements 4–5 from Section 3.1)

The Policy Handler in the Data Service Layer abstracts the policy querying process from the upper Business Logic Layer. Thus, the Policy Handler is responsible for converting domain objects to queries and replies from the Policy Agent to domain objects. Such layered architecture with separation of responsibility enables easy extension of existing security policies and adding new policy types. The use of a Policy Engine in the Supporting Services Layer ensures the scalable management and querying of the security policies. Being an integral part of the cloud ecosystem, Policy Engines such as OPA are designed with scalability in mind. Policy Engines have optimised query performance with concurrent query support and distributed deployment support to scale with the request load.

4.4   Discussion

Results demonstrate the ability of the framework to successfully integrate PaC-based application security definitions to an intelligent *Control Plane* for secure management of microservices within a multi-domain edge-cloud environment. PaC enables the flexibility for the *Control Plane* to retrieve and utilize policy data/ evaluations during different stages of the microservice management process (e.g., during the execution of *Microservice Management Algorithms* and updating Yaml resources related to microservice composition). The framework also demonstrated the ability to successfully handle multiple policy definition types with the use of *Policy Handler* component that abstracts the communication among Control Plane and *Policy Agent*. These observations and results solidify our framework's role as a foundational architectural blueprint for facilitating the development of platforms for secure microservice management, with inherent extensibility to enhance the Control Plane to handle edge-cloud use cases with diverse security requirements.

5   Conclusions and Future Work

In this paper, we present a framework for secure management of microservices during their deployment and dynamic composition within multi-domain edge-cloud environments. Our proposed framework uses PaC to define and manage granular microservice security policies. Policies are integrated into an intelligent control plane that utilises them to ensure security requirements. We created a prototype of the framework and evaluated its ability to achieve secure microservice management. The results validate the functionality of the framework, demonstrating its potential as a foundational architecture for facilitating the development of platforms for secure management of distributed microservices. In future work, we plan to improve the framework to handle security policy updates by integrating this framework into CI/CD pipelines.

Acknowledgments. This work has been supported by the Cyber Security Cooperative Research Centre Limited whose activities are partially funded by the Australian Government's Cooperative Research Centre Program.

Data Availibility Statement. As this project is funded by an industry partner, we are unable to publish the source code at this stage. To increase reproducibility, we extended an open-source framework along with open-source tools and explained the implementation of our framework in detail. High quality images of all figures used in the manuscript are available at https://doi.org/10.5281/zenodo.12524961.

References

1. Al-Doghman, F., Moustafa, N., Khalil, I., Tari, Z., Zomaya, A.: AI-enabled secure microservices in edge computing: opportunities and challenges. IEEE Trans. Serv. Comput. **16**, 1485–1504 (2022)
2. Atieh, A.T.: The next generation cloud technologies: a review on distributed cloud, fog and edge computing and their opportunities and challenges. ResearchBerg Rev. Sci. Technol. **1**(1), 1–15 (2021)
3. Chandramouli, R.: Implementation of DevSecOps for a microservices-based application with service mesh. NIST Spec. Publ. **800**, 204C (2022)
4. Deng, Q., Goudarzi, M., Buyya, R.: FogBus2: a lightweight and distributed container-based framework for integration of IoT-enabled systems with edge and cloud computing. In: Proceedings of the International Workshop on Big Data in Emergent Distributed Environments, pp. 1–8 (2021)
5. Ermolenko, D., Kilicheva, C., Muthanna, A., Khakimov, A.: Internet of things services orchestration framework based on kubernetes and edge computing. In: Proceedings of the 2021 IEEE Conference of Russian Young Researchers in Electrical and Electronic Engineering (ElConRus), pp. 12–17. IEEE (2021)
6. Farzin, P., Azizi, S., Shojafar, M., Rana, O., Singhal, M.: FLEX: a platform for scalable service placement in multi-fog and multi-cloud environments. In: Proceedings of the 2022 Australasian Computer Science Week, pp. 106–114 (2022)
7. Faticanti, F., De Pellegrini, F., Siracusa, D., Santoro, D., Cretti, S.: Throughput-aware partitioning and placement of applications in fog computing. IEEE Trans. Netw. Serv. Manage. **17**(4), 2436–2450 (2020)
8. Faticanti, F., Savi, M., De Pellegrini, F., Siracusa, D.: Locality-aware deployment of application microservices for multi-domain fog computing. Comput. Commun. **203**, 180–191 (2023)
9. https://fogatlas.fbk.eu/. Accessed January 2024
10. GitLab: What are the benefits of a microservices architecture? (2022). https://about.gitlab.com/blog/2022/09/29/what-are-the-benefits-of-a-microservices-architecture/
11. Guerrero, C., Lera, I., Juiz, C.: Evaluation and efficiency comparison of evolutionary algorithms for service placement optimization in fog architectures. Futur. Gener. Comput. Syst. **97**, 131–144 (2019)
12. Haumer, P., Pohl, K., Weidenhaupt, K.: Requirements elicitation and validation with real world scenes. IEEE Trans. Softw. Eng. **24**(12), 1036–1054 (1998)
13. IDC: Future of industry ecosystems: Shared data and insights (2021). https://blogs.idc.com/2021/01/06/future-of-industry-ecosystems-shared-data-and-insights/
14. Mahmud, R., Toosi, A.N.: Con-Pi: a distributed container-based edge and fog computing framework. IEEE Internet Things J. **9**(6), 4125–4138 (2021)
15. Marchese, A., Tomarchio, O.: Sophos: a framework for application orchestration in the cloud-to-edge continuum. In: Proceedings of the 13th International Conference on Cloud Computing and Services Science, CLOSER 2023, pp. 261–268 (2023)

16. Miller, L., Mérindol, P., Gallais, A., Pelsser, C.: Towards secure and leak-free work-flows using microservice isolation. In: 2021 IEEE 22nd International Conference on High Performance Switching and Routing (HPSR), pp. 1–5. IEEE (2021)
17. Pahl, C., El Ioini, N., Helmer, S., Lee, B.: An architecture pattern for trusted orchestration in IoT edge clouds. In: 2018 Third International Conference on Fog and Mobile Edge Computing (FMEC), pp. 63–70. IEEE (2018)
18. Pallewatta, S., Kostakos, V., Buyya, R.: Placement of microservices-based IoT applications in fog computing: a taxonomy and future directions. ACM Comput. Surv. **55**, 1–43 (2023)
19. Pallewatta, S., Kostakos, V., Buyya, R.: MicroFog: a framework for scalable place-ment of microservices-based IoT applications in federated fog environments. J. Syst. Softw. **209**, 111910 (2024)
20. Santoro, D., Zozin, D., Pizzolli, D., De Pellegrini, F., Cretti, S.: Foggy: a platform for workload orchestration in a fog computing environment. In: Proceedings of the 2017 IEEE International Conference on Cloud Computing Technology and Science (CloudCom), pp. 231–234. IEEE (2017)
21. Tuli, S., Mahmud, R., Tuli, S., Buyya, R.: FogBus: a blockchain-based lightweight framework for edge and fog computing. J. Syst. Softw. **154**, 22–36 (2019)
22. Varadharajan, V., Bansal, S.: Data security and privacy in the internet of things (IoT) environment. In: Mahmood, Z. (ed.) Connectivity Frameworks for Smart Devices. CCN, pp. 261–281. Springer, Cham (2016). https://doi.org/10.1007/978-3-319-33124-9_11
23. Wang, Z., Goudarzi, M., Aryal, J., Buyya, R.: Container orchestration in edge and fog computing environments for real-time IoT applications. In: Buyya, R., Hernandez, S.M., Kovvur, R.M.R., Sarma, T.H. (eds.) Computational Intelligence and Data Analytics. LNDECT, vol. 142, pp. 1–21. Springer, Singapore (2022). https://doi.org/10.1007/978-981-19-3391-2_1
24. Xiong, J., et al.: Enhancing privacy and availability for data clustering in intelligent electrical service of IoT. IEEE Internet Things J. **6**(2), 1530–1540 (2018)
25. Yousefpour, A., et al.: FOGPLAN: a lightweight QoS-aware dynamic fog service provisioning framework. IEEE Internet Things J. **6**(3), 5080–5096 (2019)

Electric Vehicle Fast-Charging Software: Architectural Considerations Towards Trustworthiness

Vick Dini$^{(\boxtimes)}$, Damian Andrew Tamburri, and Elisabetta Di Nitto

Politecnico di Milano, DEIB, Via Golgi 42, Milan, Italy
{vickpierce.dini,damianandrew.tamburri,elisabetta.dinitto}@polimi.it
https://www.deepse.deib.polimi.it

Abstract. The industry of electric vehicle charging stations is rapidly evolving. As more people purchase electric vehicles, the need for an efficient and reliable charging service value chain becomes paramount. In this paper we perform a literature review and topic modeling in the field to identify the most relevant topics and challenges therein as well as any standard or reference software architectures proposed to date. Subsequently, by means of ethnomethodological research and interview study with industry practitioners from an emerging Charging Point Operator (CPO) we identify major practical architectural concerns. We conclude that typical -Ops issues related to monitoring and autonomous runtime management of fast-charging software architectures are not addressed in the literature so far, and yet are the most critical for practitioners; this inspired a revised augment of state-of-the-art reference architecture frameworks—offered as the first key original contribution of this paper— for practitioners and researchers to use in the future. Finally, evaluating such framework with practitioners' insights, we elaborated a second key contribution of this work, namely, six architectural considerations to be taken into account towards trustworthy electric vehicle fast-charging software architectures of the future.

Keywords: charging station · cloud-edge computing · electric vehicle · fast charging · software architecture · software engineering

1 Introduction

Nowadays, electric vehicles (EVs) are considered as one of the main solutions to reduce the carbon footprint in highly urbanized areas in the most industrialized countries [1]. However, a critical requirement for their diffusion is the availability of fast charging services [10].

Offering such services on a large scale requires several issues to be addressed, such as the need to: (a) install the EV charging stations (EVCSs) in strategic locations, which require planning and management software services close to the demand and to electrical power sources [38]; (b) fulfill an increasing customer demand—39.5% YOY [12], projected revenue of US$623.3bn in the US

© The Author(s), under exclusive license to Springer Nature Switzerland AG 2024
M. Galster et al. (Eds.): ECSA 2024, LNCS 14889, pp. 288–304, 2024.
https://doi.org/10.1007/978-3-031-70797-1_19

in 2024 [30]—keeping into account the limitations of the electrical grid [26]; (c) support all the EV charging standards as well as the interaction protocols with multiple stakeholders, e.g. vehicle owners and manufacturers, energy distributors (DSOs), and e-mobility service providers (eMSPs) [9].

Several big players of the EV industry are developing proprietary software systems capable of addressing such challenges. For example, Tesla not only covers the electric vehicle production chain, but also creates and manages EV charging stations to provide full support to its customers [25]. Moreover, multiple companies, including Charging Point Operators (CPOs), are backing the standardization of communication protocols (e.g. OCPP, OCPI, OSCP, and ISO 15118) and software solutions, such as the Charge Point Management System (CPMS) and Energy Management System (EMS), that address the most critical aspects concerning the interoperability and efficiency of the electric vehicle charging infrastructure. This concerted effort aims to streamline the deployment and operation of charging stations across diverse platforms.

In particular, CPOs are highly relevant to our research, as they are the *last mile* of the smart grid with respect to EVs and are thus crucial to the extension of the EVCS network and consequently to the generalized adoption of EVs.

However, through a pilot study of the literature, we report that the development of a comprehensive and open architecture framework capable of addressing all aspects of EVCS management and ensuring trustworthy, orchestrated cooperation between different stakeholders and their software assets is still missing. Thus, we defined the following research questions:

- RQ1: What is the state of the art in software architectures for EVCS?
- RQ2: Is research in this context focusing on trustworthy DevOps?
- RQ3: What are the main problems perceived by EVCS operators (CPOs)?

In addition to providing answers to these questions (RQ1 in Sect. 3, RQ2 in Sect. 4, and RQ3 in Sect. 5), our objective in this paper is to contribute to the aforementioned definition by looking into what considerations shall be made from an architectural perspective over such expected research results.

2   Research Methodology

The research methodology we have adopted is based on the three steps described in the following subsections.

2.1   Literature Analysis Approach

We performed a literature review to answer RQ1 and identify the main research topics in the EVCS state of the art. Our working hypothesis is that work has already been done on trustworthy DevOps for EVCS. To elaborate on such a hypothesis, we adhere to guidelines specifically recommended for information systems research by [32]. To attain our primary studies, we ran the following query on the well-known Scopus search engine[1] on April 3rd 2024 and obtained 4,544 results in computer science.

electric AND vehicle AND charging AND stations

We did not include the keywords *software architecture* in the query because we also wanted to identify the trending topics in computer science. Following the aforementioned recommendations, we excluded the articles written in languages other than English (-148) and kept only scientific papers (-77). Then, we manually excluded the results whose titles are not related to stationary EVCS software, control, or energy management systems (-4,245). From the remaining 74 articles we excluded 9 which were inaccessible and applied the Latent Dirichlet Allocation (LDA) [4] on the remaining ones. This allowed us to identify the most relevant topics in this area. We then selected the 14 papers related to software architecture applied to the EVCS industry for further analysis.

2.2 Architectural Analysis

To address RQ2, we used literature review sources to distill an architecture overview for EVCS in industry with a twofold approach. On the one hand, we intended to distill an architecture overview which explicitly addressed trustworthiness as a primary architecture asset while, on the other hand, we aimed for such an overview to be eventually validated with the aid of practitioners involved in the matter. Even though multiple architectures for the EVCS industry have been created over time, we indeed found that none of them was designed starting from a hierarchical EVCS control model (such as the one proposed by [37]) and taking into account software feature/component deployment and orchestration, and the respective flexibility and resilience thereby obtained [13].

2.3 Distilling Architectural Considerations

To address RQ3, we conducted an industrial confirmation study featuring a six-month ethnographic study [34], which consisted of four steps: (a) daily participation in activities of our target company; (b) four unstructured interviews [6] to gauge current architectural status; (c) two semi-structured interviews [6] to understand current challenges; (d) an interview analysis [18] to distill a taxonomy of concepts. An architecture refinement approach, as well as problem detection and cause identification [36], was employed to model current baselines, together with a card-sorting exercise [39][2]. This approach allowed us to distill six key architectural considerations to be made when reflecting on the trustworthiness of electric vehicle fast-charging software; these considerations, grouped into four

---

[1] https://www.scopus.com/.
[2] For confidentiality, we excluded the generated artifacts from this paper.

categories: (1) EVCS management, (2) Data Management, (3) Service Continuity, (4) IoT & Electronics, are intended as design principles to drive architecture decision-making, as well as constraints to orchestrate architectures and their self-adaptation.

3   Electric Vehicle Charging Stations Overview

To analyse our primary literature studies, we applied the well-known LDA topic modeling approach [4] to the text of the 65 accessible articles and removed the keywords included in the query, as well as adjectives, adverbs, and verbs, and considered the top 5 words in each cluster to obtain the topics in Table 1.

Table 1. LDA Topics.

#	Topic	% of tokens	Top 5 words
1	User management	20.3	system, user, datum, service, reservation
2	Power management	18.3	power, converter, battery, voltage, system
3	System control	14.7	power, system, energy, battery, control
4	Edge scheduling	12.5	power, energy, time, datum, node
5	Event management	12.4	time, energy, event, user, system
6	Application data management	12.0	datum, system, user, application, service
7	Mobility data management	4.3	network, datum, mobility, service, node
8	Edge control	3.6	terminal, station, microcontroller, time, process
9	Energy forecast	1.8	time, energy, battery, source, traffic

We gave these topics their names by reflecting on feasible relationships among the terms associated with, or contained in each cluster. We further refined the topic names by modifying the values of the relevance metric (λ) parameter between 1.0 and 0.5, while observing the changes in the order of the top 5 words. In Table 1 the column about the percentage of tokens indicates the size of each cluster, which may also be interpreted as the frequency and relevance of each topic with respect to the contents of the papers under analysis. The intertopic distance map and ordered list of most salient terms are illustrated in Fig. 1. Additionally, according to the location of the clusters on the principal components (i.e. PC1 and PC2), we named each of the quadrants coherently as follows: Q1: Event management and power control; Q2: User management; Q3: Application and scheduling data management; Q4: Edge control and forecasting.

Following a snowballing approach [35], we added a group of publications about topics other than software to the set of articles under analysis, in addition to key standards in the EVCS context: (a) the Open Charge Alliance[3] (OCPP and OSCP), (b) EVRoaming Foundation[4] (OCPI), and (c) International Standards Organization (ISO 15118[5]), as shown in Table 2.

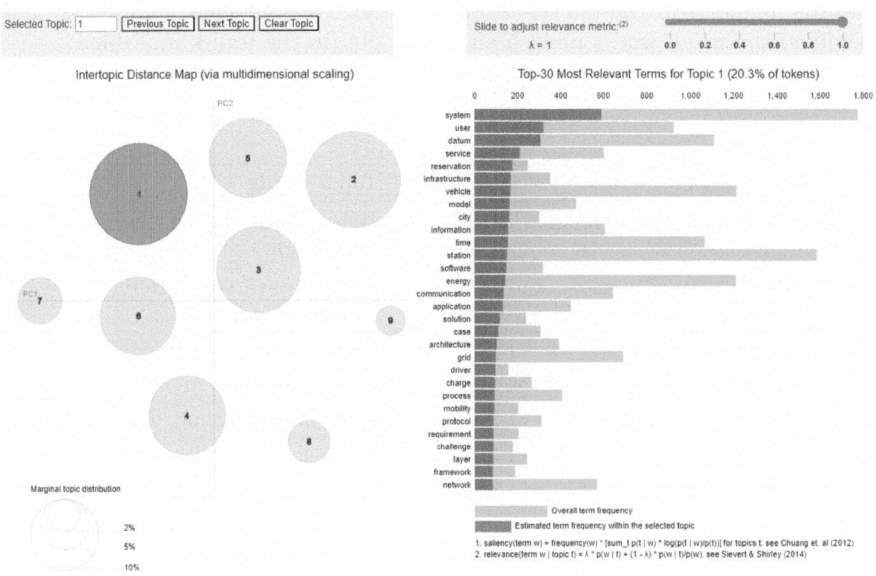

Fig. 1. Intertopic Distance Map & Most Salient Terms.

3.1   Main Stakeholders, Components, and Protocols

By analyzing the standards, we gained an overview of the key players in EVCSs and a top level perspective of the organization of each station. Figure 2 illustrates the scenario of a CPO with two EVCSs in different locations. Each station may include several charging points (EVSE: Electric Vehicle Supply Equipment), batteries to accumulate and distribute energy as needed (managed by a BESS: Battery Energy Storage System), other sources of energy (e.g. solar panels), smart meters that connect the station to the external grid, and an Energy Management

---

[3] Open Charge Alliance: https://openchargealliance.org/.
[4] EVRoaming Foundation: https://evroaming.org/.
[5] ISO 15118: https://www.iso.org/standard/77845.html.

System (EMS) in charge of controlling the energy exchange among the different sources (e.g. the external grid, batteries, and solar panels) and consumers. The CPO usually relies on a CPMS, which acts as an intermediary between EVCSs and eMSPs, enabling the CPO to communicate with its customers. The CPO also interacts with the DSOs, which are intermediaries in the energy supply market. The interactions among parties occur through the specific protocols highlighted in Fig. 2.

Table 2. Parties and Associated Interaction Protocols.

Parties	Protocol	Acronym
eMSP - CPO/CPMS	Open Charge Point Interface	OCPI
CPO - DSO	Open Smart Charging Protocol	OSCP
CPO/CPMS - CPO/EVSE	Open Charge Point Protocol	OCPP
EV - EVSE	Road vehicles – Vehicle to grid communication interface	ISO 15118

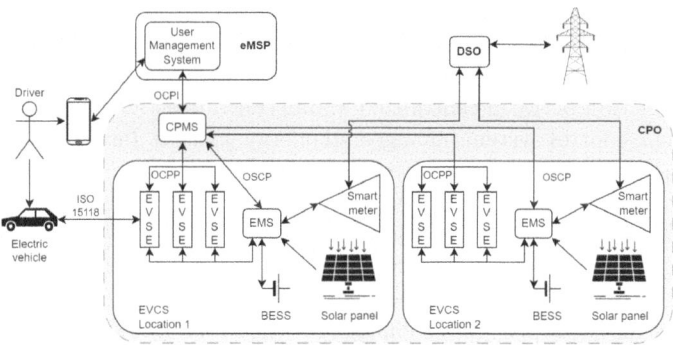

Fig. 2. EVCS Communications and Systems.

3.2 Critical EMS Service Functions

A general overview of EVCSs from an energy management perspective is provided by Wu et al. in [37]. Here the authors highlight the hierarchical EVCS control structure, present a list of the main standards and protocols used in this industry, and mention several features that a software system should include to support the EVCS control system, such as the implementation of strategies for energy allocation, trading, and demand forecasting.

The prominent role of EMSs in EVCSs is highlighted in multiple publications. [29] states that EMSs "are automation systems that collect energy measurement data from the field and make it available to users through graphics, online monitoring tools, and energy quality analyzers, thus enabling the management of energy resources." Specifically, according to [3], their features could include:

- Power Limitation: The EMS can receive power limitation signals and adjust the energy flow of connected devices, including EVSEs, accordingly. This ensures compliance with physical and grid-initiated constraints at the grid connection point, supporting stable local grid operation.
- Dynamic Pricing: The EMS can receive the rates from the DSO and adjust the overall power consumption of connected devices.
- Load Management: The EMS can safely limit the overall EV power draw on the system so that it does not exceed the available capacity. It can also efficiently balance and distribute charging power to consumers.
- EV Charging: The EVSE or EV submits its energy demand and estimated departure time to the EMS. If supported, the end user could choose CO_2 friendly charging and the EMS would take care of charging with a higher portion of solar energy, when available.

3.3   EMS Architecture

As of the EMS's architecture, [28] highlights that it is "based on different layers including management, data communication, vehicle control, electric transmission, and electric backup and storage modules." Moreover Torreglosa et al. [31] make a distinction between centralized and decentralized EMSs. In a centralized EMS, a single control system manages all energy sources. Integrating additional sources requires modifications to the EMS and installation of a communication infrastructure between the new sources and the centralized system. Instead, in a decentralized EMS, each energy source works independently, managing energy flow among the system's sources without requiring communication interfaces between sources or with a supervisory energy system.

More recently, [22] proposed a charging station management system featuring decentralized, fair, and impartial energy trading via a blockchain distributed ledger and smart contracts; as of the system's control hierarchy, the authors placed centralized control at the station level and decentralized control on the EVSEs, while the cloud server is mostly dedicated to the energy trading platform.

3.4   EVCS Service Orchestration and Monitoring

In [27] the authors present a framework for the interoperation and integration of heterogeneous systems in the industrial automation domain and evaluate its application to the EVCS industry through a case study, although it must be noted that their focus is on service orchestration (on "local clouds") and IoT, and that by service orchestrator they intend a service dispatcher or router. Later on, [21] developed a cloud-IoT monitoring system for solar panel stations, which they described as "robust, reliable, workable, and suitable."

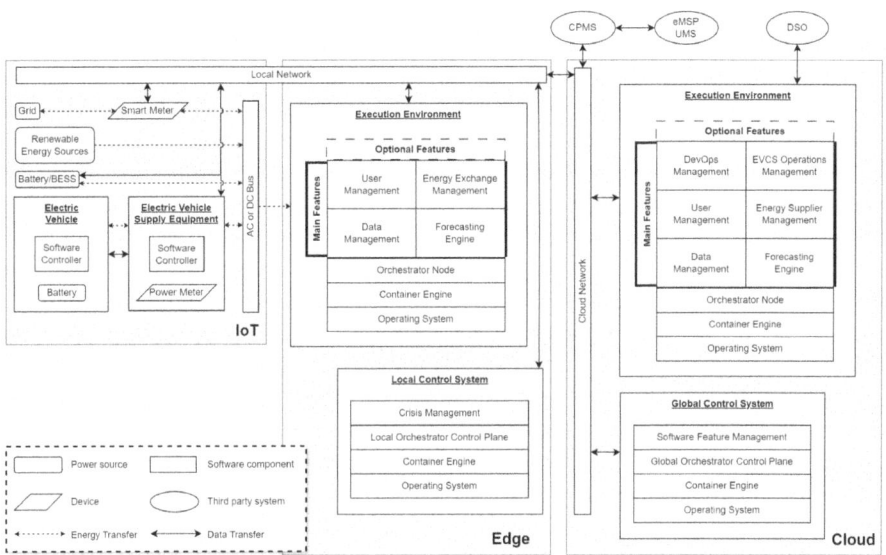

Fig. 3. Proposed Distributed Cloud-Edge Fast Charging Architecture Framework.

3.5 EMS Control Software Architecture

Some papers that emerged from the LR proposed software architectures that are similar to the OSI model, while others implemented variants of the client/server model. [33] designed an architecture comparable to the cloud-edge-IoT model. Among the others, we consider [19] to be a key paper in this research because its authors applied the Smart Grid Architecture Model (SGAM) to the e-mobility industry, focusing on interoperability and complexity management.

4 EVCS: A Revised Architecture Framework

Taking into account the findings of our literature review, we designed a distributed cloud-edge architecture framework (AF) (Fig. 3) for the fast charging domain, in which we mapped the topics that surfaced from the LDA as software components (Table 3) and allocated them in the respective execution environments. This AF should support the computing continuum and interactions from several stakeholders (multiple ownership), in addition to extending the stations' service continuity by making them more autonomous in local decision making.

Optimal EVCS management is of critical importance to EV drivers, especially those who urgently need to charge their vehicles and do not have any options nearby. In the following subsections we delve into the components of our proposed solution.

Internet of Things. From left to right, the EVCS architecture starts with the power sources, which may be renewable or fossil-based. The former refers to the power grid (to which the station connects via a smart meter) [11], whereas the latter includes wind turbines [37], train generative breaking [17], and solar panels. It should be noted that some or all of these renewable sources might not be available in all EVCSs. Regarding energy storage, EVCSs could house batteries (with their respective BESS) and electric vehicles contain their own battery cells, which, if bidirectional energy transfer is supported, may store and return energy to a common bus. In fact, all of these sources are connected to a common AC or DC bus, which provides energy to charging points and electric vehicles (EVs) [37]. As of the latter, their batteries (controlled by the BMS: battery management system) provide energy to vehicle devices and motors, and the respective processes are managed by a software controller. Charging points have microcontrollers [5] that manage their operation and react to user input and power meter data, in addition to receiving settings and transmitting data about a charging point's operation [14]. In particular, the smart meter, BESS, and EVSE are connected to the station's local network.

Table 3. Topic-Component Mapping.

Topic	Component
User Management	User Management
Power Management	Energy Exchange Management, Energy Supplier Management
System Control	EVCS Operations Management, DevOps Management
Edge Scheduling	User Management, Forecasting Engine
Event Management	Crisis Management, EVCS Operations Management
Application Data Management	Data Management, DevOps Management
Mobility Data Management	Data Management, User Management
Edge Control	Energy Exchange Management, EVCS Operations Management
Energy Forecast	Forecasting Engine

Control Systems. EVCS systems must be kept under continuous control to ensure safety of users and continuity of service. Given the presence in EVCSs of the three contexts, IoT and edge at the periphery and cloud at the center, we deem the three-level control hierarchy presented in [37] particularly relevant for the EVCS industry. We could consider the primary control level to be in charge of overseeing EVSEs, whereas the secondary control level manages EVCS devices and, in turn, communicates with the tertiary control level located in the cloud. Also, there is a bidirectional data flow that starts in the cloud, passes through the EVCS controllers, and reaches the EVSEs, and vice versa. Building on this idea, the tertiary control level should be deployed to execution environments on the cloud to better exercise the global (centralized) monitoring and control of the EVCS microgrid via industrial and single-board computers, which lie at the secondary control level and perform local (decentralized) tasks. The primary control level should manage the digital electronics of each EVCS.

Such multi-layer control system is primarily in charge of guaranteeing EVCS architectures *trustworthiness*, defined as the combination of a system's reliability and recoverability as they are described in the ISO 25010 standard[6]:

Reliability: *Capability of a product to perform specified functions under specified conditions for a specified period of time without interruptions and failures.*

Recoverability: *Capability of a product in the event of an interruption or a failure to recover the data directly affected and re-establish the desired state of the system.*

We deem that by placing container orchestrators both locally (edge) and globally (cloud) we can increase the trustworthiness of a distributed architecture, because orchestrators offer benefits such as lower mean time to repair (MTTR) and higher availability by detecting and correcting issues automatically. They improve infrastructure performance through efficient resource management and workload distribution and automate container management tasks like deployment, scaling, and operations. Containers package applications with their dependencies, ensuring consistent functionality and reducing overhead by providing a tailored minimalistic execution environment. They streamline continuous deployment through automated testing, integration, and deployment. Containers enhance security and stability by isolating applications from the underlying OS and other applications. They also enable faster deployment times and more efficient resource utilization compared to traditional virtual machines.

Execution Environments and Functional Components. We placed execution environments at the cloud and edge to run the components that emerged from our literature review and others that may arise over time. We therefore analyzed several aspects of these components to discern the most appropriate execution environment(s) for each of them, resulting in Table 4.

These components may be translated to microservices, which are "small applications with a single responsibility that can be deployed, scaled, and tested

---

[6] ISO 25010: https://www.iso.org/standard/78176.html.

Table 4. Key Architectural Drivers for EVCS: A Comparative Analysis.

Component	Cloud (Global)	Edge (Local)	Resource Intensive	Continuous Execution	Execution Environment Specific	Industry Specific
Data Management	✓	✓	✓			
Forecasting Engine	✓	✓	✓			
User Mgmt	✓	✓				
Energy Exchange Mgmt		✓		✓	✓	✓
Energy Supplier Mgmt	✓				✓	✓
DevOps Mgmt	✓			✓	✓	
EVCS Operations Mgmt	✓			✓	✓	✓
Crisis Mgmt		✓		✓	✓	

independently" [20], to accommodate new features, address emerging challenges, and optimize performance without compromising reliability. However, designing an architecture based on this principle has several challenges, such as lower performance (vs. monolithic applications), the increased complexity in monitoring operations and debugging code, and how to guarantee data consistency [23]. Nevertheless, applications designed for our architecture framework could dynamically scale by leveraging the expansive capabilities of the cloud [24], while ensuring that latency-sensitive tasks are executed at the edge [8]. This dynamic scalability accommodates varying workloads and provides a cost-effective solution by optimizing resource allocation [16].

As of their functionalities, the energy exchange and energy supplier management components should focus on automated dynamic energy procurement and sales among DSOs, CPOs, and drivers, in addition to adaptive energy storage and allocation, depending on changing market conditions. Similarly, the DevOps and EVCS operations management components should cover the DevOps life cycle and the centralized monitoring and control of charging stations.

Meanwhile, we placed the forecasting engine and the data and user management components in both the cloud and edge execution environments, because they could offer functionalities that are useful, and even necessary, at both ends, considering the limitations of the underlying hardware.

Finally, we designed the crisis management component to handle all unforeseen adversities and events at the stations and guarantee a certain level of service continuity while the system recovers or adapts to the new circumstances.

5   Trustworthy EVCS: Architectural Considerations

5.1   Ethnographic Study About the Proposed Architecture

For industrial grounding and confirmation of the aforementioned architecture framework, we interviewed practitioners who work for a CPO that is a key player in the Southern European market of fast-charging stations, powered by renewables, energy storage, and vehicle-grid integration. They intend to deploy

5,000 remotely managed fast charging points by 2025 and more than 35,000 by 2030. Additionally, they outsource certain services and systems to focus on their core business. To delve into the architectural design, implementation, and maintenance of an EMS infrastructure, we carried out four unstructured interviews at the company, each one lasting one work day, which helped us perform an initial validation of our theoretical EVCS model. Then, we held two semi-structured interviews focused on the software aspects of EVCSs and problems in this industry that might be solved with software. Next, we analyzed the interviews by reviewing the recordings and transcripts through content analysis. Afterwards, we categorized the synthesized data according to each question and produced conceptualizations to ground the software architecture to the industry. Finally, we elicited the most significant problems by reviewing the interviews and diagrams with the practitioners. Specifically, a thematic analysis [18] of the interview data revealed recurring challenges and issues raised by participants, providing qualitative insights into critical areas of concern in this industry. These are discussed in the following subsections.

5.2 EVCS Software Architecture Management

Neither the state of the art nor the insights from practitioner interviews provide clear guidelines on how to divide EVCS management software into components and which execution environment is more convenient to converge towards a computing continuum. Yet, this is reported as a critical aspect for the stations to function in a highly autonomous fashion. The following consideration is made:

1: Computing Continuum of EVCSs is a First-Class Citizen.
The plurality and heterogeneity of devices and stakeholders that compose and interact with an EVCS lead to a greater relevance of the continuum in this industry. It is thus crucial to determine which features shall be allocated to which layer of the computing continuum.

Subsequently, we report emerging needs for the underlying software architecture to offer a high degree of flexibility. For example, the deployment of updates to the industrial computers that control each EVCS involves a certain degree of complexity. These devices may run a variety of component containers, serverless functions, native OS programs, or even virtual machines, and thus require architectural flexibility based on the capabilities of each computer and its components. Additionally, in such a continuously evolving industry, the underlying technologies could change anytime. The following consideration therefore stands:

2: DevOps-oriented Architecture Flexibility.
Metrics shall be defined to measure the extent of flexibility (e.g. modifiability analysis [2]) of EVCS architectures and how they exert continuity over the

---

surrounding DevOps pipelines. The use of abstraction layers should help increase system flexibility[7].

Finally, the absence of established guidelines, standards, or universally acknowledged best practices in this realm increases the degree of complexity of the architectural decision-making process.

5.3 Data Management

Regarding data management, the interviews helped us quantify the data volume transmitted and received daily by each EVCS ($\sim$100 MB). This empirical data is significant for further (e.g., dataflow) architecture-based experimentation given its absence within existing literature and the related lack of ad-hoc experimentation on the matter. The following consideration is made:

3: Controlled Architectural Experimentation Shall Reflect Dataflow Modelling & Analysis.
Many approaches that feature dataflow modelling and analysis may be leveraged to better define the constraints and execution policies under which EVCS management design can be performed.

At the same time, disparities between the data stored in the cloud and the edge can emerge, resulting in *data divergence* [15]. This phenomenon introduces a significant challenge for architectural decision-making processes, as insights derived from cloud-based data may no longer accurately reflect the real-time conditions and dynamics occurring at the edge and vice versa. Consequently, decisions based on outdated or divergent data may prove inadequate or even counterproductive, leading to sub-optimal outcomes and operational inefficiencies. The following consideration is made:

4: EVCS Architecture Decision-Making is a Continuous Architecting Exercise.
Fostering closer alignment between cloud and edge data entails leveraging advanced analytics and machine learning algorithms to detect and reconcile discrepancies in architectural decision-making in real-time, reflecting a continuity in such decisions; new adaptation mechanisms shall hence assume such decision continuity and operate decision-making in account of it.

Thus, organizations shall proactively identify cases of data discrepancy and implement corrective measures to re-align cloud and edge data streams, e.g. by continuously monitoring data integrity and consistency across distributed nodes.

5.4   Service Continuity

During the interviews, the participants elucidated various factors contributing to service interruptions within the EVCS infrastructure, including that:

Fault logging is a whole issue on its own, because typically when the system fails, it either brings down components or even the entire site [i.e. EVCS]. *Therefore, understanding how to ensure the collection and retention of information, and the ability to recover when the site is down can save you from physically going there to conduct diagnostics.*

Hence, the potential for industrial hardware to experience system halts or encounter bugs within its software is a notable concern, as it may result in operational disruptions, especially because usually no one is present at the EVCSs. In the event of a system hang-up in a containerized environment, an orchestrator mechanism can swiftly detect and rectify the issue by initiating the deployment of a new container while decommissioning the malfunctioning one. Similarly, in the case of bugs affecting serverless functions, the system's automated processes can seamlessly execute error-handling protocols to ensure uninterrupted service delivery. Thus, the following consideration is made:

5: EVCS Architecture Management is an Orchestration-centric Exercise.

By adopting containerization or serverless architectures, EVCS operators can enhance system robustness and minimize the likelihood of prolonged service disruptions. These deployment strategies not only facilitate rapid fault recovery but also afford greater scalability and agility in adapting to evolving operational requirements.

The interviewees also highlighted the risk of communication disruptions in charging stations and charging points. This stresses the criticality of maintaining seamless connectivity to ensure uninterrupted service delivery. Therefore, future experimentation must be directed towards implementing robust redundancy measures over said orchestration exercises to mitigate the impact of potential communication misalignment or other failures.

5.5   IoT and Electronics

The practitioners gave particular importance to phenomena concerning charging-points malfunctioning due to firmware bugs. Unlike industrial computers, IoT devices typically lack the resources to run containerized environments or lambda functions. However, the imperative to swiftly detect and address charging point failures is crucial in ensuring seamless service continuity and enhancing user experience. In light of this challenge, the following consideration is made:

> **6: EVCS Architectures Require Fault Detection by Design.**
> Proactive measures such as automated system alerts and remote troubleshooting protocols enable operators to expedite the resolution of firmware-related issues, thus maximizing service uptime and availability.

While IoT devices may not inherently support containerization or serverless computing paradigms, alternative architecture approaches—and the connected fault-tolerance analysis—can be employed to mitigate the impact of firmware bugs on EVCS operations. For instance, leveraging edge computing capabilities, such as edge analytics and edge processing, can empower IoT devices to perform localized fault detection and remediation tasks autonomously. By distributing computational resources closer to the point of data generation, edge computing frameworks expedite response times and reduce reliance on centralized infrastructure for critical decision-making processes.

6   Conclusion

We consider this paper to be a preliminary document on software architecture principles applied to the electric vehicle charging stations industry, upon which we expect more authors to elaborate and develop ad hoc solutions. Throughout its sections, we applied qualitative research methods to gather relevant information that helped us model a cloud-edge fast charging architecture framework and identify current challenges in the industry, thereby grounding the theory to the practice and with hopes of developing an industry-actionable solution. In addition to providing answers to the research questions (RQ1 in Sect. 3, RQ2 in Sect. 4, and RQ3 in Sect. 5), we hope to have contributed to the scientific community by delving into important aspects of EVCS management from a software architecture perspective. In the future we plan to follow up on our considerations with an experimental approach, in search of quantitative guidelines to compound each consideration.

Data Availability and Acknowledgements. The Scopus search results, the filtered articles, the LDA Jupyter notebook, and the interview questionnaire are openly available on our online repository (https://zenodo.org/doi/10.5281/zenodo.11060482). This work is partially supported by the Italian Ministry of University and Research under the PNRR program (financed by the EU, NextGenerationEU), Ministerial Decree n. 352, and Atlante SRL.

References

1. Asimow, N.G., Turner, A.J., Cohen, R.C.: Sustained reductions of bay area CO2 emissions 2018–2022. Environ. Sci. Technol. **58**, 6586–6594 (2024)
2. Bengtsson, P., Lassing, N., Bosch, J., van Vliet, H.: Architecture-level modifiability analysis (ALMA). J. Syst. Softw. **69**(1–2), 129–147 (2004)

3. Bhattacharjee, S., Nandi, C.: Advanced energy management system (A-EMS) design of a grid-integrated hybrid system. Iran J. Sci. Technol. Trans. Electr. Eng. **47**(3), 1021–1044 (2023)

4. Blei, D.M., Ng, A.Y., Jordan, M.I.: Latent Dirichlet allocation. J. Mach. Learn. Res. **3**, 993–1022 (2003)

5. Bose, P., Sivraj, P.: Smart charging infrastructure for electric vehicles in a charging station. In: ICICCS, pp. 186–192. IEEE (2020)

6. Brinkmann, S.: Unstructured and semi-structured interviewing. Oxf. Handb. Qual. Res. **2**, 277–299 (2014)

7. Broniatowski, D.A.: Flexibility due to abstraction and decomposition. Syst. Eng. **20**(2), 98–117 (2017)

8. Carvalho, G., Cabral, B., Pereira, V., Bernardino, J.: Edge computing: current trends, research challenges and future directions. Computing **103**, 993–1023 (2021)

9. Chamberlain, K., Al-Majeed, S.: Standardisation of UK electric vehicle charging protocol, payment and charge point connection. World Electr. Veh. J. **12**(2), 63 (2021)

10. Chatterjee, P., Hermwille, M.: Tackling the challenges of electric vehicle fast charging. ATZelectron. Worldwide **15**(10), 18–22 (2020)

11. Das, H.S., Li, S., Chowdhury, M.F.F., Ghosh, T.: Game theoretical energy management of EV fast charging station with V2G capability. In: PECon, pp. 344–349. IEEE (2020)

12. Evans, G.: 2024 EV forecast: the supply chain, charging network, and battery materials market (2023)

13. Flora, J., Gonçalves, P., Teixeira, M., Antunes, N.: A study on the aging and fault tolerance of microservices in kubernetes. IEEE Access **10**, 132786–132799 (2022)

14. Germanà, R., De Santis, E., Liberati, F., Di Giorgio, A.: On the participation of charging point operators to the frequency regulation service using plug-in electric vehicles and 5G communications. In: EEEIC/I&CPS Europe, pp. 1–6. IEEE (2021)

15. Gonçalves, A.M.A.: Estimating data divergence in cloud computing storage systems. Ph.D. thesis, Universidade Nova de Lisboa (2013)

16. Gonzalez, N.M., Carvalho, T.C.M.D.B., Miers, C.C.: Cloud resource management: towards efficient execution of large-scale scientific applications and workflows on complex infrastructures. J. Cloud Comput. **6**(1), 1–20 (2017)

17. Hernandez, J., Sutil, F.: Electric vehicle charging stations fed by renewables: PV and train regenerative braking. IEEE Lat. Am. Trans. **14**(7), 3262–3269 (2016)

18. Hsieh, H.F., Shannon, S.E.: Three approaches to qualitative content analysis. Qual. Health Res. **15**(9), 1277–1288 (2005)

19. Kirpes, B., Danner, P., Basmadjian, R., Meer, H.D., Becker, C.: E-mobility systems architecture: a model-based framework for managing complexity and interoperability. Energy Inform. **2**, 1–31 (2019)

20. Larrucea, X., Santamaria, I., Colomo-Palacios, R., Ebert, C.: Microservices. IEEE Softw. **35**(3), 96–100 (2018)

21. Lee, H.C., Liu, H.Y., Lin, T.C., Lee, C.Y.: A customized energy management system for distributed PV, energy storage units, and charging stations on Kinmen Island of Taiwan. Sensors **23**(11), 5286 (2023)

22. Lin, Y.J., Chen, Y.C., Zheng, J.Y., Shao, D.W., Chu, D., Yang, H.T.: Blockchain-based intelligent charging station management system platform. IEEE Access **10**, 101936–101956 (2022)

23. Liu, G., Huang, B., Liang, Z., Qin, M., Zhou, H., Li, Z.: Microservices: architecture, container, and challenges. In: QRS-C, pp. 629–635. IEEE (2020)

24. Mahmood, Z.: Cloud computing for enterprise architectures: concepts, principles and approaches. In: Mahmood, Z., Hill, R. (eds.) Cloud Computing for Enterprise Architectures. Computer Communications and Networks, pp. 3–19. Springer, London (2011). https://doi.org/10.1007/978-1-4471-2236-4_1

25. Naor, M., Coman, A., Wiznizer, A.: Vertically integrated supply chain of batteries, electric vehicles, and charging infrastructure: a review of three milestone projects from theory of constraints perspective. Sustainability **13**(7), 3632 (2021)

26. Oliinyk, M., Džmura, J., et al.: Impact of electric vehicles and demand management systems on electrical distribution networks. Electr. Eng. **104**(2), 667–680 (2022)

27. Péceli, B., Singler, G., Theisz, Z., Hegedűs, C., Szepessy, Z., Varga, P.: Integrated infrastructure for electro mobility powered by the arrowhead framework. IEEE Access **6**, 73210–73222 (2018)

28. Qureshi, K.N., Alhudhaif, A., Jeon, G.: Electric-vehicle energy management and charging scheduling system in sustainable cities and society. Sustain. Urban Areas **71**, 102990 (2021)

29. Segatto, M.E.V., de Oliveira Rocha, H.R., Silva, J.A.L., Paiva, M.H.M., do Rosário Santos Cruz, M.A.: 14 - telecommunication technologies for smart grids: total cost optimization. In: Yahyaoui, I. (ed.) Advances in Renewable Energies and Power Technologies, pp. 451–478. Elsevier (2018)

30. Statista: Electric Vehicles. https://www.statista.com/outlook/mmo/electric-vehicles/worldwide

31. Torreglosa, J.P., García-Triviño, P., Fernández-Ramirez, L.M., Jurado, F.: Decentralized energy management strategy based on predictive controllers for a medium voltage direct current photovoltaic electric vehicle charging station. Energy Convers. Manage. **108**, 1–13 (2016)

32. Tranfield, D., Denyer, D., Smart, P.: Towards a methodology for developing evidence-informed management knowledge by means of systematic review. Br. J. Manag. **14**(3), 207–222 (2003)

33. Vujasinović, J., Savić, G., Prokin, M.: Terminal for remote control of renewable energy sources powered station for electric vehicles charging. In: MECO, pp. 1–6. IEEE (2021)

34. Wilson, W.J., Chaddha, A.: The role of theory in ethnographic research. Ethnography **10**(4), 549–564 (2009)

35. Wohlin, C.: Guidelines for snowballing in systematic literature studies and a replication in software engineering. In: EASE. ACM, New York (2014)

36. Wong, K.C.: Using an Ishikawa diagram as a tool to assist memory and retrieval of relevant medical cases from the medical literature (2011)

37. Wu, Y., Wang, Z., Huangfu, Y., Ravey, A., Chrenko, D., Gao, F.: Hierarchical operation of electric vehicle charging station in smart grid integration applications-an overview. Int. J. Electr. Power Energy Syst. **139**, 108005 (2022)

38. Xiong, Y., An, B., Kraus, S.: Electric vehicle charging strategy study and the application on charging station placement. Auton. Agent. Multi-Agent Syst. **35**, 1–19 (2021)

39. Zimmermann, T.: Card-sorting. In: Menzies, T., Williams, L.A., Zimmermann, T. (eds.) Perspectives on Data Science for Software Engineering, pp. 137–141. Academic Press (2016)

Architecture Decision Making

Exploring Architectural Design Decisions in Mailing Lists and Their Traceability to Issue Trackers

Mohamed Soliman[1,2](✉) (iD)

[1] Universität Paderborn, Paderborn, Germany
mohamed.soliman@uni-paderborn.de
[2] University of Groningen (RUG), Groningen, The Netherlands

Abstract. Software developers commonly use mailing lists to communicate and discuss issues. Some of the emails discuss architectural design decisions (ADDs) to design the components of a system, or achieve quality attributes. In this paper, we explore (a) The types of ADDs in mailing lists, and their co-occurrences, and (b) Methods used by practitioners to discuss ADDs using both mailing lists and issue trackers (e.g. Jira). We applied qualitative methods, as well as deep learning classification to classify ADDs in mailing lists. Furthermore, we applied similarity algorithms to detect related ADDs between emails and issues. Our findings support empirically-grounded approaches to identify and classify ADDs in mailing lists and issue trackers for future system design.

Keywords: Software architecture · Architectural design decisions · Architectural knowledge · Mailing lists · Issue tracker · Software engineering

1 Introduction

Software engineers make different types of *architectural design decisions* (ADDs) [10]. For instance, Kruchten et al. [10] classified ADDs into three types: 1) *Existence ADDs* on the components structure and behavior, such as ADDs on the layers or microservices of a system and their interactions. 2) *Property ADDs* are concerned with cross-cutting quality attributes (e.g. performance or security), such as ADDs on performance or security tactics [1]. 3) *Executive ADDs* are enforced ADDs due to contextual constraints, such as *process ADDs* on organizational aspects to implement ADDs, such as testing strategies, and *technology ADDs* on selecting technologies, such as libraries or APIs.

ADDs of different types are often interrelated and do not exist in isolation; instead, they frequently *co-occur* with one another [24]. For instance, when a software engineer makes a technology ADD, such as selecting a library, it typically entails making additional property ADDs to address any quality drawbacks associated with that technology. Moreover, the process of technology selection may necessitate modifications to various components of the system, thereby require making existence ADDs as well.

While software engineers make different types of ADDs, empirical studies show that software engineers do not formally document ADDs [5]. They rather discuss ADDs spontaneously without documentation as part of their daily work. For instance, when they create an issue in issue tracking systems [2], or send an email to a mailing list [4,8]. However, these informal communications tend to be lost within large software repositories, due to the massive amount of data, and the scarcity of discussions on ADDs compared to other discussions on testing or bug fixing. This leads to losing the knowledge on ADDs in a phenomenon known as *architectural knowledge* (AK) vaporization [10].

To overcome the problem of AK vaporization, researchers aim to develop approaches that can find and structure ADDs from software repositories, such as issue tracking systems and mailing lists. In the past decade, researchers (e.g., [2]) explored ADDs in issue tracking systems, and created datasets and tools to find ADDs in issue trackers. In contrast, mailing lists received less attention from researchers in spite of their importance as the most commonly used method to discuss ADDs [16]. Only few recent studies [4,8] explored general decisions (e.g., requirements and design decisions), and AK (e.g. rationale of ADDs) in mailing lists of one or two open source systems. However, these studies did not explore the types of ADDs in mailing lists, and their co-occurrences (i.e., which ADD types occur together in the same email). Furthermore, it is unknown how software engineers share their ADDs among mailing lists and issue trackers. Such knowledge is important for researchers and practitioners to find ADDs in both platforms. Therefore, in this paper, we **aim** to *explore the types and co-occurrences of ADDs in mailing lists, and the methods used by software engineers to discuss ADDs using both mailing lists and issue trackers.*

To achieve our aim, we explored the mailing lists of six open source projects. We then used information retrieval to find emails that discuss ADDs, and classified 3155 emails manually based on their contained ADDs using qualitative content analysis. Using our curated dataset of emails, we developed deep learning models to automatically classify ADDs in 40,604 emails in our dataset. Finally, we applied similarity algorithms and qualitative analysis on pairs of emails and issues to explore ADDs in mailing lists and issue trackers. In summary, we provide the following **contributions**:

- *Types and co-occurrences of ADDs in mailing lists*, and *methods of discussing ADDs using both mailing lists and issue trackers*. These contributions provide an overview about ADDs in mailing lists, and clarify how software engineers discuss ADDs in software repositories using both mailing lists and issue trackers. This can guide researchers and practitioners to find ADDs in mailing lists, and connect them with ADDs in issue trackers.
- *A dataset of 3155 manually curated emails, and another 40,604 automatically classified emails according to the types of ADDs from six open-source projects*, as well as *a tool to download, annotate, and search for emails in mailing lists*, and *deep learning models to classify ADDs in emails*. The dataset, tool and

deep learning models are available for researchers and practitioners to be directly used or extended. These contributions provide a useful addition to the infrastructure for software architecture researchers.

The paper is structured as follows: In Sect. 2, we explain our research questions and steps. We then present our results in Sects. 3 and 4, which are subsequently discussed in Sect. 5. The threats to validity and related work are discussed in Sects. 6 and 7, while the paper is concluded in Sect. 8.

2   Study Design

To achieve our aim in Sect. 1, we ask the following research questions (**RQs**):

(**RQ1**) *What are the types and co-occurrences of ADDs in mailing lists?*
Software engineers discuss different types of ADDs, such as existence and property ADDs. Some of these are related and discussed together to develop a functionality or achieve a system quality. In mailing lists, software engineers can discuss different ADDs in the same email or thread (i.e., sequences of emails), where each thread can involve emails with different types of ADDs. Thus, we ask this RQ to determine the types of ADDs and their co-occurrences within emails and threads, which can help practitioners and researchers to find and connect ADDs about existing systems from their emails and threads within mailing lists.
(**RQ2**) *How are ADDs discussed using both mailing lists and issue trackers?*
In large software systems, software engineers can use both mailing lists and issue trackers to discuss ADDs. For instance, if software engineers started a discussion on an ADD in an email thread, and then created an issue to continue the discussion on the same ADD. In this case, the discussion on this ADD is scattered among email threads and issues, which make it challenging for software engineers to find and integrate knowledge on ADDs. Thus, we ask this RQ to determine possible methods used by practitioners to discuss ADDs using both platforms: mailing lists and issue trackers. The answer to this RQ can help researchers and practitioners to trace discussions on ADDs between mailing lists and issue trackers.

To answer the aforementioned RQs, we followed the five steps depicted in Fig. 1, and explained in the following sub-sections.

Step 1 - Select Mailing Lists, and Pre-process Emails and Threads

To explore ADDs in mailing lists, we need a sample of mailing lists to be explored. We selected the development mailing lists from six Apache (sub)-projects: Cassandra, Tajo, Hadoop Common, HDFS, Map-Reduce, and Yarn. We have decided on these projects as they are among the largest Apache projects. Furthermore, in our previous research [13,14], we explored the types of ADDs in issue trackers

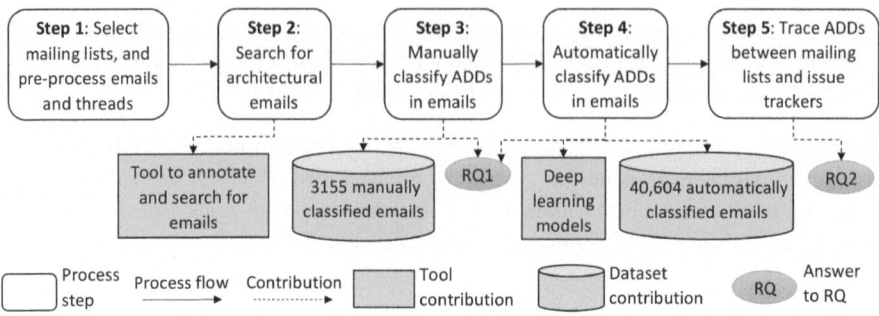

Fig. 1. Research process

of these projects, and created a dataset[1], which will be used to answer RQ2 (see Step 5).

To explore ADDs in the mailing lists of the selected projects, we need to download the emails of these projects from their mailing lists. Apache projects offer an archive of all emails sent to their mailing lists. We downloaded in total 246,673 emails from the mailing lists of the six selected Apache (sub)-projects. However, by checking a sample of emails, we found out that many of the emails come from bots or spams. Thus, we filtered-out emails generated by bots (e.g., emails about commits), as well as spam emails (e.g., call for papers). We did this automatically using regular expressions. This filtering step reduced the number of emails significantly to 43,759 emails. Another problem in the archives of mailing lists, that emails are stored separately, not whole threads of emails. As we aim to analyze ADDs and their co-occurrences, it was essential to build-up the discussion threads from their individual emails to understand full discussions on ADDs. Thus, we associated individual emails to their parent emails in the same thread using their IDs. Moreover, we stored the emails and threads in a database (online[2]), and developed a tool to facilitate their analysis.

Step 2: Search for Architectural Emails

To explore ADDs in emails, we need to find emails that discuss ADDs. However, the vast majority of emails discuss other topics such as decisions on requirements, management and testing [11]. Thus, manually searching for emails that discuss ADDs is ineffective due to the small percentage of these emails compared to other topics. Therefore, we have decided to search for ADDs in emails using keywords searches (i.e., information retrieval). Thus, we implemented keywords searches in our tool (see footnote 2) using Apache Lucene[3]. The tool allows users to search with keywords, browse and tag emails in their hierarchical threads, and facilitates analyzing ADDs in a whole thread rather than isolated emails.

---

[1] https://github.com/mining-design-decisions/Maestro.
[2] https://doi.org/10.5281/zenodo.12516827.
[3] https://lucene.apache.org/.

To search for ADDs in emails, we carefully gathered the most common keywords from prominent architectural literature on ADDs (e.g., [1,10,24]), as well as developer communities (e.g., Stack Overflow [3]). These keywords have been used in previous experiments to find ADDs in issue tracking systems, and showed their effectiveness [13,14]. Some of the search keywords refer to decision factors (e.g., "constraint", "performance", "security", "time", "market"), other search keywords refer to components and connectors (e.g., "broker", "call", "architecture", "interface"), other search keywords refer to patterns or tactics (e.g., "authentication", "concurrency", "heartbeat"), and other search keywords refer to the rationale of ADDs (e.g., "advantage", "risk", "tradeoff"). The full list of keywords are available online (see footnote 2). We executed the search using these keywords, and retrieved a ranked-lists of threads with emails that contain terms with the highest similarity to the search keywords.

Step 3: Manually Classify ADDs in Emails

To find and classify ADDs in emails, we manually analyzed emails in threads at the highest ranking from the ranked list of threads (from step 2). We classified in total 3155 emails from 309 threads. To classify emails, we followed a deductive category assignment content analysis as defined by Mayring [17]:

1) Define category system: We decided to classify emails based on the types of ADDs according to Kruchten et al. [10], because they are comprehensive, and have been previously used to classify issues [2].

2) Develop coding rules: We need to specify rules that determine the types of ADDs. To achieve this, the first author of this paper and an independent researcher classified emails in the top 20 threads in the ranked list of threads. Both researchers discussed their classifications to agree upon the coding rules. At the beginning, we aimed to classify all emails in the threads. However, by going through threads, we realized that some threads are significantly long (contain more than 40 emails), with less discussions on ADDs. These could consume lots of time, but have less benefit. To effectively analyze threads for ADDs, we set two rules: First, we only analyze threads that start with an ADD. Second, we decided to stop our analysis for certain threads at emails, when software engineers change the topic of discussion away from ADDs.

3) Revision of categories: During our initial analysis of emails, we realized that some types of ADDs from Kruchten et al. are explicitly discussed in separate emails (e.g., process ADDs), while others are always discussed together in one email (e.g. structural and behavioral ADDs). Thus, we limited our categories of ADDs to the following four types of ADDs: existence, property, technology and process ADDs (see Sect. 1). These types of ADDs could be explicitly found in emails. By the end of this step, we created the first version of the coding book (available online (see footnote 2)), which define emails that discuss each type of ADD. We repeated this step in nine iterations. In each iteration, we discussed at least 50% of the emails to refine the coding book. By the last iterations, we reached high agreement in our classification.

4) Classify the rest of emails: After settling on the coding rules and categories, we analyzed the rest of the emails iteratively to analyze in total 309 threads. After each iteration, researchers took a sample to ensure agreement, and updated the coding book if needed. Furthermore, for each iteration, we re-calculated the precision of the keywords search approach to find ADDs. We stopped the analysis as the precision degraded below 0.3, as it became ineffective to find ADDs. The full dataset, coding book, and tool are available online (see footnote 2).

Using the dataset of manually classified emails, we answered RQ1 and calculated $\tilde{\chi}^2$ to determine significant co-occurrences among the types of ADDs. We note that we answered RQ1 again in Step 4 using an automatically classified dataset to strengthen the validity of our results.

Step 4: Automatically Classify ADDs in Emails

In Step 3, we applied qualitative analysis to explore the types of ADDs in mailing lists, and curated a dataset of 3155 emails. To explore ADDs in a wider scope, and strengthen the external validity of our results, we trained deep learning models on the dataset of manually classified emails (from Step 3), and then automatically classified the types of ADDs in the rest of the 40,604 emails from the six projects. We kept our focus on these six projects as the accuracy of the classifiers degrade, if we classify ADDs in other projects. Below, we provide brief explanation of the developed classifiers. Further technical details are provided online (see footnote 2). In summary, we followed these steps:

1) Pre-process emails: We removed noise from emails that confuse classifiers, and do not involve ADDs. This includes HTML tags, special characters, stop words and numbers. Furthermore, we identified URLs and class paths, and replaced them with special tokens. Finally, we performed lemmatization to unify different variations of the same word (e.g., scalability and scalable became scale).

2) Create features vectors and train deep learning classifiers: Deep learning classifiers require numerical vectors as input. Thus, emails need to be transformed to such vectors, we applied two techniques: Word embedding (WE) using Skip-Gram [18], and BERT transformer [6] using the DistilBERT [20]. We applied these approaches as they showed success to classify ADDs in issue trackers [13,14], and have been commonly applied in software engineering tasks (e.g., [12]). Using the feature vectors, we trained three deep learning classifiers: 1) A WE layer followed by a four layer convolutional neural network. 2) A WE layer followed by a four layer recurrent neural network. 3) A BERT layer followed by a two layer forward neural network. Further details on the classifiers are online (see footnote 2).

3) Evaluate classifiers: We evaluated the classifiers using 10-fold cross validation, where 80% of the emails are used as a *training* set, 10% of the emails are used as a *validation* set to tune the classifiers, and 10% of the emails are used to *test* the accuracy of the classifiers. In each of the 10 iterations, the training, validation and test sets are selected using stratified random sampling. For each iteration, we train the classifiers using the training set, stop the training based on the validation set, and calculate the following metrics on the test set: Precision, recall, and F1-score. Based on our evaluation, BERT achieved the best results with an F1-score of 0.6. We note that multi-class classification is a challenging task. A single email can involve multiple types of ADDs, which confuses the classifiers between the different types of ADDs. In Sect. 6, we discuss the impact of the accuracy of the classifiers on the validity of our results. Furthermore, we provide online (see footnote 2) further technical details and results.

4) Predict ADDs: Using the trained BERT model (with the best F1-score), we automatically classified all 40,604 emails in our dataset of the six projects, which allow us to determine the amount of emails for each type of ADD, and answer RQ1 on a bigger number of automatically classified emails.

Step 5: Trace ADDs Between Mailing Lists and Issue Trackers

This step aims to answer RQ2, and find ADDs that are discussed using both issue trackers and mailing lists. Due to the big amount of issues and emails, we applied text similarity algorithms to effectively find pairs of issues and emails that discuss the same ADDs. We experimented and combined two algorithms: 1) *Cosine similarity* focus on the textual similarity between terms, and has been commonly applied in software engineering problems [12], and 2) *Sentence Bert (SBERT)* [19] is a recent algorithm based on Bert large language model, which focus on the semantic similarity. We explain below the steps of applying them:

1) Execute similarity algorithms: We applied the similarity algorithms on two combinations of issue-email pairs: 1) On 2,642 issues with ADDs from our issues dataset (see footnote 3), with all 43,759 emails from our dataset from Step 1. 2) 1,155 emails with ADDs from Step 3 with all issues of the six projects (69,637 issues). In this way, we guarantee that at least an issue or email of each pair is manually verified to contain ADDs. This step produced two ranked lists of pairs (one for each algorithm). The higher the rank, the bigger the similarity score.

2) Improve effectiveness: The first author and an independent researcher conducted a pilot analysis for a sample of 100 issue-email pairs with the highest similarity in both lists, and calculated the precision of finding issue-email pairs that discuss the same ADDs. We noticed that the precision of both similarity algorithms were low ($\approx$0.5@100). To improve the precision, we performed the following: 1) We filtered out pairs of issue-email created in two distant points of time (>500 days), because more than 98% of the correctly identified pairs in our sample were created in closer points of time. 2) We filtered out pairs that involve emails with short text (<50 words), because more than 88% of the correctly identified pairs in our sample have longer text. 3) We filtered out duplicate pairs, with emails in the same thread and the same issue (including parents of issues). 4) We took the average of the Cosine similarity and SBert similarity, and created a unified ranked list based on the average values. We did this because both algorithms have the same precision, but each has its own advantage, either lexical or semantic. Based on this step we have one filtered list of issue-email pairs with higher precision ($\approx$0.7@100). We provide online (see footnote 2) further details.

3) Qualitative analysis of issue-email pairs: To answer RQ2, we analyzed the lists of issue-email pairs using steps from the grounded theory. Specifically, we followed open coding and selective coding from the classical grounded theory variation [22]. *In open coding,* the first author and an independent researcher checked each email thread and issue to determine the correct flow of discussion on ADDs. For each issue-email pair, we first verified that the issue and email-thread discuss *the same design issues*. We note that some pairs discuss related design issues; these are not in our scope, and are part of our future work. For each pair that discuss the same design issues, we checked *when* and *where* the discussion on ADDs started and ended. A discussion on ADDs can start or end in an issue or in an email thread. Furthermore, We determined the purpose of discussion, and why it started and ended in an issue or an email. In this, we checked issue descriptions, issue comments, as well as all emails in the respective thread. Based on this information, we annotated an issue-email thread pair with a *method*. We analyzed in total 200 unique issue-email pairs in four iterations. In each iteration, both researchers discussed and refined their categories in a coding book (available online(see footnote 2)). We stopped after 200 pairs as we did not find new methods from the last 100 pairs, which indicated theoretical saturation. In *selective coding,* we refined categories by conducting constant comparison of their respective pairs. We refined and merged categories that rarely occurred, and improved the definition of categories to provide clear differences between them. We present the methods of discussing ADDs in Sect. 4.

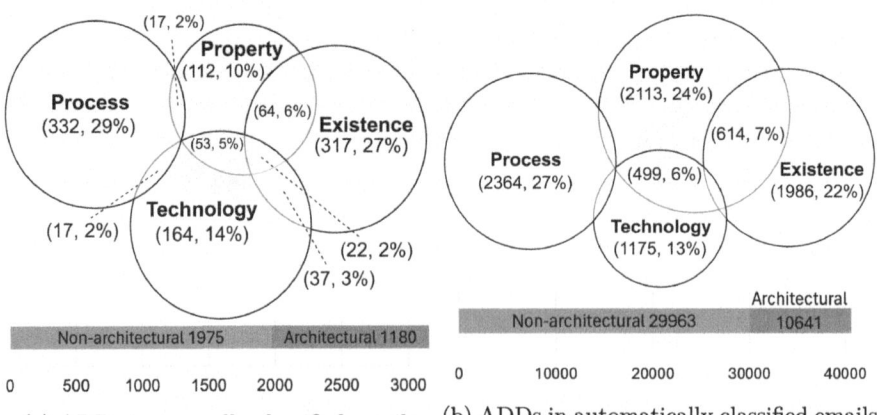

(a) ADDs in manually classified emails   (b) ADDs in automatically classified emails

Fig. 2. The amounts and percentages of ADDs in emails, and their significant co-occurrences based on $\tilde{\chi}^2 > 10$.

3   RQ1: Types and Co-occurrences of Design Decisions

Figure 2 illustrates the distribution of architectural emails across the different types of ADDs and their co-occurrences in our manually analyzed dataset (Fig. 2a) and automatically classified dataset (Fig. 2b). As shown in the horizontal bar chart in Fig. 2, emails that discuss ADDs account for 37% of the manually classified dataset and 23% of the automatically classified dataset. *This verifies the significance of mailing lists as a platform for sharing and discussing ADDs.*

Based on our analysis, we found that software engineers discuss four types of ADDs in emails: Process, existence, property, and technology. We explain below each type of ADD and support them with examples from our dataset. We also write their percentages in our manual and automatic dataset in this format (manual %, automatic %).

– *Process ADDs* (33%, 27%) discuss organisational decisions, such as development process, strategies to achieve code quality, testing strategies, code review process, and releasing strategies. For example, in thread[4], developers discuss code quality *"What principles should the community have in place about code quality and ensuring its long term productivity?"*.
– *Existence ADDs* (38%, 29%) discuss decisions on changing components structure and behavior. For example, in thread[5], developers discuss behavioral ADDs about Hadoop's single-sign-on authentication flow *"client presents credentials to an authentication service endpoint exposed by the SSO server (AS) and receives a token...client then presents the identity token...to the token endpoint exposed by the SSO server (TGS) to request an access token to a particular Hadoop service..."*.
– *Property ADDs* (21%, 37%) are concerned with achieving cross cutting quality attributes such as performance and security. For example, in thread[6] developers discuss improving the security of Cassandra in a cloud environment *"I've been recently looking into how we could improve security in Cassandra...There are very interesting projects out there, such as Vault. Wouldn't it be cool to have automated, build-in certificate management instead?"*.
– *Technology ADDs* (24%, 19%) are concerned with selecting specific technologies such as libraries or development tools. For example, in thread[7] developers discuss the pros and cons of tools to test Cassandra *"The primary tradeoffs as I understand them...Pros: 1. Better debugging functionality 2. Integration with simulator 3. More deterministic runtime...Cons: 1. The framework is not as mature as the python dtest framework..."*.

---

[4] www.mail-archive.com/dev@cassandra.apache.org/msg10843.html.
[5] www.mail-archive.com/common-dev@hadoop.apache.org/msg09911.html.
[6] www.mail-archive.com/dev@cassandra.apache.org/msg11686.html.
[7] www.mail-archive.com/dev@cassandra.apache.org/msg18548.html.

Table 1. Number of co-occurring ADDs in the same thread, but different emails.

Co-occuring types of ADDs in the same thread	Count in manually classified emails	Count in automatically classified emails
Existence ↔ Property	166	982
Technology ↔ Property	155	660
Technology ↔ Existence	138	356
Existence ↔ Process	87	243
Process ↔ Property	61	284
Process ↔ Technology	70	157

Our analysis also show that software engineers discuss different types of ADDs in the same email and in the same thread. Figure 2 (Venn diagrams) shows the amount and percentages of co-occurring types of ADDs in the same email, while Table 1 shows the number of co-occurring ADDs in the same thread. We can observe that the majority of emails (80% in Fig. 2a, and 86% in Fig. 2b) discuss one type of ADD in the same email. But 14–20% of the emails discuss two or more types of ADDs in the same email. Process ADDs are discussed in separate and dedicated emails and threads, with few intersections with the other types of ADDs. Thus, software engineers discuss organisational decisions such as strategies for development process and testing separate from other types of ADDs. In contrast, property ADDs are commonly discussed with other types of ADDs in the same email and thread, especially with technology and existence ADDs. For example, in thread[8], software developers discuss improving the performance of Cassandra through improving the components behavior *"This is a new memtable solution...developed with the following objectives: lowering the on-heap complexity...leveraging byte order and a trie structure to lower the memory footprint and improve mutation and lookup performance"*.

RQ1 key takeaways:
• Software engineers discuss four types of ADDs in mailing lists: process, existence, property, and technology ADDs.
• Existence and process ADDs present around two third of the ADDs in mailing lists.
• Property ADDs are discussed with existence and technology ADDs in the same emails or threads, while process ADDs are discussed separately.

---

[8] www.mail-archive.com/dev@cassandra.apache.org/msg18146.html.

4   RQ2: Methods to Discuss Design Decisions Across Mailing Lists and Issue Trackers

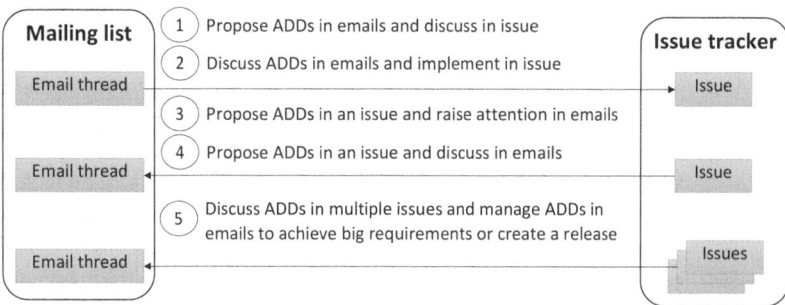

Fig. 3. Methods to discuss design decisions across mailing lists and issue trackers. The arrow direction presents the chronological order.

Based on our analysis of email-issue pairs from Step 5 in Sect. 2, we identified five distinct methods (see Fig. 3) between discussions about ADDs in mailing lists and their corresponding issues in issue trackers. Some of the design discussions start in issues, while others start in mailing lists. We explain the five methods in the following paragraphs. For each method, we present the percentage of pairs (%) from our analysis and specify whether the pairs discuss the same or different types of ADDs across issues and emails. Based on a Fisher exact significance test, none of the methods achieved significance with specific types of ADDs (from RQ1). However, we provide examples of ADDs from our analysis for each method.

(1) **Propose ADDs in emails and discuss in issue (12%):** ADDs are first proposed in an email to ask for feedback. For example, a property ADD is proposed in an email[9] *"I'm working on some performance improvements...What do you think?"*. Issue[10] [CASSANDRA-14448] is created with a property and an existence ADD. The ADDs are then further discussed in issue comments *"It's a tradeoff though, not a clear cut optimization..."*. 60% of our analyzed pairs discuss different types of ADDs across issues and emails. An example for this is to discuss property ADDs and technology ADDs in emails, and then focus on existence ADDs in issues.

(2) **Discuss ADDs in emails and implement in issue (16%):** ADDs are first discussed in an email. For example, an executive ADD is proposed and discussed in an email[11] *"Replacing ivy with maven-ant-tasks...It will make deploying to central easier"*. After finishing the design discussion in emails, an issue

---

[9] www.mail-archive.com/dev@cassandra.apache.org/msg12600.html.

[10] Concatenate issue ID to https://issues.apache.org/jira/browse/.

[11] www.mail-archive.com/dev@cassandra.apache.org/msg01591.html.

(see footnote 10) [CASSANDRA-2017] is created to track the implementation of ADDs. 56% of our analyzed pairs discuss the same types of ADDs across issues and emails. In this case, the ADDs are replicated in issues to facilitate the implementation.

③ **Propose ADDs in an issue and raise attention in emails (20%):** A software engineer proposes an ADD in an issue (see footnote 10) [HADOOP-8803], then an email[12] is sent to pull the attention for this ADD. Then, the design discussion continues further in issue comments. 64% of our analyzed pairs discussed the same types of ADDs across issues and emails. In this case, software engineers replicate ADDs in emails to raise attention.

④ **Propose ADDs in an issue and discuss in emails (7%):** ADDs are first proposed in an issue. For example, property and existence ADDs are proposed in issue(see footnote 10) [HADOOP-8803] to refactor a component *"This is an umbrella jira for improving the current JSP-based HDFS Web UI...one task of this jira is to modernize the UIs"*. Then, an email[13] is sent to discuss this ADD *"I have a few concerns about removing the old UI...I think the new web UI has to have the same level of unit testing"*. 60% of our analyzed pairs discuss different types of ADDs across issues and emails. For example, when software engineers discuss existence ADDs for components design in an issue, and then discuss quality attributes for property ADDs in emails.

⑤ **Discuss ADDs in multiple issues and manage ADDs in emails to achieve big requirements or create a release (45%):** Multiple issues with different ADDs have been discussed. These ADDs are discussed together in an email thread. The goal of the discussion is to manage these ADDs to ensure consistency and integration among them to achieve certain requirements: either new functionalities or quality attributes. Another goal of the discussion is about creating a release that contains the implementation of these ADDs. For example, in the email[14] *"I'd like to open a discussion on merging the Timeline Service v.2 feature to trunk (YARN-2928 and MAPREDUCE-6331)...my concern to release it in 3.0.0-alpha...is we haven't provide any security support in ATS v2 yet."*. 71% of our analyzed pairs discussed different types of ADDs across issues and emails. For example, when software engineers discuss process and property ADDs in mailing lists to manage the creation of a release.

RQ2 key takeaways:
- There are five different methods used by practitioners to discuss ADDs using both mailing lists and issue trackers.
- The majority (72%) of discussions on ADDs start in issues then emails.
- Nearly half of the design discussions in mailing lists focus on a single issue, and the other half involve multiple issues.
- Nearly half of related email threads and issues discuss different types of ADDs, and the other half discuss the same types of ADDs.

---

[12] www.mail-archive.com/common-dev@hadoop.apache.org/msg07358.html.

[13] www.mail-archive.com/hdfs-dev@hadoop.apache.org/msg11691.html.

[14] www.mail-archive.com/yarn-dev@hadoop.apache.org/msg23808.html.

5   Discussion

5.1   Implications for Researchers

Table 2. Comparison on the % of ADDs in mailing lists and issue trackers.

	Mailing lists		**Issue trackers**	
	Manual	Automatic	Manual	Automatic
ADDs in the whole dataset	37%	23%	13%	15%
% Existence ADDs in all ADDs	38%	29%	73%	45%
% of Property ADDs in all ADDs	21%	37%	50%	17%
% of Executive ADDs in all ADDs	52%	46%	30%	52%
% of co-occurences between ADDs from all ADDs	18% per email 42% per thread	13% per email 17% per thread	45%	14%

The results of RQ1 show that the amounts and co-occurrences of ADDs in mailing lists are different from issue trackers. In Table 2, we compare the amounts of ADDs in mailing lists (based on RQ1) and issue trackers (based on [7]). From which, we can conclude the following:

- *Mailing lists have nearly double the percentage of ADDs, compared to issue trackers.* This should motivate researchers to explore ADDs in mailing lists, because the chance is higher to find ADDs compared to issue trackers.
- *Existence and property ADDs are more discussed in issue trackers than mailing lists, while executive ADDs tend to be discussed in mailing lists more than issue trackers.* This should guide researchers when developing general approaches to find ADDs in both mailing lists and issue trackers. One can focus on existence and property ADDs in issue trackers, and executive ADDs in mailing lists.
- *The majority of executive ADDs in issue trackers pertain to technology ADDs, while both process and technology ADDs are discussed in mailing lists.* This should derive researchers to specifically consider process ADDs in mailing lists as they are not commonly discussed in issue trackers.
- *The co-occurrences of ADDs in an issue description happen more often (45%) compared to individual emails (max. 18%). In contrast, the co-occurrences of ADDs in mailing lists happen mostly on the thread level (max. 42%).* This result can help to extract related ADDs. In issue trackers, related ADDs should be extracted from paragraphs of issue description, while in mailing lists, related ADDs should be extracted from multiple emails in the same thread.

The results of RQ2 provide the first conceptual foundation on how software engineers discuss ADDs between mailing lists and issue trackers. Based on this, researchers could develop traceability approaches to manage the discussions on ADDs among mailing lists and issue tracker. Specifically, the methods in Sect. 4 can provide heuristics for traceability approaches to accurately trace the discussions on ADDs. One example for a heuristic is to assume that ADDs start in issues and then emails because most practitioners start with creating an issue as shown in our results in Sect. 4.

Our datasets, searching tool, and classifiers provide researchers with the required infrastructure to explore ADDs in different projects. For instance, researchers can use the dataset, expand it with emails from other projects, and train classifiers to achieve better accuracy to classify ADDs in mailing lists.

5.2   Implications for Practitioners

The results of RQ1 set the expectations of practitioners regarding the types of ADDs in mailing lists and their co-occurrences, which can motivate practitioners to search for certain types of ADDs in mailing lists, and guide them during the search. From Fig. 2, process ADDs are one of the most discussed ADDs, and they are usually discussed in separate emails and threads. This should guide practitioners to search for process ADDs in mailing lists, and find its AK. In contrast, property ADDs are usually discussed with existence and technology ADDs. Thus, to find property ADDs, it is helpful to find other ADDs as well.

The results of RQ2 can guide practitioners on how to discuss and trace ADDs using both issue trackers and mailing lists. In one situation, practitioners can decide to propose ADDs in an issue, as this is the most commonly used method by practitioners as presented in Sect. 4, and follow the issue with an email to discuss ADDs. In another situation, if a practitioner reads an issue with ADDs in its description. However, the issue does not contain sufficient details about the ADDs. This might mean that the design discussions are actually in mailing lists. Thus, practitioners can search in mailing list archive (e.g., using our tool (see footnote 2)) to find emails related to this issue.

Our dataset, searching tool, and classifiers support practitioners with tools to search and classify ADDs. All our tools are open source (see footnote 2) and available for practitioners to be directly used. Furthermore, the searching tool has an easy to use user-interface, which facilitate searching and filtering emails.

6   Threats to Validity

External Validity: Like several other studies (e.g., [4,8]), our investigation relies on choosing mailing lists from open-source projects. This approach could pose challenges when generalizing the findings to industrial projects or different ecosystems. Additionally, our analysis is constrained by a limited set of projects and emails, potentially affecting the broader generalizability of our results. Nonetheless, our study encompasses a larger number of projects and emails than other closely related works [4,8]. Furthermore, we aim to enhance the generalizability by analyzing a big amount of emails through deep learning techniques.

Construct Validity: The accuracy of the deep learning classifier caused wrong classifications of ADDs, and accordingly affect the results of RQ1. However, the results of the deep learning classifier to answer RQ1 were comparable to the qualitative approach (see Sect. 3). Furthermore, the methods of discussing ADDs in Sect. 4 might not be exhaustive. However, during our qualitative analysis (see Sect. 2), we reached theoretical saturation, and thus covered most of the used methods to discuss ADDs using both emails and issues.

Reliability: The consensus on the types of ADDs in RQ1 and the identified methods in RQ2 may pose reliability concerns. To address this, two researchers collaborated to address both research questions, and we evaluated the agreement between them for both RQs. For RQ1, we achieved a kappa value of 0.67 on the types of ADDs. For RQ2, all pairs have been fully agreed by both researchers. Additionally, we developed a coding guide with precise definitions and illustrative examples to ensure consistency among researchers. Furthermore, we offer our dataset along with a specialized tool for email browsing and annotation to support the replication of the study.

7 Related Work

Early research on ADDs (e.g., [10,23,24]) provide its foundational concepts. Recent research efforts explore ADDs in software repositories, such as in issue trackers [2], mailing lists [4,8], and pull requests [15], as well as on the Web, such as in forums [3], technology documentation [9], and blogs [21]. We discuss below the most related work to this paper.

Bhat et al. [2] propose a machine learning approach to classify issues in issue trackers, which contain existence ADDs. The approach focuses solely on existence ADDs in issue trackers. In contrast, our study explores all types of ADDs in mailing lists, and methods to discuss ADDs with issue trackers.

Fu et al. [8] propose a machine learning approach to classify different decisions from the mailing list of an open source project. Specifically, Fu et al. classify requirements, design, management, construction and testing decisions in mailing lists. The goal of this paper is different, we classify types of design decisions, their co-occurrences and methods to discuss ADD with issue trackers.

Bi et al. [4] is the closest work to this paper. Bi et al. explored the mailing lists of two open source projects, and determined the type of architectural information, such as components, dependencies, and quality requirements. The authors also determined who use mailing lists, and when it is used. Bi et al. applied qualitative methods to analyze emails. In contrast, this paper focuses on the types of ADDs in mailing lists, their co-occurrences, and their methods to discuss ADDs with issue trackers. Both qualitative methods and deep learning have been applied in this paper. Thus, we argue that our paper is different from Bi et al. in its contributions and methodology. Both contributions complement each other to understand the architectural knowledge in mailing lists.

8   Conclusions

Our aim of this study is to explore ADDs in mailing lists and the methods used to discuss ADD with issue trackers. Our results cover the types of ADDs and their co-occurrences in mailing lists, as well as methods to discuss ADDs in issue trackers. These support determining how to find and trace ADDs using both mailing lists and issue trackers. Our future work aims to expand our study to explore relationships between ADDs, and to develop approaches to find and trace ADDs across mailing lists and issue trackers.

Data Availibility Statement. All the data and artifacts of this paper are available online (see footnote 2).

References

1. Bass, L., Clements, P., Kazman, R.: Software Architecture in Practice, 3rd edn. Addison-Wesley (2012)
2. Bhat, M., Shumaiev, K., Biesdorf, A., Hohenstein, U., Matthes, F.: Automatic extraction of design decisions from issue management systems: a machine learning based approach. In: Lopes, A., de Lemos, R. (eds.) ECSA 2017. LNCS, vol. 10475, pp. 138–154. Springer, Cham (2017). https://doi.org/10.1007/978-3-319-65831-5_10
3. Bi, T., Liang, P., Tang, A., Xia, X.: Mining architecture tactics and quality attributes knowledge in stack overflow. J. Syst. Softw. 111005 (2021). https://doi.org/10.1016/j.jss.2021.111005
4. Bi, T., Ding, W., Liang, P., Tang, A.: Architecture information communication in two OSS projects: the why, who, when, and what. J. Syst. Softw. **181**, 111035 (2021). https://doi.org/10.1016/J.JSS.2021.111035
5. Capilla, R., Jansen, A., Tang, A., Avgeriou, P., Babar, M.A.: 10 years of software architecture knowledge management. J. Syst. Softw. **116** (2016). https://doi.org/10.1016/j.jss.2015.08.054, http://dx.doi.org/10.1016/j.jss.2015.08.054
6. Devlin, J., Chang, M., Lee, K., Google, K.T., Language, A.I.: BERT: pre-training of deep bidirectional transformers for language understanding. In: Proceedings of the Conference of the North American Chapter of the Association for Computational Linguistics, vol. 1 (2019). https://doi.org/10.18653/V1/N19-1423
7. Druyts, S.: Exploring design decisions in issue tracking systems for projects in different software domains. Ph.D. thesis (2023). https://fse.studenttheses.ub.rug.nl/31767/
8. Fu, L., Liang, P., Li, X., Yang, C.: A machine learning based ensemble method for automatic multiclass classification of decisions. In: ACM International Conference on Evaluation and Assessment in Software Engineering (EASE 2021), pp. 40–49 (2021). https://doi.org/10.1145/3463274.3463325
9. Gorton, I., Xu, R., Yang, Y., Liu, H., Zheng, G.: Experiments in curation: towards machine-assisted construction of software architecture knowledge bases. In: IEEE/IFIP ICSA 2017, pp. 79–88 (2017)
10. Kruchten, P., Lago, P., van Vliet, H.: Building up and reasoning about architectural knowledge. In: Hofmeister, C., Crnkovic, I., Reussner, R. (eds.) QoSA 2006. LNCS, vol. 4214, pp. 43–58. Springer, Heidelberg (2006). https://doi.org/10.1007/11921998_8

11. Li, X., Liang, P., Li, Z.: Automatic identification of decisions from the hibernate developer mailing list. In: ACM International Conference on the Evaluation and Assessment in Software Engineering (EASE 2020), pp. 51–60 (2020). https://doi.org/10.1145/3383219.3383225
12. Li, Y., Soliman, M., Avgeriou, P.: Automatically identifying relations between self-admitted technical debt across different sources. In: Proceedings - 2023 ACM/IEEE International Conference on Technical Debt, TechDebt 2023, pp. 11–21 (2023). https://doi.org/10.1109/TECHDEBT59074.2023.00008
13. Maarleveld, J., Dekker, A.: Developing deep learning approaches to find and classify architectural design decisions in issue tracking systems. MSc. thesis (2023). https://fse.studenttheses.ub.rug.nl/31368/
14. Maarleveld, J., Dekker, A., Druyts, S., Soliman, M.: Maestro: a deep learning based tool to find and explore architectural design decisions in issue tracking systems. In: Tekinerdoğan, B., Spalazzese, R., Sözer, H., Bonfanti, S., Weyns, D. (eds.) ECSA 2023. LNCS, vol. 14590, pp. 390–405. Springer, Cham (2023). https://doi.org/10.1007/978-3-031-66326-0_24
15. Mahadi, A., Ernst, N.A., Tongay, K.: Conclusion stability for natural language based mining of design discussions. Empir. Softw. Eng. (2022). https://doi.org/10.1007/S10664-021-10009-1/TABLES/7
16. Mannan, U.A., Ahmed, I., Jensen, C., Sarma, A.: On the relationship between design discussions and design quality: a case study of Apache projects. FSE (2020). https://doi.org/10.1145/3368089.3409707
17. Mayring, P.: Qualitative content analysis. In: Forum Qualitative Sozialforschung/Forum: Qualitative Social Research, vol. 1 (2000)
18. Mikolov, T., Chen, K., Corrado, G., Dean, J.: Efficient estimation of word representations in vector space. In: 1st International Conference on Learning Representations, ICLR 2013 - Workshop Track Proceedings (2013). https://arxiv.org/abs/1301.3781v3
19. Reimers, N., Gurevych, I.: Sentence-BERT: sentence embeddings using siamese BERT-networks. In: EMNLP-IJCNLP 2019 - 2019 Conference on Empirical Methods in Natural Language Processing and 9th International Joint Conference on Natural Language Processing, Proceedings of the Conference, pp. 3982–3992 (2019). https://doi.org/10.18653/v1/d19-1410, https://arxiv.org/abs/1908.10084v1
20. Sanh, V., Debut, L., Chaumond, J., Wolf, T.: DistilBERT, a distilled version of BERT: smaller, faster, cheaper and lighter (2019). https://arxiv.org/abs/1910.01108v4
21. Soliman, M., Gericke, K., Avgeriou, P.: Where and what do software architects blog?: an exploratory study on architectural knowledge in blogs, and their relevance to design steps. In: ICSA (2023). https://doi.org/10.1109/ICSA56044.2023.00020
22. Stol, K.J., Ralph, P., Fitzgerald, B.: Grounded theory in software engineering research: a critical review and guidelines. In: Proceedings - International Conference on Software Engineering, 14–22 May 2016, pp. 120–131. IEEE Computer Society (2016). https://doi.org/10.1145/2884781.2884833
23. Tang, A., Jin, Y., Han, J.: A rationale-based architecture model for design traceability and reasoning. J. Syst. Softw. **80**(6), 918–934 (2007). https://doi.org/10.1016/j.jss.2006.08.040
24. Zimmermann, O., Koehler, J., Leymann, F., Polley, R., Schuster, N.: Managing architectural decision models with dependency relations, integrity constraints, and production rules. J. Syst. Softw. **82**(8), 1249–1267 (2009). https://doi.org/10.1016/j.jss.2009.01.039

Helping Novice Architects to Make Quality Design Decisions Using an LLM-Based Assistant

J. Andrés Díaz-Pace[1]([✉]) [iD], Antonela Tommasel[1] [iD], and Rafael Capilla[2] [iD]

[1] ISISTAN, CONICET/UNICEN University, Tandil, Buenos Aires, Argentina
{andres.diazpace,antonela.tommasel}@isistan.unicen.edu.ar
[2] Rey Juan Carlos University, Madrid, Spain
rafael.capilla@urjc.es

Abstract. Architectural knowledge and specifically design decisions have become first-class entities to be captured routinely in a design process. However, the quality of the decisions captured is often low. Part of the problem is that reflections intended to criticize, and thus improve, the decisions are seldom made in architecture teams, particularly when involving novice architects. To improve reflective practices and capture better decisions, we propose an design assistant approach based on generative AI techniques. Our assistant, called ArchMind, relies on two information sources: architectural knowledge about patterns, and information about the system under design. Furthermore, the assistant takes advantage of LLMs to progressively aid users in selecting and assessing alternative decisions, until capturing them using an Architecture Decision Record format. ArchMind mainly targets novice architects as discussed in an initial experiment using the assistant off-line for a classroom project. The generated ADRs were concrete and well-justified in their design rationale, although they tended to miss system-specific details.

Keywords: Design decisions · Architectural knowledge · Assistant · Large Language Models · Reflection · Architecture Decision Records

1 Introduction

For over 20 years, the software architecture community has emphasized the need to document key design decisions in software projects. Diverse tools and techniques have been developed to capture, manage, and share architectural knowledge (AK) in academia and industry. However, architectural decision-making involves more than capturing AK. Architects often start by deriving decisions for software requirements and then evaluate alternative choices based on their pros and cons. Once discussed, decisions can be documented using templates such as the Architecture Decision Record (ADR).

Related to the deliberative process in which decisions are discussed, Razavian et al. [6] proposed the *Mind1-Mind2* model, in which some architects make

© The Author(s), under exclusive license to Springer Nature Switzerland AG 2024
M. Galster et al. (Eds.): ECSA 2024, LNCS 14889, pp. 324–332, 2024.
https://doi.org/10.1007/978-3-031-70797-1_21

decisions (*Mind1* stage) while others criticize those decisions (*Mind2* stage). The latter refers to as *cognitive architects* and their reflective tasks aim at validating decisions, identify overlooked aspects, and enhance the overall quality of the decisions. In practice, less experienced architects tend to struggle with reflection about design decisions. Some challenges include identifying good alternative decisions (e.g., different architectural styles or design patterns), weighing quality-attribute trade-offs, and effective comparison of design alternatives. These situations also appear frequently in architecture training courses. In this context, we argue that generative AI techniques, like Large Language Models (LLMs) can assist by suggesting alternatives and evaluating their pros and cons. Additionally, once contextual information and potential decisions are gathered, an LLM could generate preliminary ADRs [3].

This papers investigates an approach to support novice architects in decision-making activities by relying on a set of LLM-based agents, called *copilots*, that act as assistive cognitive architects within the *Mind1-Mind2* model. We present a prototype called `ArchMind` that uses two information sources: *existing AK on software patterns, and the system under design* (e.g., context, requirements, past decisions). Using LLM techniques like prompting and Retrieval-Augmented Generation (RAG) [4], `ArchMind` assists architects with tasks such as suggesting alternative patterns for a requirement, evaluating and ranking them based on pros and cons, assessing about chosen decisions, and structuring the information using an ADR template. Our approach enables architects' interventions to adjust the inputs for the tasks managed by the copilots. This process builds an ADR progressively, enabling the architect and copilots to focus on different aspects of the final decision document. We discuss initial findings from comparing ADRs generated by `ArchMind` with those produced in an architecture classroom.

2 Background

The *Mind1-Mind2* model has been used in literature to study the dynamics of the software architecture process [6]. Capilla et al. [2] implemented a model with undergraduate students in a software design course. The students assumed one of three roles: senior architects making decisions (*Mind1*), cognitive architects providing critique (*Mind2*), and junior architects documenting decisions. A typical exercise involved considering design patterns to meet a functional requirement The study showed that reflective tasks somehow improved the final architecture, and the architects' interactions contributed to a thorough decision analysis. However, in certain cases the students' bias may have limited the exploration of alternatives, thus affecting decision quality. This factor motivated our proposal to use generative AI to enhance the reflective practices.

AI-assisted Decision-making. In software engineering, AI assistants and LLMs have been commonly used for implementation activities [5], but their application to architecture design remains limited [3]. For instance, White et al. [7] introduced a prompt pattern catalog for activities such as requirement elicitation, system design, and refactoring. Ozkaya [5] discusses how generative AI can help develop or refine system architectures, aligning with our goal of using an LLM tool to support better decision-making. However, crafting suitable prompts and other LLM abstractions for architectural tasks is not simple and depends on the knowledge base provided to the tool. Moreover, human oversight is crucial to mitigate "hallucinations" in the LLM outputs.

LLMs and Retrieval Augmented Generation. In recent years, LLMs have been trained on extensive datasets and fine-tuned with human instructions [1], enabling them to understand and respond to natural language questions. LLMs are particularly useful for in-context learning. Interacting with LLMs typically involves instructions (prompts), which combine user-specific details and relevant context for the LLM to generate responses. In the architecture domain, a prompt might include functional requirements, system context, or preferences for solutions from a pattern catalog (e.g., the Gang-of-Four book). By processing this information, an LLM can infer a design decision and document it using an ADR.

LLMs often have limitations for dealing with factual information in tasks that require specific domain expertise. A promising technique here is retrieval-augmented generation (RAG), which integrates external domain-specific sources into the LLM generative process to improve response accuracy and relevance. In essence, a RAG helps ground the LLM, making it focus on specific contents for its response [4]. For instance, in our architectural example, the Gang-of-Four catalog could serve as a foundation for an architecture-centric RAG framework.

3   Study Design

This work evaluates whether generative AI can enhance human decision-making and reflective practices based on prior results of novice architects. In this context, we investigated the quality of ADRs generated by an LLM-based assistant. Unlike previous research for generating ADRs in one single step [3], we adopted an *incremental* approach. Our method integrates various information pieces (e.g., architectural patterns, system details, assessment results) and allows architects to interact with the assistant to refine the final decision. The assistant consists of a set of *copilots*, each responsible for a specific design task, as depicted in Fig. 1. This modular schema facilitates reasoning by breaking the ADR generation process into simpler steps.

As an initial benchmark for human design decisions, we used a student project from an undergraduate architecture course at a Spanish university. In this project, a team of students captured their decisions as ADRs. For the sake of the study, these students were considered as novice architects. We developed a prototype called `ArchMind` to support five tasks including reflective practices. Once `ArchMind` was loaded with information from the students' project, we carried out design sessions based on the *Mind1-Mind2* model. In these sessions, a human (acting as a senior architect) submits a requirement, and a copilot (acting as a cognitive architect) provides an initial decision. This decision is progressively refined by other copilots, with human interventions if needed until the

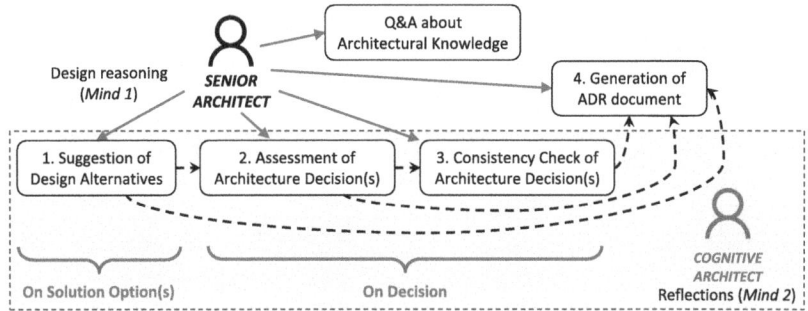

Fig. 1. Reflective tasks supporting architect's in decision-making. The dotted arrows show the information dependencies between the tasks.

final ADR is generated. We evaluated the resulting ADRs and compared them qualitatively to the students' decisions.

We used both zero-shot and RAG strategies for our experiment. Zero-shot learning involves performing a task without prior exposure to related training examples. On the other hand, RAG [4] is a practical way of adding knowledge to an LLM, even if the knowledge might be already known to the model (e.g., the pattern catalogs in our domain). First, RAG ensures that the AK being retrieved is relevant to the specific context or query, instead of relying on the model's generalized knowledge, which might be too broad or inaccurate if the question is ambiguous. Second, in RAG, the information sources can be directly identified, providing greater traceability. Third, RAG directs the LLM to key pieces of information, improving the relevance and accuracy of the outputs.

3.1   Approach

To conduct a design session, we followed a four-step protocol involving a senior architect (the user[1]) and a cognitive architect (the copilots). Figure 2 provides an example of the questions each copilot addresses. The user can intervene at specific points in the process to provide input or refine the copilots' partial outputs. Initially, we fed the prototype with AK including architectural styles, design patterns, and microservice patterns from well-known sources[2]. This AK is required for the RAG but not for the zero-shot strategy. Additionally, we provided information about the system under design and its functional requirements. Given the system context S and a subset of requirements $\{R\}$ (decision drivers), our goal was to yield a decision D through a series of interactions with the copilots,

---

[1] In our experiment, this role was played by two of the authors.

[2] We included *Design Patterns: Elements of Reusable Object-Oriented Software* (Gamma, Helm, Johnson, Vlissides), *Microservices Patterns: With examples in Java* (Richardson), *Design It!: From Programmer to Software Architect* (Keeling) and *Software Architecture in Practice* (Bass, Clements, Kazman, Safari).

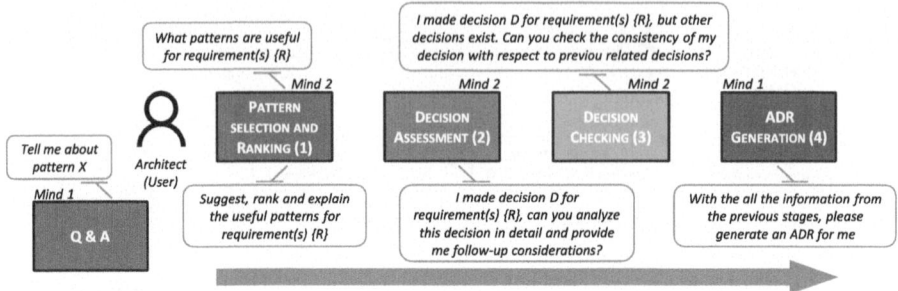

Fig. 2. Example of an architect's journey and questions supported by each copilot (numbered boxes). The copilots have a one-to-one mapping with the tasks.

until producing an ADR for D. This process proceeded in iterations until all requirements were met. Each iteration involved 4 tasks:

Task 1. The user asks the *PatternSuggestion&Ranking* copilot to generate a list of design alternatives for $\{R\}$ using any available patterns. Patterns are sourced either from the RAG knowledge base or the LLM (zero-shot). Afterwards, the copilot evaluates each alternative weighting its pros and cons, ranks the alternatives from best to worst (for S and $\{R\}$), and discards any irrelevant pattern. The alternative ranked first is suggested as a candidate decision D for $\{R\}$.

Task 2. The *DecisionAssessment* copilot analyzes the previous D in more detail, focusing on quality attributes, risks, and subsequent decisions, among others.

Task 3. The *DecisionChecking* copilot expands the analysis of D by ensuring its consistency with other related or committed decisions for $\{R\}$.

Task 4. The *ADRGeneration* copilot consolidates the analyses from tasks $1-3$ and produces an ADR. Although one could attempt ADR generation after tasks 1 or 2, the copilot may lack enough information for a comprehensive ADR.

Finally, we included a *Q&A* task which allows users to ask questions about any architectural topic at any time. This task aims to assist about doubts or clarifications, but it is not mandatory for the generation of the ADR.

3.2   Target System

In our study, we chose a student project about a system whose requirements and decisions provide enough detail to compare it with **ArchMind**'s results. Details of the project materials and tool are provided in Sect. 5. The student team was composed by 6 participants, according to the two roles of the *Mind1-Mind2* model plus the role of *junior architect*, who was responsible for modeling the architecture in UML based on the decisions resulting from the senior and cognitive architects. Before making their decisions, participants elicited a list of functional requirements from the problem description. The target problem was about the migration of a monolithic system to a microservices solution. Subsequently, participants identified design issues hinted by the requirements and explored decisions for them. The decisions, originally written in Spanish, were translated into English to align with the AK sources and also for anonymization.

3.3 The ArchMind Tool

The assistant comprises five copilots: *Q&A*, *PatternSuggestion&Ranking*, *DecisionAssessment*, *DecisionChecking* and *ADRGeneration*. Each copilot is accessible as a separate tab in the tool's GUI, allowing users to navigate between them freely. Intermediate results are shared across tabs. Each copilot uses parametrized prompts for its LLM, summarized in Table 1. The tool is implemented using *Langchain*, *OpenAI GPT-3.5*, *Chroma* and *Streamlit*.

Table 1. Prompt instructions and parameters for the LLM in each copilot.

	Copilot	Summary of instructions	Parameters
	Q&A	*You are an architecture expert, answer question Q concisely*	Q: system-agnostic question
1.	Pattern Suggestion	*Given a system context S and requirements R, provide a list of candidate patterns for S and R. For each pattern P, assess pros and cons. Create a ranking of the patterns and remove irrelevant patterns. For the first ranked pattern P, suggest decision D = P and instantiate it for requirements R and system context S*	S: system context. R: target requirement(s). List of candidate patterns (intermediate result)
2.	Decision Assessment	*I want to better understand the implications of my decision D for requirements R in the provided context S. Please assess: appropriateness of D, clarifying questions, follow-up decisions, assumptions and constraints, consequences on quality attributes, potential risks and trade-offs*	S: system context. R: requirement(s). D: decision being assessed
3.	Decision Checking	*Search for related decisions for my decision D for requirement(s) R, and check if past/related decisions lead to conflicts or inconsistencies*	S: System context. R: requirement(s). D: decision being assessed. Related decisions
4.	ADR Generation	*Generate ADR for requirement(s) R based on initial decision D and using a MADR template. Leverage on prior assessment and consistency checking of D.*	S: system context. R: requirement(s). D: decision being assessed. Assessment of candidates (pros & cons). Assessment of consistency of D with related decisions

The tool allows configuring a RAG strategy for the first three copilots. In RAG, a copilot first uses similarity search to match its prompts to pieces from the knowledge base, called vectors, and then asks the LLM to generate a response. If no vectors can be retrieved for a prompt, the LLM call is skipped, and a default answer is returned. Conversely, with a zero-shot strategy, all prompts are directly sent to the LLM for responses. ArchMind has two repositories: one with pattern catalogs as the AK for the RAG and another containing information about the target system. To ingest the patterns as documents, they are divided into chunks and converted into embeddings, which are numeric vector representations of the text. These embeddings enable similarity search across documents in the RAG strategy. Users can select a specific pattern catalog or opt for a zero-shot mode.

The system repository provides descriptions (e.g., requirements, decisions) as input for the copilots. For example, in *Task 1*, a user can select a requirement and ask the *PatternSuggestion&Rank* copilot to explore and rank design options based on their pros and cons. A system summary provides context for the LLM to identify candidate options. Once a list of possible patterns is returned, the top-ranked pattern is suggested as the initial decision. The user can then change the patterns, the decision, or even write a new one if they disagree with the suggestion. Once an initial decision for the requirement is made, the user can ask the *DecisionAssessment* copilot for a more in-depth assessment. In *Task 2*, this task includes identifying assumptions, constraints, consequences over quality attributes, risks, trade-offs, and clarifying or follow-up questions for implementation. In *Task 3*, the *DecisionChecking* copilot retrieves semantically similar prior decisions from the system repository to check for potential conflicts or inconsistencies. Finally, the user can use the *ADRGeneration* copilot to request an ADR. In *Task 4*, the LLM employs a zero-shot strategy, combining the outputs from the previous copilots into an ADR document.

4   Initial Evaluation

In this research, our goal was to analyze the characteristics of the generated ADRs and the influence of the copilots' pipeline on them. As `ArchMind` is still experimental, we did not aim to conduct a systematic comparison of prompting techniques or user studies. We started by exercising the tool on individual tasks to ensure each copilot could produce useful outputs, adjusting the prompts as needed. Once the AK repository was ready, we tested the *Q&A* copilot with various architectural questions. For the other copilots, we tested end-to-end generation scenarios for selected requirements. On the other hand, we implemented a batch procedure to take all the requirements from the target system through the copilot pipeline to generate 16 ADRs, using the RAG and zero-shot strategies.

We then analyzed the quality of these ADRs offline, evaluating the relevancy of the generated content compared to the ADRs produced by students[3]. We manually examined a sample of 5 key ADRs to identify similarities and differences. Based on our expertise, we judged most ADRs as correct in addressing decision drivers, with only a few misleading decisions. These issues arose because the LLM lacked sufficient system context compared to humans. For instance, students often derived a "problem statement" or design issue from the requirement before evaluating design options, which the current copilots did not capture. Regarding the search of the design options, a positive aspect of the copilots is that they suggested and justified more candidate patterns than humans. However, the copilots struggled to recognize when a functional module or a vendor-specific technology can be considered a convenient option.

For the RAG strategy, the LLM often provided no answers due to insufficient context, and when it did, the responses were specific to concrete patterns or system-related aspects. As a result, the ADRs from `ArchMind` were

---

[3] This does not mean that the ADRs produced by the students were considered as the ground truth but rather as a reference solution for comparison purposes.

sometimes concise and stylistically different from the more verbose but less specific decisions and options considered by the students. In contrast, the zero-shot strategy always produced an output but with general comments. For example, several responses mentioned the broad "Microservices" pattern, which was not specific enough for our target system. Consequently, the generated ADRs had few design options and lacked detailed pros and cons of the patterns. One key lesson from this study is that the RAG strategy effectively focuses the design work on specific topics, such as those from our AK base. Thus, this strategy can benefit novice architects by improving their decision-making and assessment skills, and it can even be useful in real-life project scenarios. Our approach highlights a trade-off between specific (but constrained) copilots and more general ones, presenting a research challenge for using LLMs in the software architecture field. Another challenge involves accurately representing the system context and requirements as inputs for the copilots. During our experiments, a brief system context (about 10 lines of text) and short functional requirements (1–2 lines) were provided. However, further evaluation with more complex problems and human interactions is needed, considering variations in context (e.g., acceptance criteria, quality-attribute concerns) and AK sources.

5   Conclusions and Future Work

In this paper, we proposed an LLM-based approach and tool to assist architects in creating ADRs using zero-shot and RAG strategies. An initial offline study comparing human-generated and copilot-generated ADRs showed that a RAG strategy can yield more precise results than using a zero-shot approach. We conjecture that this advantage arises from using intermediate tasks and reflections, as facilitated by our copilots' modular design. This approach holds potential for both architecture training and assessing design alternatives. We believe that LLM-based assistants can significantly enhance software architecture education. They can also automate decision-making and documentation tasks while also expanding human capabilities in exploring large spaces of design alternatives.

Related to our work, Dhar et al. [3] experimented with LLMs to generate design decisions. We differ from this approach in how we generate the ADRs, focusing on interactive and step-wise activities rather than on a single-shot effort. We provide more elaborate prompts and address a different research question.

For future work, we plan to conduct empirical experiments using ADRs for different projects and perform a deeper evaluation of the tool. This will include a qualitative, fine-grained analysis of human answers to refine the copilot inputs and prompts for the ADR sections. We will also investigate how copilots perform with alternative LLMs including multi-modal capabilities. Another research topic is the representation of AK for the RAG strategy.

Data Availibility Statement. We provide a Github site with the source code and data used in the paper: https://github.com/tommantonela/archmind. A version of **ArchMind** can be run via *Streamlit*: https://archmind.streamlit.app/.

References

1. Brown, T., et al.: Language models are few-shot learners. Adv. Neural Inf. Process. Syst. **33**, 1877–1901 (2020)
2. Capilla, R., Zimmermann, O., Carrillo, C., Astudillo, H.: Teaching students software architecture decision making. In: Jansen, A., Malavolta, I., Muccini, H., Ozkaya, I., Zimmermann, O. (eds.) ECSA 2020. LNCS, vol. 12292, pp. 231–246. Springer, Cham (2020). https://doi.org/10.1007/978-3-030-58923-3_16
3. Dhar, R., Vaidhyanathan, K., Varma, V.: Can llms generate architectural design decisions?-An exploratory empirical study. arXiv preprint arXiv:2403.01709 (2024)
4. Lewis, P., et al.: Retrieval-augmented generation for knowledge-intensive NLP tasks. Adv. Neural Inf. Process. Syst. **33**, 9459–9474 (2020)
5. Ozkaya, I.: Can architecture knowledge guide software development with generative AI? IEEE Softw. **40**(5), 4–8 (2023)
6. Razavian, M., Tang, A., Capilla, R., Lago, P.: In two minds: how reflections influence software design thinking. J. Softw. Evol. Process **28**(6), 394–426 (2016)
7. White, J., Hays, S., Fu, Q., Spencer-Smith, J., Schmidt, D.C.: Chatgpt prompt patterns for improving code quality, refactoring, requirements elicitation, and software design. CoRR abs/2303.07839 (2023)

Architecture Decision Records in Practice: An Action Research Study

Bardha Ahmeti, Maja Linder, Raffaela Groner$^{(\boxtimes)}$, and Rebekka Wohlrab

Department of Computer Science and Engineering, Chalmers | University of
Gothenburg, Gothenburg, Sweden
{gusahmeba,guskalmas}@student.gu.se, {raffaela,wohlrab}@chalmers.se

Abstract. To establish good architectural practice and knowledge sharing during software development and maintenance, architectural design decisions need to be documented. While a lot of research has been done on documenting architectural decisions, there exist few approaches with empirical evidence on their applicability and usefulness in practice. Architecture Decision Records are a popular documentation approach in industry, but there is a lack of research focusing on how Architecture Decision Records can be introduced in different company contexts. To tackle this shortcoming, we performed an action research study in cooperation with a company that develops a microservice-based system without proper architecture design decision documentation. We performed seven interviews to identify the challenges faced by the developers of the system. Afterward, we introduced Architecture Decision Records as a means of documentation. Over the course of three months, we observed whether this markdown-based documentation approach addresses the identified challenges. Our results show that practitioners face challenges related to the documentation culture, knowledge transfer, prioritization of information to be documented, as well as handling documentation for shared and distributed components. The first three types of challenges are well addressed by Architecture Decision Records. However, challenges arising from developing distributed systems remain open. Thus, there is a need for further research that helps to document design decisions for distributed systems effectively. We also compiled a list of lessons learned from our study. We found that cooperation among the teams was improved after the introduction of Architecture Decision Records. At the same time, the decision on where documentation is stored has a massive influence on its perceived usefulness. Practitioners should carefully consider what information to store centrally and what information to store in local repositories.

Keywords: Action Research · Architecture Decision Records ·
Architecture Decision Documentation · Challenges · Software
Architecture

M. Galster et al. (Eds.): ECSA 2024, LNCS 14889, pp. 333–349, 2024.
https://doi.org/10.1007/978-3-031-70797-1_22

1 Introduction

A selected software architecture influences a software system in all stages of the software life cycle. Therefore, its documentation is an important part of maintaining a system efficiently for years, since missing or inadequate documentation can delay maintenance or the onboarding process of new employees.

In particular, the documentation of architecture design decisions (ADDs) plays an important role. Each decision is based on an analysis of the advantages and disadvantages as well as a discussion of the technical options. Documenting this information helps later to understand why a system has been implemented in a certain way. It also helps to support future decisions and thus to avoid the repetition of wrong decisions.

A lot of research has examined ADDs. For example, several tools facilitate ADDs, like the Architecture Design Decision Support System (ADDSS) [4] tool and Knowledge Architect (KA) [14]. Additionally, there are several approaches to document ADDs, e.g., Architecture Decision Records (ADRs) [12,17] or UML.

While ADRs are popular in industry [10], there is a lack of empirical evidence on how ADRs can be introduced in companies as a means of documenting ADDs. Most empirical studies focus on architectural documentation in general. For example, Rost and Robillard [7] examined the extent to which the format of architectural documentation affects comprehension. Manteuffel et al. [15] explored whether architectural decisions from previous projects can be reused in new projects. Finally, Buchgeher et al. [3] mined open-source repositories at GitHub to analyze the current use of ADRs.

However, no study addresses the challenges that practitioners face if they do not have any proper documentation approach and how introducing ADRs can overcome these challenges from a culture, tooling, and knowledge management perspective. Therefore, we address the following research questions in our work:

RQ1: What challenges do practitioners face while developing microservices without documentation of ADDs?

RQ2: To what extent are these challenges addressed when introducing ADRs as a means of documenting ADDs?

To answer our two research questions, we performed action research in cooperation with a company to investigate whether ADRs are suitable as a means of documenting ADDs in practice. The system under consideration is based on microservices and is developed and maintained by various agile teams.

Our results show that different types of challenges are more or less well addressed by ADRs. For example, challenges related to documentation culture or knowledge transfer are well addressed by ADRs. However, challenges arising from the distributed development of partially dependent components are insurmountable using only ADRs and require further development. Also, a clear definition of documentation guidelines is crucial for implementing a documentation approach.

2 Related Work

In this section, we discuss a selection of related works. Firstly, we present work that presents tools or other documentation approaches to document ADDs. Secondly, we describe studies that deal with architecture documentation in practice.

Documentation of ADDs: Several works deal with the documentation of ADDs. In contrast to our work, however, no challenges are mentioned in these works and these works do not consider microservice-based systems.

For example, Shahin et al. [20] evaluated nine different ADD models and compare different tools that work with these models. Many papers, such as Lee and Kruchten [13], Tyree and Akerman [22], Heesch et al. [23], and Kopp et al. [12], present tools, views, or models for documenting ADDs. Often, these approaches are similar to ADRs or use ADRs. Heesch et al. [23], for example, described various views on documentation, one of which is very similar to the ADRs we use. Kopp et al. [12] presented their tool for documenting ADDs with the help of ADRs and evaluate their tool with several participants.

Haselböck et al. [9] analyzed different applications of decision models in microservice-based architectures and developed their own decision models. They introduced these models as part of a technical action research process in several companies to identify relevant stakeholders and use cases for their models. In this phase, the authors also collected requirements for their models. Like our work, Haselböck et al. [9] also considered the documentation of ADDs in the context of microservice-based systems. In comparison to their work, we use action research to identify challenges that arise in such systems without proper documentation and observe how the introduction of ADRs affects them.

Alexeeva et al. [2] analyzed 96 publications covering 58 approaches to document ADDs and obtained a taxonomy for ADDs. The authors have also compiled a list of possible factors for the rare use of ADDs in practice, such as the resulting documentation overhead not being taken into account and the lack of integration in commercial tools. This list includes system-independent challenges about ADDs. In comparison, our work focuses on the challenges in the context of microservice-based systems.

Studies on Architecture Documentation in Practice: Several studies have focused on the challenges of documenting the architecture of a system or ADDs. Dasanayake et al. [6] performed a case study to examine the approaches developers use to make an ADD and what challenges they face during the decision-making. Rost et al. [18] also conducted a study in which they interviewed 147 developers from the industry. Their study aimed to explore the challenges developers face when realizing an architecture based on its documentation. In comparison to these studies, we include different groups of stakeholders, such as developers, architects, and team leaders, and examine the challenges they face without proper ADD documentation. We also analyze how those challenges can be mitigated by introducing ADRs.

Forward and Lethbridge [8] surveyed 32 developers about their documentation, with a focus on the tools used and the maintenance of the documentation.

Manteuffel et al. [16] performed a case study at a company. During this study, the authors identified several challenges the developers face while they document ADDs. For instance, the developers lack tool support and guidelines for documenting ADDs. Based on their observations, a tool was developed that implements the Decision Documentation Framework [23].

Although the studies by Forward and Lethbridge [8] and Manteuffel et al. [16] presented challenges related to documentation and ADDs, they did not explicitly focus on distributed microservice-based systems. In addition, they do not include how the introduction of a documentation approach addresses these challenges.

The publication by Kleehaus and Matthes [11] is an example of a work that deals with the documentation of microservice-based systems. The authors presented various challenges that developers face when documenting such a distributed system. In comparison to this work, we focus specifically on the documentation challenges in a real-world context, the introduction of ADRs, and observations regarding to what extent the observed challenges were mitigated.

3   Architecture Decision Records

In this section, we provide some background on Architecture Decision Records and introduce the template we use in this work.

Architecture Decision Records (ADRs) were introduced by Nygard [17]. The idea is to make decisions comprehensible with the help of lightweight documentation in a Markdown language. Nygard proposes a five-part structure, starting with the *title* of the document. The *context* section summarizes the factors influencing a decision, such as technical, social, or political factors. The *decision* section describes the responses to the influencing factors from the context section. The *status* section documents whether a decision is proposed, accepted, deprecated, or superseded. Finally, the resulting *consequences* are summarized [17].

Listing 1.1 shows a simplified version of an ADR that we provided to the study participants as an initial example. The structure of this ADR is mainly based on the structure presented by Nygard. However, our template also contains a section to document other options considered and the reasoning for why they were discarded. We also included a section for tags to label an ADR. Before the introduction of ADRs at the company at which we conducted our study, we held a workshop on ADRs. At this workshop, the practitioners provided feedback that they would like to include these two sections in the ADR template.

```
# ADR 000 Using ADRs

## Context
Architecture decisions are currently not documented in a
structured and consistant way.

## Decision
We will document every architecturally significant
```

```
decisions as ADRs. They will be stored in Markdown format.

## Considered Options
--

## Status
Accepted

## Consequences
1. Team members should write an ADR and submit it for
review before implementing an approach to any architectural
decision.
2. We will have a visible history of the decisions
evolution through version control.

## Tags
Documentation, Architecture Decision
```

Listing 1.1. Simplified Architecture Decision Record.

4 Study Context

We performed our study in cooperation with a Swedish engineering and consulting company that offers engineering, design, digital, and advisory services in different areas. Two teams participated in our study, both of which applied the principles of agile software development. The teams already used tools, e.g., a wiki page, and visualizations to document parts of their system but lacked an approach to document ADDs in an effective and structured way. Both teams were responsible for developing and maintaining a microservice-based system that was used by the 19,000 employees of the company.

One team ($team_{Sys}$) consisted of eight employees, namely one business analyst, one tester, and seven developers/architects. $Team_{Sys}$ was responsible for the overall architecture and platform design of the system considered in our study. The other team ($team_{App}$) consisted of six employees, namely one team leader, three developers, one tester, and one business analyst. $Team_{App}$ was responsible for a few specific application services consumed by the system considered.

The teams regularly communicated architecture decisions through a developer forum, where all the teams met, presented, and discussed the decisions taken. However, the lack of centralization and accessibility of documentation led to confusion and slowed down the decision-making process.

5 Research Method

In this section, we describe the research method we used to answer our two research questions. All supplementary material can be found online [1].

Since we want to learn the challenges that practitioners face while developing microservice-based systems without proper documentation of architecture design decisions and how architecture decision records tackle these challenges as a means of documenting ADDs in practice, we selected a research method that focuses on the study context, learning, and improvement of the practice. Our study was conducted over the course of three months. We applied action research [21], which consists of five phases, namely *Diagnosing*, *Action Planning*, *Action Taking*, *Evaluation*, and *Learning* [21]. In the following, we describe our study design that implements these five phases.

Diagnosing: The first phase of action research was dedicated to identifying the problem to be addressed [21]. We combined semi-structured interviews, a survey, and participant observations to collect data in this phase and answer **RQ1** [5]. A total of seven employees from both teams took part in the interviews. We transcribed the recordings of the interviews and analyzed them with the help of open coding using descriptive codes [19]. We provide information on the codes used during our analysis at [1]. Six employees participated in our survey, and we analyzed the results using descriptive statistics. For the observations, we accompanied the participants to various daily meetings to observe the team members in their usual activities. The observations were carried out in a non-intrusive manner using notes to collect data about the daily work of the teams. We present the results of our *Diagnosing* phase in Sect. 6.1.

Action Planning: This phase was dedicated to planning our study based on the results from the *Diagnosing* phase [21]. Based on the result from the previous phase, we designed a plan and guidelines to introduce Architecture Decision Records (ADRs) as a means to document ADDs. The plan elaborated by us consisted of three stages. First, a workshop was held to introduce ADRs. We also discussed the structure, storage, visibility, and handling of ADRs with the participants at this workshop. Second, in collaboration with the participants, we introduced seven guidelines on how the team members should manage ADRs. We assigned each guideline a unique ID using the naming schema **G<number>** These guidelines were also published on a central wiki page and are as follows:

G1 ADRs affecting several microservices are stored in a central repository.

G2 ADRs are referred to in commit messages and user stories.

G3 ADRs are checked in with the code fragments connected to them in the corresponding repository.

G4 ADRs are reviewed by another developer and are published after a successful review.

G5 ADRs are tagged with the quality attributes, components, or features they are related to.

G6 ADRs are labeled with a status to document whether an ADR is proposed, accepted, rejected, deprecated, or superseded.

G7 New ADRs are discussed in biweekly developer forums.

Finally, the team members used ADRs in their daily work. To make this step easier, we provided the teams with an example ADR. A simplified version of this ADR is shown in Listing 1.1. During this stage, we collected data using semi-structured interviews, a survey, and observation of the study participants.

Action Taking: During this phase, the plan designed during the *Action Planning* phase was performed [21]. During this stage of our study, we spent four weeks at the company to collect data.

Evaluation: In this phase, the aim was to understand the influence of the previously performed action [21]. As already mentioned, we analyzed our data qualitatively with the help of coding.

Learning: The final phase in the cycle dealt with the lessons learned from the previous phases [21]. Based on the combination of our evaluation of the results of the *Action Taking* phase and our results from the *Diagnosing* phase, we obtained different results for this phase to answer **RQ2**. We present the results of this phase and our lessons learned in more detail in Sect. 6.2.

6   Findings

In this section, we present our findings and present our lessons learned. In Sect. 6.1, we present the challenges the teams face without proper documentation of ADDs. In Sect. 6.2, we present the observations we made after ADRs were used for documentation and map them to the previously identified challenges. Subsequently, we discuss and summarize our lessons learned.

6.1   Challenges Due to Insufficient Documentation (RQ1)

In this section, we focus on the challenges that practitioners face without proper ADD documentation. First, we present insights that show how current team members perceived the practice of documenting ADDs before introducing ADRs. Afterward, we present the seven challenges that practitioners face without proper documentation of ADDs.

Figures 1 and 2 show the results of the survey we performed during the *Diagnosing* phase. Both plots consist of two parts: The part on the right shows the number of participants who answered "Don't know". The part on the left shows the number of participants who answered the respective question. The bars in the left part of the plot grow from the center to the left to present the negative answers of the participants. To represent the positive answers the respective bar grows from the center to the right. The annotations show the percentage share of the six participants who answered the respective question positively or negatively.

In Figs. 1 and 2, it is noticeable that at the beginning of our study, 83% of respondents stated that ADDs are only rarely or occasionally documented and that there are no clear guidelines for documenting ADDs. In addition, half of the survey participants are dissatisfied with the way ADDs are documented.

Based on the analysis of our qualitative data, we identified seven challenges that can be categorized into four main categories, namely **Documentation**

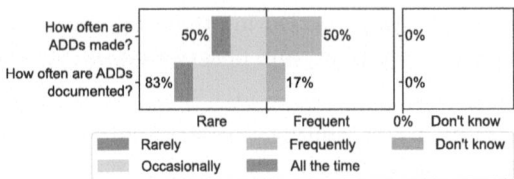

Fig. 1. Frequency of ADDs and their documentation before introducing ADRs

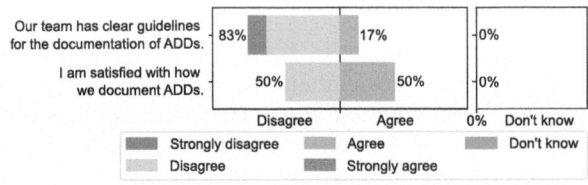

Fig. 2. Assessment of the current ADD documentation before introducing ADRs

Culture, Shared/Distributed System Parts, Knowledge Transfer, and **Prioritization**. To enable traceability, we assign each challenge a unique ID using the naming schema **C<number>**.

Documentation Culture: The first challenge is related to the problems of fostering a documentation culture in an agile team (**C1**). Figure 2 shows the dissatisfaction of the practitioners with the currently available documentation. Additionally, some team members elaborated on their dissatisfaction due to missing documentation during our interviews. Based on the participants' statements, we identified one challenge related to the documentation culture. For example, an architect from team$_{Sys}$ mentioned: *"It seems to be challenging also when you work with agile and you have to continuously deliver. It seems like stopping for documentation is a challenge for teams across the industry so yeah, documentation is hard."*

Shared/Distributed System Parts: Different teams are involved in developing the microservice-based system at the company we cooperated with. Thus, the corresponding code is stored in different repositories. In this context, some participants mentioned that the placement of documents can affect their usefulness. One challenge relates to identifying and receiving information about ADDs that affect all or multiple components of a system (**C2**). This challenge is supported by the statement from an architect from team$_{Sys}$: *"We have a lot of different teams that are working with the different parts of our platforms. But sometimes, there is stuff that is common for all the teams, and then finding that decision and finding why we did something or haven't done something, it is hard."*

Another challenge relates to the location where the documentation is stored because the documentation of ADDs is currently scattered across different tools and platforms (**C3**).

Knowledge Transfer: Documentation is a means of knowledge transfer. We identified two challenges arising from a lack of/insufficient documentation in knowledge transfer.

The lack of documentation leads to relying on individual team members to maintain the context and history of ADDs and recall them from memory (**C4**). Relying on the memory of individual team members as a means of documentation is not expedient. On the one hand, essential knowledge gets lost when team members leave the company. On the other hand, individual team members are burdened with a high mental load and responsibility, as they are expected to remember all decisions correctly.

Another challenge is related to onboarding new team members. Due to outdated and missing documentation, onboarding depends on meetings with other team members (**C5**). As a result, the employees' productivity decreases during the onboarding of new team members. On the one hand, senior employees are busy with onboarding tasks that could be covered by providing documentation. On the other hand, the onboarding of a new employee depends on the availability of other team members. This problem can also be seen in the statement of a developer from Team$_{App}$: *"[...] it kind of takes time from developers to share the information every time a new person starts."*

Prioritization: Besides the issue of where the documentation should be stored, another important aspect is deciding on what should be documented. One challenge deals with missing prioritization of information to be documented (**C6**). A lack of prioritization might lead to too much documentation, which overwhelms the team members due to the resulting workload and excessive documentation.

Another challenge is partly related to the previously mentioned challenge, as it encompasses the lack of guidelines for prioritizing information to be documented (**C7**). This lack of guidelines is also reflected by the answers to our survey during the *Diagnosing* phase. As shown in Fig. 2, four of the six survey participants stated that they disagree with the statement *"Our team has clear guidelines for the documentation of ADDs."* Additionally, one participant stated that they strongly disagreed with the statement.

Answer to RQ1: The majority of the challenges we identified are independent of the developed system. For example, the lack of guidelines for documentation or the loss of expertise due to a lack of documentation are challenges that exist independently of the system architecture. Nevertheless, we identified two system-specific challenges. The first challenge is concerned with identifying and receiving information that is relevant for multiple microservices. The second one is concerned with documentation scattered across different tools.

6.2 Observations After Introducing ADRs (RQ2)

After introducing ADRs as a means of documenting architectural design decisions, we observed several changes in the teams that are related to the seven challenges we presented in Sect. 6.1.

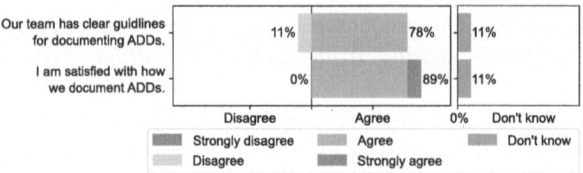

Fig. 3. Assessment of the ADD documentation using ADRs

Figure 3 shows an overview of the team members' perceptions after introducing ADRs. It can be seen that most participants considered their team to have clear guidelines and that they were satisfied with their documentation approach. If we compare Fig. 3 with Fig. 2, we can see a more positive assessment compared to before the introduction of ADRs for documentation. If we assign numerical values to the nominal scale with "strongly disagree" = 1 to "strongly agree" = 4 and exclude the two "don't know" responses, we see that the average agreement values also increased. The average agreement with the statement about guidelines increased from 2 to 2.9. The average agreement increased from 2.5 to 3.1.

In this section, we describe our observations and map them to their related challenges using our identifier schema: **C<number>**. In the following, we assign each observation an identifier using the naming schema **O<number>**.

In total, we gathered nine observations that show to what extent the challenges were tackled by the introduction of ADRs to document ADDs. An overview of our findings can be seen in Table 1 which summarizes the identified challenges and maps them to our observations. In this section, since we map the previously identified challenges to observations, we reuse our four main categories (**Documentation Culture**, **Shared/Distributed System Parts**, **Knowledge Transfer**, and **Prioritization**) to classify our observations.

Documentation Culture: C1 is related to the difficulties in maintaining documentation. During our study, we observed different changes related to this challenge: First, writing ADRs led to team members discussing the importance of documentation (**O1**). Team$_{App}$ introduced a review process for ADRs and integrated ADRs into its daily meetings. This also encouraged collaboration between developers in the creation of ADRs (**O2**). Furthermore, team members mentioned that ADRs were easy to create (**O3**). For example, an architect from team$_{Sys}$ mentioned "*I think it's really easy to write them actually. They are basic enough, as they should be.*"

Shared/Distributed System Parts: C2 is concerned with retrieving information about decisions that affect multiple parts of the system. To deal with this challenge, for this study we decided to document such decisions in a separate repository for all teams (cf. **G1** in Sect. 5). Seven of nine participants of our second survey stated that they used the documentation (**O4**). In addition to the guidelines we provided, some participants expressed the wish to integrate the documented ADRs into the rest of the system documentation (**O5**).

Table 1. Overview of the identified challenges and the associated observations.

Challenges		Observations	
ID	Description	ID	Description
C1	Foster and maintain a documentation culture in an agile team	O1	Writing ADRs sparked discussions about the importance of documentation
		O2	ADRs are perceived as encouraging collaboration
		O3	ADRs are perceived as an easy way to document ADDs
C2	Identifying and receiving information about ADDs that affect all or multiple components of a system	O4	Participants used the documentation of overarching ADRs
		O5	Participants want to include ADRs in the system documentation
C3	The documentation of ADDs is scattered across different tools and platforms	O6	Participants suggest standardized tags to filter ADRs, independently of the repositories they are stored in
C4	Relying on individual team members to maintain the context and history of ADDs	O7	By using ADRs, team members no longer need to memorize ADDs
C5	Onboarding depends on meetings with other team members	O8	Participants rate it as likely that they will integrate ADRs into the onboarding process
C6	Lack of prioritization of the information to be documented	O9	Provided guidelines helped the participants, but they need further improvement
C7	Lack of guidelines for prioritizing		

The practitioners proposed a tag-based system to address the challenge of scattered ADD documentation across different tools and platforms (**C3**). Their idea was that ADRs should be categorized using standardized tags and then all ADRs could be filtered using these categories (**O6**). The ADR template we used in our study already provides a section for tags, but these tags were not standardized and we did not provide a search mechanism.

Knowledge Transfer: C4 is concerned with relying on individual team members to remember and correctly recall ADDs. We observed that this challenge is eliminated by using ADRs as a means of documenting ADDs (**O7**). This is also reflected in the statement of a developer from Team$_{App}$: *"What I like most about ADRs is that I do not need to remember these things further, I can write them down here and now and then be sure that I won't lose this information."*

The timeframe of our study was too short to determine whether the use of ADRs makes the onboarding process less dependent on meetings with experienced team members (**C5**). In our second survey, we asked participants using a 5-point scale (1="very unlikely" to 5="very likely") about the likeliness of including ADRs in the onboarding process. Two participants rated the likeliness with 3, five participants rated it with 4, and two rated the likeliness with 5. The rating indicates that the participants are very positive about integrating ADRs into the onboarding process (**O8**).

Prioritization: In terms of prioritization, we were able to identify two challenges. Firstly, what should be documented (**C6**), and secondly, what guidelines should be applied for prioritization (**C7**). Related to these two challenges, we observed that the seven guidelines we provided are already helping the team members. Nevertheless, the participants noted that the guidelines need to be further improved (**O9**). For example, some team members struggled to decide whether a change should be considered as an ADD and thus needs to be documented using an ADR. This is also reflected in the statement of an architect from team$_{Sys}$: *"The hardest part might be to actually decide whether it's a change that is made, whether it's to be considered as an ADR, or if it's just part of normal maintenance."*

Answer to RQ2: Our results show that challenges related to system-independent categories like documentation culture or knowledge transfer are addressed well using ADRs. Challenges that practitioners face due to the distributed nature of the microservice-based system are not addressed and need further research. To tackle this issue, the participants made some suggestions like implementing standardized tags and a filtering mechanism that allows them to search for ADRs across all tools and repositories.

6.3   Discussion and Lessons Learned

Based on our findings, we have compiled a list of lessons learned, which we discuss and summarize in this section.

Documentation Culture: We have learned that introducing ADRs has led to stronger cooperation among the teams (O2).

Previously, ADDs were discussed and communicated in wiki pages and discussions in development forums or via teams. After the introduction of ADRs, the developers wrote ADRs cooperatively. Additionally, there was a stronger focus and more discussions in the teams on how, where, and what should be documented. Despite these positive effects, some effort is needed to integrate ADRs into existing processes. For example, some developers had difficulties integrating the creation of ADRs into their daily workflow.

It is not always the case that the documentation of decisions strengthens cooperation within a team. For example, Dasanayake's [6] survey shows that cooperation depends on the respective teams. This is also shown by the case study of Forward and Lethbridge [8], in which decisions are often made in cooperation, regardless of whether they are documented.

Therefore, we assume that this lesson learned is independent of the method of documentation used. However, we expect the corporate culture and social dynamics within a development team to have a strong influence. The teams in our study use agile software development, which might have also influenced this lesson learned.

Guidelines and Prioritization: We have learned that clear guidelines on the content to be documented are crucial for implementing a documentation approach successfully (C7).

During our study, practitioners sometimes struggled to decide whether a decision was an architecturally significant decision and whether they should create an ADR. Furthermore, practitioners need additional guidelines to help them prioritize the information to be documented, as they only have limited time available due to the short development cycles in agile software development. Our participants have prioritized their documentation based on the complexity of the logic involved, the impact of changes on the system, and the number of developers or teams affected. Some practitioners suggested prioritizing based on the criticality of the system or component or the frequency of changes.

The problem of a lack of guidelines for documentation is also mentioned by the participants in the evaluation of the documentation framework by Manteuffel et al. [16]. In this work, the participants state that they document "major" decisions without a clear definition of what "major" means in a given context.

Our seven guidelines from Sect. 5 are context-independent. Thus, they can applied to other teams without any adjustments. However, we expect the two shortcomings related to explicitness and prioritization mentioned above to be context-independent. Thus, these shortcomings might also occur in other teams.

Shared and Distributed System Parts: We have learned that the storage location of the documentation has an extreme impact on its usefulness and accessibility (C3).

On the one hand, the documentation of shared system components should be accessible to all teams for which the corresponding ADRs are relevant. On the other hand, the practitioners prefer to store the documentation with the corresponding system part/code. One developer commented on this way of organizing documentation. They stated that the documentation for a service might be stored in its repository because the developers do not know that their ADD affects several other services as well. Thus, developers of a service need to know which services depend on their service and whether their ADD affects other services to decide where to store the documentation.

One approach is to store the documentation locally with the automated publication of the ADRs on a central wiki page or similar. In our study, the practitioners also noted that ADRs should be provided with standardized tags. These tags should be used to filter the ADRs and identify relevant ADRs across all platforms and tools used.

Regardless of whether a system is distributed or not, practitioners report that important information is often scattered strongly [18]. For example, developers use different tools for documentation [8]. Practitioners also mention difficulties in navigating existing documentation [18] or revisiting the design rationale [6]. In the Forward and Lethbridge survey [8], participants mentioned that centralized documentation beyond individual projects would help them.

This lesson applies to all distributed systems, especially when ADDs affect multiple system components. However, we also see in other works that teams have problems organizing documentation in a consistent and easy-to-find way, regardless of the system developed. Whether the solution proposed by the practitioners solves this issue needs to be explored in more detail.

7   Threats to Validity

In the following, we discuss threats to the validity of our study.

Conclusion Validity: Our results might be only a small subset of possible challenges and observations since we performed our study in the context of one company and used mainly data collection methods that produce qualitative data. However, with the help of action research in the context of one specific company, we can directly observe the impact of introducing ADRs to document ADDs in practice. Additionally, the qualitative data we obtained provides us with more insights into the challenges that practitioners face without proper documentation and what changes arise in the company due to introducing a systematic way to document ADDs.

Internal Validity: Interviewing participants can always lead to incorrect data being collected due to the interviewee's bias or inaccuracies. Therefore, we applied multiple methods, as this allows us to triangulate data from interviews, surveys, and participant observation.

The interviews were coded by two researchers independently and their results were subsequently compared to avoid researcher bias and mistakes.

Construct Validity: We performed pilot interviews with two students and a software developer to test the interview procedure and identify ambiguous questions.

The implementation of data collection based on observations is always prone to participants behaving differently under observation than in real life. We tried to observe the participants as non-intrusively as possible. However, we cannot say to what extent our presence influenced the participants' behavior.

Terms like "decision" can be interpreted in different ways by different participants. To ensure that a common meaning is used, those constructs were briefly discussed at the beginning of the interviews, when introducing the topic.

External Validity: We cannot say to what extent our results apply to other companies. However, we expect the identified challenges to apply to other companies with a similar working setup as the company in our study. There might be further challenges, e.g., related to the storage and distribution of ADD documentation, the size of the development teams, or the complexity of the microservice-based system. The behavior of the participants, on the other hand, may differ from our observations even in an identical company, as this depends on various factors such as team dynamics, personality, and company culture.

We used ADRs for the documentation of ADDs. We cannot say to what extent this influenced our results. We assume that the use of other documentation approaches will produce slightly different results. For example, if our study is repeated using a documentation approach that provides dedicated tool support, the usability of such a tool is an additional, decisive factor influencing the results.

Reliability: We published our study material at [1] to increase the reliability of our study. However, we cannot publish the raw data due to confidentiality. It should be noted, that our results are based on experience in a company. Thus, we cannot guarantee that researchers produce the same results if they repeat our study in another company.

8   Conclusion

This research was motivated by a lack of studies on how introducing ADRs can overcome documentation challenges from a culture, tooling, and knowledge management perspective. To close this gap, this action research study has examined the impact of ADRs on the challenges associated with documentation practices during the distributed development of a microservice-based system.

Our results show that challenges can be divided into different categories. Challenges that belong to categories that can also occur in non-distributed systems, such as knowledge transfer, are already well addressed by ADRs in combination with clear guidelines for documentation. However, for challenges arising from the distributed system developed, there is currently no suitable solution. This lack of support in documenting ADDs for distributed systems is also evident from our lessons learned. Therefore, we see a need for future work, especially on the effective management of ADD documentation in distributed systems.

Acknowledgments. This work was partially supported by the Wallenberg AI, Autonomous Systems and Software Program (WASP) funded by the Knut and Alice Wallenberg Foundation.

Data Availibility Statement. We published our study material consisting of two questionnaires, our interview guides, and information on the codes used during our analysis at [1]. We cannot publish the collected raw data due to confidentiality.

References

1. Ahmeti, B., Linder, M., Groner, R., Wohlrab, R.: Supplementary material (2024). https://doi.org/10.5281/zenodo.11635100
2. Alexeeva, Z., Perez-Palacin, D., Mirandola, R.: Design decision documentation: a literature overview. In: Tekinerdogan, B., Zdun, U., Babar, A. (eds.) ECSA 2016. LNCS, vol. 9839, pp. 84–101. Springer, Cham (2016). https://doi.org/10.1007/978-3-319-48992-6_6
3. Buchgeher, G., Schöberl, S., Geist, V., Dorninger, B., Haindl, P., Weinreich, R.: Using architecture decision records in open source projects-an MSR study on GitHub. IEEE Access **11**, 63725–63740 (2023). https://doi.org/10.1109/ACCESS.2023.3287654
4. Capilla, R., Nava, F., Montes, J., Carrillo, C., et al.: ADDSS: architecture design decision support system tool (2010)
5. Clark, V.L.P., Ivankova, N.V.: Mixed Methods Research: A Guide to the Field, vol. 3. Sage Publications, Thousand Oaks (2015)
6. Dasanayake, S., Markkula, J., Aaramaa, S., Oivo, M.: Software architecture decision-making practices and challenges: an industrial case study. In: Proceedings of the 2015 24th Australasian Software Engineering Conference, pp. 88–97 (2015). https://doi.org/10.1109/ASWEC.2015.20
7. Ernst, N.A., Robillard, M.P.: A study of documentation for software architecture. Empir. Softw. Eng. **28**(5), 122 (2023)
8. Forward, A., Lethbridge, T.C.: The relevance of software documentation, tools and technologies: a survey. In: Proceedings of the 2002 ACM Symposium on Document Engineering, pp. 26–33. DocEng '02, Association for Computing Machinery, New York, NY, USA (2002). https://doi.org/10.1145/585058.585065
9. Haselböck, S., Weinreich, R., Buchgeher, G.: Decision models for microservices: design areas, stakeholders, use cases, and requirements. In: Lopes, A., de Lemos, R. (eds.) ECSA 2017. LNCS, vol. 10475, pp. 155–170. Springer, Cham (2017). https://doi.org/10.1007/978-3-319-65831-5_11
10. Keeling, M.: The psychology of architecture decision records. IEEE Softw. **39**(6), 114–117 (2022). https://doi.org/10.1109/MS.2022.3198195
11. Kleehaus, M., Matthes, F.: Challenges in documenting microservice-based it landscape: a survey from an enterprise architecture management perspective. In: Proceedings of the 2019 IEEE 23rd International Enterprise Distributed Object Computing Conference (EDOC), pp. 11–20 (2019).https://doi.org/10.1109/EDOC.2019.00012
12. Kopp, O., Armbruster, A., Zimmermann, O.: Markdown architectural decision records: format and tool support. In: ZEUS, pp. 55–62 (2018)
13. Lee, L., Kruchten, P.: A tool to visualize architectural design decisions. In: Becker, S., Plasil, F., Reussner, R. (eds.) QoSA 2008. LNCS, vol. 5281, pp. 43–54. Springer, Heidelberg (2008). https://doi.org/10.1007/978-3-540-87879-7_3
14. Liang, P., Jansen, A., Avgeriou, P.: Knowledge Architect: A Tool Suite for Managing Software Architecture Knowledge. University of Groningen, Johann Bernoulli Institute for Mathematics and Computer Science (2009). relation: http://www.rug.nl/informatica/organisatie/overorganisatie/iwi Rights: University of Groningen, Research Institute for Mathematics and Computing Science (IWI)
15. Manteuffel, C., Avgeriou, P., Hamberg, R.: An exploratory case study on reusing architecture decisions in software-intensive system projects. J. Syst. Softw. **144**, 60–83 (2018). https://doi.org/10.1016/j.jss.2018.05.064

16. Manteuffel, C., Tofan, D., Koziolek, H., Goldschmidt, T., Avgeriou, P.: Industrial implementation of a documentation framework for architectural decisions. In: Proceedings of the 2014 IEEE/IFIP Conference on Software Architecture, pp. 225–234. IEEE (2014)
17. Nygard, M.: Documenting architecture decisions (2011). https://www.cognitect.com/blog/2011/11/15/documenting-architecture-decisions. Accessed 12 Apr 2024
18. Rost, D., Naab, M., Lima, C., von Flach Garcia Chavez, C.: Software architecture documentation for developers: a survey. In: Drira, K. (ed.) ECSA 2013. LNCS, vol. 7957, pp. 72–88. Springer, Heidelberg (2013). https://doi.org/10.1007/978-3-642-39031-9_7
19. Saldaña, J.: The coding manual for qualitative researchers (2013)
20. Shahin, M., Liang, P., Khayyambashi, M.R.: Architectural design decision: existing models and tools. In: Proceedings of the 2009 Joint Working IEEE/IFIP Conference on Software Architecture & European Conference on Software Architecture, pp. 293–296 (2009). https://doi.org/10.1109/WICSA.2009.5290823
21. Staron, M.: Action Research in Software Engineering. Springer, Berlin, Heidelberg (2020). https://doi.org/10.1007/978-3-030-32610-4
22. Tyree, J., Akerman, A.: Architecture decisions: demystifying architecture. IEEE Softw. **22**(2), 19–27 (2005). https://doi.org/10.1109/MS.2005.27
23. van Heesch, U., Avgeriou, P., Hilliard, R.: A documentation framework for architecture decisions. J. Syst. Softw. **85**(4), 795–820 (2012). https://doi.org/10.1016/j.jss.2011.10.017

Towards Teamwise Informed Decisions On Microservice Security Smells

Francisco Ponce[1]([✉])[iD], Jacopo Soldani[1][iD], Hernán Astudillo[2][iD],
and Antonio Brogi[1][iD]

[1] University of Pisa, Pisa, Italy
francisco.ponce@di.unipi.it, {jacopo.soldani,antonio.brogi}@unipi.it
[2] ITiSB, Universidad Andrés Bello, Viña del Mar, Chile
hernan.astudillo@unab.cl

Abstract. Security smells, i.e., possible symptoms of bad security decisions, can occur in microservice-based applications, potentially resulting in violations of key security properties. The decision of whether or not to refactor a service to mitigate the potential effects of security smells is complex, considering the distributed responsibility of services across teams and the possible impact on their development schedules. In this work-in-progress paper, we propose a team-centric approach that provides insights into the effects of refactorings on quality attributes, the urgency and effort of a refactoring, and its implications for other teams. The ultimate goal is to support teams in making decisions in the context of microservice-based application security and to improve the scheduling of the refactorings that mitigate the potential effects of microservice security smell instances.

Keywords: microservices · teamwise · security smells · refactoring

1 Introduction

Microservice-based applications (MSAs) enable building *cloud-native* applications [7], namely applications that can fully exploit the capabilities of cloud computing, exhibiting distributed, dynamic, and fault-resilient behavior [22]. Along with their benefits, MSAs also introduce new security challenges, including the so-called *security smells*, i.e. symptoms of bad security decisions that may impact key security properties [16]. Security smells must be carefully checked, and if possible resolved by refactoring the application or the services therein, to preempt security issues.

When multiple instances of different security smells appear in an MSA, it is necessary to identify the refactorings that mitigate the effects of the security smell instances that demand urgent attention [17]. Subsequently, decisions must be made regarding the implementation of these refactorings, e.g., which ones should be implemented. To this end, teams must consider, among other things, the urgency of each refactoring, its impact on quality attributes (QAs), and the effort required for its implementation.

On the other hand, given the nature of MSAs, the responsibility for the services composing an MSA is distributed among different teams. Refactoring a specific service to mitigate the effects of security smells could impact how the service interacts with others. Consequently, the decision of refactoring a service may also impact the decisions and development schedule of another team responsible for a different service.

In this work-in-progress paper, inspired by ATAM utility trees [10], we aim at supporting teams in making decisions in the context of MSA security and to improve the scheduling of the refactorings that mitigate the potential effects of microservice security smell instances. In concrete, we propose a team-centric approach that provides insights into the relevance of each assigned service to the business, the importance of its key quality attributes, and the refactorings that a team needs to implement. This view also provides information about the implementation effort of each refactoring and the urgency of its associated security smell instance [17].

The rest of this paper is organized as follows. Section 2 provides a motivating scenario based on the Lakeside Mutual MSA. Section 3 introduces the team-centric approach and illustrates it using the motivating scenario. Finally, Sects. 4 and 5 discuss related work and draw some concluding remarks, respectively.

2   Motivating Scenario

To illustrate our proposal, we will use the Lakeside Mutual MSA[1], a third-party MSA that simulates an insurance company. Figure 1 shows the services composing the architecture of the Lakeside Mutual MSA. The colors in the figure represent the four development teams that are part of the insurance company. The Yellow, Purple, Red, and Blue Teams are then responsible for the services of the Lakeside Mutual MSA. For our motivating scenario, we will focus on Purple Team and its two assigned services, the *Policy Management Frontend* and *Policy Management Backend* services.

For our motivating scenario, we will rely on the 32 security smell instances affecting the Lakeside Mutual MSA identified in our previous work [17]. Table 1 contains the 8 security smell instances affecting the *Policy Management Frontend* and *Policy Management Backend* services.

Now, suppose that the Product Owner of the Lakeside Mutual insurance company specifies the relevance of each service and the importance of its quality attributes. Table 2 indicates the information specified for the Purple Team services, i.e., the relevance of each service for the business and the importance of its key quality attributes.

With all the information indicated above, the Purple Team can start identifying the urgency of the 8 security smell instances currently affecting its services,

---

[1] https://github.com/Microservice-API-Patterns/LakesideMutual.

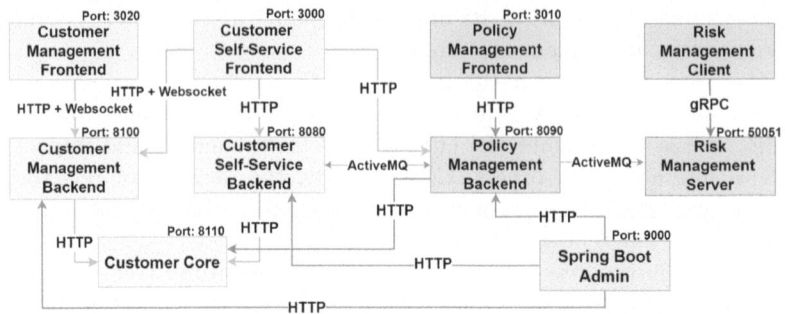

Fig. 1. Lakeside Mutual MSA - Colors represent the company Teams and their services.

Table 1. The 8 security smell instances affecting Purple Team services.

Security smell	Affected services
Publicly accessible microservices	*Policy Management Backend,* *Policy Management Frontend*
Insufficient access control	*Policy Management Backend*
Unauthenticated traffic	*Policy Management Backend,* *Policy Management Frontend*
Hardcoded secrets	*Policy Management Backend*
Non-secured service-to-service communications	*Policy Management Backend,* *Policy Management Frontend*

using, e.g., the TriSS method [17]. However, even when the urgency of security smell instances is known, it remains the open question of which security smell instance to start first. The choice is indeed not only based on the urgency of the smell instance itself, but driven also by other aspects, e.g., the effort needed to apply the refactoring for resolving the smell instance, whether interactions with other teams are needed, and whether a team has enough time and resources to deal with such effort and interactions.

3 Team-Centric Approach

We propose a team-centric approach in response to the question raised in Fig. 2. Our approach places every service under the purview of the team, offering insights into the relevance of each service to the overall business (provided by the Product Owner), the importance of its key QAs, and the list of refactorings that the team needs to implement. Additionally, it provides information on the effort required for the implementation and the urgency of the associated smell instance for each refactoring. Lastly, it considers the potential impact on other teams resulting from the decision to implement each specific refactoring.

Table 2. Relevance and quality attributes importance for Purple Team services.

Service Name	Relevance	Importance of QAs
Policy Management Backend	*High*	*High*: integrity, authenticity. *Medium*: confidentiality. *Low*: analyzability
Policy Management Frontend	*Medium*	*High*: integrity, authenticity. *Medium*: confidentiality. *Low*: analyzability

The proposed approach aims to enable teams to make more informed decisions and, consequently, better scheduling of the development timelines. To illustrate this we will retake our motivating scenario. Figure 2 showcases the Purple Team view containing the information described above. Each of the teams that are part of the Lakeside Mutual insurance company will have their own view available, which will contain similar information as to that in Fig. 2, but adapted to the services for which they are responsible.

As we can observe in Fig. 2, the left part of the view shows the services for which the Purple Team is responsible, i.e., the *Policy Management Frontend* and *Policy Management Backend* services. In the center of the view, we have the key quality attributes and their importance defined by the Product Owner as indicated in Table 2. Finally, the right part of the view shows the refactorings that the Purple Team needs to implement, together with the urgency of the security smell instances that they mitigate (as the outcome of our approach proposed in [17]) and the effort needed for its implementation. Each refactoring is associated with the key quality attributes that it helps to achieve.

The Purple Team now has information about the refactorings that should be applied to its services, the urgency of the security smell instances for which the refactorings are needed, and the interaction with other teams that should be enacted to apply these refactorings, if any.

For example, the refactorings needed to resolve the smells affecting the *Policy Management Backend* service impact other services owned respectively by the Yellow and Blue Teams. The view in Fig. 2 also displays that the Yellow Team is also planning to implement some of the suggested refactorings (indicated by the yellow lines), whilst the Blue Team is not planning to implement any of the displayed refactorings. In any case, this means that an interaction with the Yellow or Blue Team is required to implement the refactorings needed to resolve the smells affecting the *Policy Management Backend* service.

Suppose now that – for the next development sprint – the Purple Team can implement a refactoring up to Medium effort, and that it does not have enough time and resources to interact and coordinate with other teams. If this is the case, even if the most urgent refactorings are adding OpenID Connect and Mutual TLS to the *Policy Management Backend* service, the Purple Team

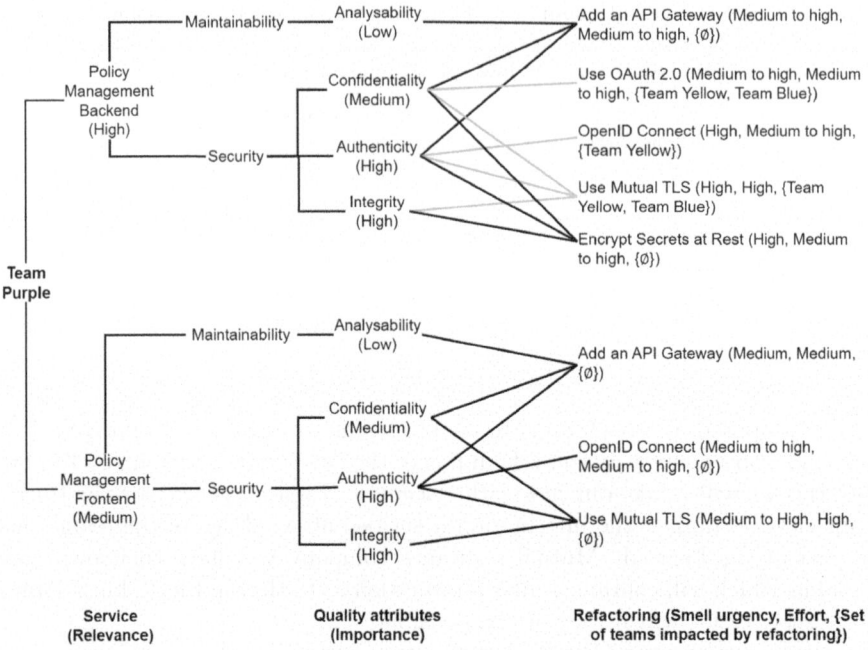

Fig. 2. Example of view for the Purple Team.

may decide to postpone them to follow up sprints, when the required time and resources get available.

The Purple Team could then focus on the bottom part of the view in Fig. 2, which links the *Policy Management Frontend* service with the refactorings it needs to resolve the smells it is affected by. Such refactorings are impacting only the services owned by the Purple Team, and therefore do not require setting up interactions with other teams. Given that the Purple Team – according to our assumptions – can only devote a Medium effort to the next development sprint, the choice of which refactoring to apply naturally falls on adding an API gateway for the *Policy Management Frontend* service.

The above described decision process is that envisioned by our proposal, and it was instantiated based on a given context for the running example. Of course, different contexts may end up with different decisions taken by different teams. The idea here is to support such teams to come to the "most convenient" decision for the context, namely finding a trade-off among, e.g., the effort needed to apply the refactoring for resolving the smell instance, whether interactions with other teams are needed, and whether a team has enough time and resources to deal with such effort and interactions.

4   Related Work

Refactoring MSAs to resolve the issues therein is crucial nowadays [4,11]. However, despite there exist solutions for detecting microservice security smell instances and reasoning about their resolution, e.g., [6,23], the problem of deciding which refactorings to apply first to resolve the detected security smell instances is still open [5]. Indeed, several approaches have been developed for prioritizing refactorings when enacting issue resolution, but, to the best of our knowledge, our proposal is the first providing a team-centric support to decide which refactorings to apply to resolve the security smell instances affecting the microservices owned by a team.

The existing approaches differ from our proposal as they typically focus on prioritizing refactorings with the ultimate goal of resolving *code smells* [20], instead we focus on the *MSA security smells* identified in [16]. For example, [9] ranks code smells by analyzing execution traces obtained with runtime tests, whereas [8] and [13] exploit different static analysis tools to estimate the severity of code smells in object-oriented applications. The above approaches differ from ours as they focus on a different type of smells affecting object-oriented applications, and because they are not intended to feature a team-centric view like that in our proposal.

Other existing approaches prioritize refactorings for code smell resolution by proposing source code analysis methods, e.g., [1,2], or by using machine learning models to rank the code smells affecting an application, e.g., [3,12,14]. Closer to our proposal, other approaches prioritize the refactoring of code smell instances by considering the context in which they occur, with priorities that may differ (even for instances of the same code smell) depending on the affected application components. In particular, [18,19] compute a context-relevant index by processing the information available in the issue tracking system of the affected component, whereas [21] ranks code smells using a combination of three criteria, namely past modifications of the affected components, their actual modifiability, and the known relevance of the smells themselves. Differently from our proposal, the methods mentioned above focus on refactoring code smells for component-based applications rather than security smells in MSAs, and they are not intended to feature a team-centric view like that in our proposal.

Finally, the closest to our proposal is perhaps the trade-off analysis in [15], which first considered the possible impacts of MSA security smells on quality attributes for affected services, by proposing to visualize such impacts as softgoal-interdepency graphs. However, the proposal in [15] differs from ours as it considers the impacts of security smells in *general*, without considering the contextual information given by security smell *instances*. We complement the proposal in [15] by providing a support for teams to decide how to prioritize the refactoring of the MSA security smell instances affecting the services that they own, while also considering the refactorings' urgency, the effort needed, and the possible interactions with other teams.

5   Conclusions

In this work-in-progress paper, we have introduced a team-centric approach to support teams in the decision-making processes related to MSA security smells [16], and the scheduling of refactorings to mitigate the effects of security smell instances. Our approach places every service under the purview of each team providing information on the effort needed to implement each refactoring and the urgency of each security smell instance. Furthermore, it provides information on the implications of these decisions on other teams. We have also showcased our approach by applying it to the Lakeside Mutual MSA, which was used as a reference architecture in our motivating scenario.

For future work, we envision making the proposed team-centric approach concrete by devising decision models that enable implementing it. Additionally, we plan to validate these decision models through empirical studies, e.g., controlled experiments with practitioners, or an exploratory study with security architects, which should allow us to collect information regarding the usefulness and applicability of the decision models. Finally, we also plan to collect information regarding the general appreciation of the effort needed to implement each of the refactorings indicated in [16].

Acknowledgments. This work was supported by the project FREEDA (CUP: I53D23003550006), funded by the frameworks PRIN (MUR, Italy) and Next Generation EU.

References

1. Alshammari, T., Alshayeb, M.: Toward a software bad smell prioritization model for software maintainability. Arabian J. Sci. Eng. **46** (2021). https://doi.org/10.1007/s13369-021-05766-6
2. Arcelli Fontana, F., Ferme, V., Zanoni, M., Roveda, R.: Towards a prioritization of code debt: a code smell intensity index. In: 2015 IEEE 7th International Workshop on Managing Technical Debt (MTD), pp. 16–24 (2015). https://doi.org/10.1109/MTD.2015.7332620
3. Aung, T.W.W., Wan, Y., Huo, H., Sui, Y.: Multi-triage: a multi-task learning framework for bug triage. J. Syst. Softw. **184**, 111133 (2022). https://doi.org/10.1016/j.jss.2021.111133
4. Besker, T., Martini, A., Bosch, J.: Technical debt triage in backlog management. In: 2019 IEEE/ACM International Conference on Technical Debt (TechDebt), pp. 13–22 (2019). https://doi.org/10.1109/TechDebt.2019.00010
5. Cerny, T., Abdelfattah, A.S., Maruf, A.A., Janes, A., Taibi, D.: Catalog and detection techniques of microservice anti-patterns and bad smells: a tertiary study. JSS **206**, 111829 (2023). https://doi.org/10.1016/j.jss.2023.111829
6. Dell'Immagine, G., Soldani, J., Brogi, A.: Kubehound: detecting microservices' security smells in Kubernetes deployments. Future Internet **15**(7) (2023). https://doi.org/10.3390/fi15070228
7. Gannon, D., Barga, R., Sundaresan, N.: Cloud-native applications. IEEE Cloud Comput. 4(5), 16–21 (2017). https://doi.org/10.1109/MCC.2017.4250939

8. Gupta, A., Chauhan, N.K.: A severity-based classification assessment of code smells in kotlin and java application. Arabian J. Sci. Eng. **47** (2022). https://doi.org/10.1007/s13369-021-06077-6

9. Haendler, T., Sobernig, S., Strembeck, M.: Towards triaging code-smell candidates via runtime scenarios and method-call dependencies. In: Proceedings of the XP2017 Scientific Workshops. XP '17, ACM (2017). https://doi.org/10.1145/3120459.3120468

10. Kazman, R., Klein, M., Barbacci, M., Longstaff, T., Lipson, H., Carriere, J.: The architecture tradeoff analysis method. In: Proceedings. Fourth IEEE International Conference on Engineering of Complex Computer Systems (Cat. No.98EX193), pp. 68–78 (1998). https://doi.org/10.1109/ICECCS.1998.706657

11. Lenarduzzi, V., Besker, T., Taibi, D., Martini, A., Arcelli Fontana, F.: A systematic literature review on technical debt prioritization: strategies, processes, factors, and tools. J. Syst. Softw. **171**, 110827 (2021). https://doi.org/10.1016/j.jss.2020.110827

12. Lim, S., Zaidi, S., Woo, H., Lee, C.G.: Toward an effective bug triage system using transformers to add new developers. J. Sens. **2022** (2022). https://doi.org/10.1155/2022/4347004

13. Malhotra, R., Singh, P.: Exploiting bad-smells and object-oriented characteristics to prioritize classes for refactoring. Int. J. Syst. Assur. Eng. Manag. **11** (2020). https://doi.org/10.1007/s13198-020-01001-x

14. Pecorelli, F., Palomba, F., Khomh, F., De Lucia, A.: Developer-driven code smell prioritization. In: Proceedings of the 17th International Conference on Mining Software Repositories, pp. 220–231. MSR '20, ACM (2020). https://doi.org/10.1145/3379597.3387457

15. Ponce, F., Soldani, J., Astudillo, H., Brogi, A.: Should microservice security smells stay or be refactored? Towards a trade-off analysis. In: Gerostathopoulos, I., Lewis, G., Batista, T., Bureš, T. (eds.) Software Architecture. ECSA 2022. LNCS, vol. 13444, pp. 131–139. Springer, Cham (2022). https://doi.org/10.1007/978-3-031-16697-6_9

16. Ponce, F., Soldani, J., Astudillo, H., Brogi, A.: Smells and refactorings for microservices security: a multivocal literature review. J. Syst. Softw. **192**, 111393 (2022). https://doi.org/10.1016/j.jss.2022.111393

17. Ponce, F., Soldani, J., Taramasco, C., Astudillo, H., Brogi, A.: Triaging microservice security smells, with TriSS. In: Proceedings of the 28th International Conference on Evaluation and Assessment in Software Engineering, pp. 698–706. EASE '24, Association for Computing Machinery, New York, NY, USA (2024). https://doi.org/10.1145/3661167.3661282

18. Sae-Lim, N., Hayashi, S., Saeki, M.: Context-based code smells prioritization for prefactoring. In: 2016 IEEE 24th International Conference on Program Comprehension (ICPC), pp. 1–10. IEEE (2016). https://doi.org/10.1109/ICPC.2016.7503705

19. Sae-Lim, N., Hayashi, S., Saeki, M.: Revisiting context-based code smells prioritization: on supporting referred context. In: Proceedings of the XP2017 Scientific Workshops. XP '17, Association for Computing Machinery, New York, NY, USA (2017). https://doi.org/10.1145/3120459.3120463

20. Verma, R., Kumar, K., Verma, H.K.: Code smell prioritization in object-oriented software systems: a systematic literature review. J. Softw. Evolut. Process **35**(12), e2536 (2023). https://doi.org/10.1002/smr.2536

21. Vidal, S.A., Marcos, C., Díaz-Pace, J.A.: An approach to prioritize code smells for refactoring. Autom. Softw. Eng. **23**, 501–532 (2016). https://doi.org/10.1007/s10515-014-0175-x

22. Wang, Y., Kadiyala, H., Rubin, J.: Promises and challenges of microservices: an exploratory study. Empir. Softw. Eng. **26**(4), 63 (2021). https://doi.org/10.1007/s10664-020-09910-y
23. Wizenty, P., et al.: Towards resolving security smells in microservices, model-driven. In: 18th International Conference on Software Technologies (ICSOFT), pp. 15–26. INSTICC, SciTePress (2023). https://doi.org/10.5220/0012049800003538

Automated Quality Concerns Extraction from User Stories and Acceptance Criteria for Early Architectural Decisions

Khubaib Amjad Alam[1,2](✉), Hira Asif[2], Irum Inayat[2],
and Saif-Ur-Rehman Khan[3]

[1] College of Engineering, Al Ain University, Abu Dhabi, UAE
khubaib.amjad@nu.edu.pk
[2] National University of Computer and Emerging Sciences (FAST-NUCES),
Islamabad, Pakistan
{i220735,irum.inayat}@nu.edu.pk
[3] Department of Computing, Shifa Tameer-e-Millat University (STMU) Park Road
Campus, Islamabad, Pakistan
saif-rehman.ssc@stmu.edu.pk

Abstract. User stories serve as a fundamental tool in agile software development methodologies, articulating the functional requirements of a system from an end-user perspective. However, while user stories are crucial for capturing the desired features and functionalities, they frequently overlook the non-functional aspects critical to the system's success. Despite their paramount importance, these quality concerns often remain implicit or underrepresented in user stories, necessitating a deliberate effort to extract them during the elicitation and architectural analysis phases. Failure to address these quality concerns upfront may lead to poor architectural decisions. Consequently, this oversight may result in sub-optimal system designs, increased development costs, delayed time-to-market, diminished user satisfaction, and increased operational risks. This paper presents an ISO-25010 compliant Transfer Learning approach for automated quality concerns extraction from user stories and corresponding acceptance criteria. The proposed solution is constructed upon the Transformer-based RoBERTa-Large model, leveraging and extending its pre-trained capabilities. This approach proficiently classifies user stories and acceptance criteria into 5 most critical user quality concerns including Usability, Performance, Reliability, Security, and Compatibility. This process involves cleaning and preprocessing the dataset followed by fine-tuning the pre-trained models on the refined data set. A comparative analysis of the Three mainstream BERT variants including RoBERTa-base, DistilBERT and XLNET is also provided. Considering the non-availability of public data sets in this scope, a dataset of approximately 1000 user stories with acceptance criteria was compiled from diverse sources and real-world projects. This dataset was subsequently labeled through an extensive labeling activity. The findings suggest that the RoBERTa-Large fine-tuned variant achieves an impressive level of performance in terms of accuracy, precision, recall and Avg F1 score.

© The Author(s), under exclusive license to Springer Nature Switzerland AG 2024
M. Galster et al. (Eds.): ECSA 2024, LNCS 14889, pp. 359–367, 2024.
https://doi.org/10.1007/978-3-031-70797-1_24

Keywords: User stories · BERT · Language Models · ISO-25010 · Quality Attributes · NLP

1 Introduction

User stories have become a prevalent and integral component of modern software development methodologies and have been found to provide benefits compared to previous methods of managing requirements [1–4]. User stories are a popular artifact in agile software development containing a wealth of information, aspects and user concerns. Natural language processing (NLP) techniques can be used to extract information from user stories. NLP techniques, such as part-of-speech chunking, Named Entity Recognition (NER), and dependency parsing, can help identify the aspects of who, what, and why in user stories. Part-of-speech (POS) tags are the most prolific approach for information extraction from USs. However, semantic approaches such as Topic Modeling and machine learning have recently gained momentum in this domain [5]. NLP research on user stories has primarily focused on extracting conceptual models, defects, roles, and scenarios, and tracing links between models and user stories [6]. Additionally, NLP can be used to automatically extract classes from user stories, improving the requirements analysis and design stages of software development [7]. Computational linguistic methods, including recurrent neural networks, have been employed to detect and handle user stories extracted from problem reports within software quality tracking systems [8]. Extracting quality concerns from user stories assumes paramount significance as it facilitates a comprehensive comprehension of the system's non-functional requirements. Furthermore, this process enables architects to prioritize and delineate the non-functional requirements that are most salient to stakeholders and align with the overarching objectives of the project. Through meticulous analysis and interpretation of user stories, architects can distill implicit quality concerns and translate them into actionable architectural decisions. In essence, the extraction of quality concerns from user stories serves as a foundational step in the architectural analysis and design process, facilitating the synthesis of functional and non-functional requirements to inform informed decision-making. By bridging the gap between user-centric narratives and architectural imperatives, this approach fosters the development of robust, scalable, secure, reliable, compatible, and user-friendly systems that align with stakeholder expectations and project goals. In this paper, we proposed an approach for automated quality concerns extraction according to the ISO 25010 quality model from user stories and acceptance criteria. We fine-tuned a Transformer-based pre-trained BERT variants to predict quality concerns present in user stories and acceptance criteria.

2 Related Work

Extracting quality concerns from user stories is crucial for making informed architectural decisions because it helps in understanding the non-functional requirements of the system. Gilson et al. Ahmed et al. [9] proposed a conceptual framework that contains two components i.e. QA extractor and QA prioritiser. This framework is used for extracting and prioritizing quality attributes using regular expressions.

Gilson et al. [10] suggested automatically analyzing user stories to find quality aspects that may be implicitly stated in them, thus reducing the need to rely on their intuition or potential bias from prior experiences. Architects may use this identification as a starting point for their decision-making. A machine learning method for identifying quality qualities as outlined in the ISO/IEC 25010 quality models was presented in this paper. To validate their method, they employed a publicly available dataset of 1,675 user stories that they had manually labeled. They had a relatively small dataset to test a machine learning classifier, the results were acceptable, particularly stories related to security, compatibility, and performance, achieving 74% precision. We conducted an extensive literature review to ascertain the type of information contained within user stories and the methodologies employed for their extraction. Figure 1 shows the artifacts that have been extracted from user stories.

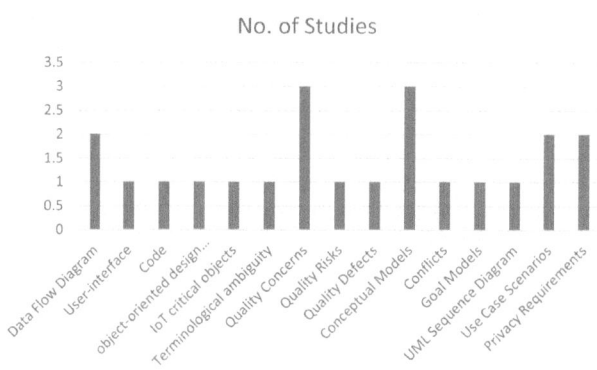

Fig. 1. Information extracted from User Stories

3 Proposed Method

The objective of our approach is to identify user stories that encompass quality concerns, thereby aiding in the enhancement of architectural design decisions. The suggested approach's working mechanism is explained in detail in this section. We experimented with our dataset using RoBBERTa-base, RoBERTa-large, BERT-large, XLNET and DistilBERT. This chapter is structured into five subsections delineating the text preprocessing methodology employed, followed by an exposition on the chosen model and the subsequent process of model training. Figure 2 illustrates the sequence of activities undertaken within the proposed framework.

3.1   Research Questions

The central research focus of this study revolves around identifying the transformer-based models demonstrating the highest proficiency.We are using transformer-based models to predict quality concerns present in user stories and acceptance criteria. The research questions formulated for this research work are as follow:

RQ 1: Are transformer-based models a better approach than traditional deep learning models for the quality concerns extraction?

RQ 2: How effectively does our method classify user stories with acceptance criteria to different categories of quality concerns according to the ISO-25010 quality model using transformer-based models?

3.2   Data Collection

Publicly accessible datasets including both user stories and acceptance criteria are non-existent. Only a single publicly available dataset comprising user stories is available, encompassing 21 software projects. However, it lacks associated acceptance criteria. Given this limitation, we compiled a new dataset by sourcing user stories and acceptance criteria from various outlets, including publicly accessible software project documentation, real-world industrial projects, GitHub repositories, and blogs. This compilation effort yielded approximately 3000 user stories paired with their respective acceptance criteria. However, manual labeling (according to the standard quality model ISO 25010) of the dataset revealed that not all user stories with acceptance criteria addressed quality concerns. Subsequently, we identified approximately 1000 user stories with acceptance criteria that contained at least one quality concern. However, the frequency of occurrence of quality concerns is not uniform. Usability and reliability were the most prolific quality concerns in the collected data. In contrast, compatibility and security are rather scarce in the collected data with limited training samples. This imbalance highlights an important future need of curated and balanced datasets.

3.3   Text Pre-processing

Text pre-processing involves removing any special characters, symbols, or numerical values from the text. This helps reduce noise and ensures that the model focuses on meaningful information. Subsequently, we converted all text to lowercase to ensure uniformity and prevent the model from treating the same word in different cases as different entities.

3.4   Model Training and Evaluation

We fine-tuned RoBERTa-base, RoBERTa-large, BERT-large, DistilBERT and XLNET transformer-based models to the most appropriate hyperparameters.

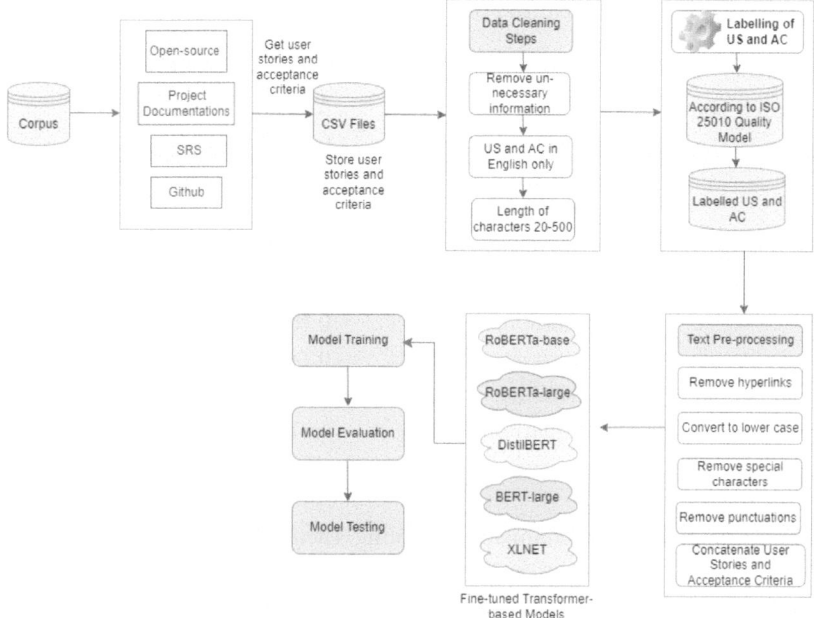

Fig. 2. Architecture of the Proposed approach

Hyperparameters for RoBERTa-Large include Learning Rate 7e-6, Batch size 4 for training, Epoch 32, and we also used an optimizer (Adam).

Standard evaluation metrics, namely Accuracy, Precision, Recall, and F1-score, were employed for assessing the performance of the models. After scrutinizing the evaluation outcomes, the model exhibiting the highest performance is selected for subsequent investigation and deliberation.

In our experimental setup, we employed RoBBERTa-base, RoBERTa-large, BERT-large, XLNET and DistilBERT models. Subsequently, these models underwent fine-tuning using our dataset. Extensive hyperparameter tuning was also performed to get the desired level of performance from the selected models.

4   Results and Discussions

This section discusses the results of the evaluation process according to the defined research questions.

RQ 1: Transformer-based models offer several advantages over traditional deep learning architectures when applied to tasks involving natural language processing (NLP), such as the analysis of user stories and acceptance criteria in software engineering datasets.

Firstly, traditional deep learning models often struggle to capture long-range dependencies in the textual data [11]. In contrast, Transformer-based models

Table 1. Comparison of CNN and RoBERTa-large Model

Model	Label	Accuracy	Precision	Recall	F1-score
CNN [10]	Usability	–	0.61	0.20	0.29
	Performance	–	0.57	0.31	0.39
	Security	–	0.66	0.50	0.57
	Reliability	–	0.29	0.14	0.18
	Compatibility	–	0.77	0.55	0.63
RoBERTa-large	Usability	0.77	0.78	0.77	0.77
	Performance	0.74	0.70	0.74	0.72
	Security	0.77	0.78	0.77	0.77
	Reliability	0.70	0.74	0.70	0.72
	Compatibility	0.83	0.77	0.83	0.79

leverage self-attention mechanisms, enabling them to efficiently capture relationships between tokens across the entire input sequence.

Secondly, Transformers offer inherent parallelization capabilities, allowing them to process tokens in parallel rather than sequentially [11].

Furthermore, Transformer-based models excel at capturing contextual information from text, a critical aspect for understanding the nuanced relationships between different elements within user stories [12].

We conducted a comparative analysis of our results obtained using RoBERTa-large against those from a model trained using a convolutional neural network (CNN) [10]. The analysis revealed that our model outperformed the conventional deep learning model in terms of accuracy as shown in Table 1. Gilson et al. [10] used a dataset exclusively composed of user stories, in contrast to our approach where we utilized a dataset that included both user stories and their acceptance criteria. Our findings indicate that the inclusion of acceptance criteria in user stories contributes to the identification of a greater number of quality concerns.

RQ 2: To answer this research question, We formulated multiple solution pipelines by fine-tuning several mainstream BERT variants including RoBERTa-base, RoBERTa-large, BERT-large, DistilBERT and XLNET on 5 quality concerns. Subsequently, we evaluated the performance of these models using Accuracy, Precision, Recall and F1-score evaluation metrics. Table 1 shows the comparative results of different BERT variants in terms of Accuracy and enlists the results of precision, Recall and F1 scores for each class. The findings indicate that RoBERTa-large consistently achieves the highest levels of Accuracy, Precision, Recall and F1 (Table 2).

Table 2. Comparative performance of the Proposed approach with different BERT variants

Model	Label	LR	BS	Acc.	Precision	Recall	F1
XLNET	Usability	7e−6	4	0.53535	0.52724	0.53535	0.53084
	Performance	7e−6	4	0.68686	0.60741	0.68686	0.64342
	Security	7e−6	4	0.80303	0.76509	0.80303	0.78059
	Reliability	7e−6	4	0.57575	0.71116	0.57575	0.51597
	Compatibility	7e−6	4	0.83333	0.72594	0.83333	0.77594
Overall	−	−	−	0.68686	0.66736	0.68686	0.64935
BERT-large	Usability	5e−6	4	0.58585	0.53586	0.58585	0.54032
	Performance	7e−6	4	0.70202	0.59787	0.70202	0.64577
	Security	7e−6	4	0.76262	0.73060	0.76262	0.74558
	Reliability	7e−6	4	0.39898	0.67513	0.39898	0.43222
	Compatibility	7e−6	4	0.83333	0.80617	0.83333	0.81645
Overall	−	−	−	0.65656	0.66912	0.65656	0.63606
DistilBERT	Usability	7e−6	4	0.63636	0.60461	0.63636	0.60470
	Performance	7e−6	4	0.61111	0.65970	0.61111	0.63192
	Security	7e−6	4	0.75757	0.76382	0.75757	0.76063
	Reliability	7e−6	4	0.60606	0.67119	0.60606	0.63288
	Compatibility	7e−6	4	0.83838	0.77472	0.83838	0.79393
Overall	−	−	−	0.68989	0.69480	0.68989	0.68481
RoBERTa-base	Usability	7e−6	4	0.67272	0.68148	0.67272	0.67694
	Performance	7e−6	4	0.70707	0.59888	0.70707	0.64849
	Security	7e−6	4	0.75757	0.75757	0.75757	0.75757
	Reliability	7e−6	4	0.75252	0.67354	0.75252	0.70157
	Compatibility	7e−6	4	0.82828	0.74933	0.82828	0.78106
Overall	−	−	−	0.74363	0.69216	0.74363	0.71312
RoBERTa-large	**Usability**	7e−6	4	**0.77272**	0.78148	0.77272	0.77694
	Performance	7e−6	4	**0.74141**	0.70305	0.74141	0.72126
	Security	7e−6	4	**0.77272**	0.78148	0.77272	0.77694
	Reliability	7e−6	4	0.70101	0.74752	0.70101	0.72185
	Compatibility	7e−6	4	**0.83333**	0.77943	0.83333	0.79717
Overall	−	−	−	**0.76423**	**0.75859**	**0.76423**	**0.75883**

5 Conclusion and Future Work

Extracting quality concerns from user stories assumes paramount significance as it facilitates a comprehensive comprehension of the system's non-functional requirements. Furthermore, this process enables architects to prioritize and delineate the non-functional requirements that are most salient to stakeholders and align with the overarching objectives of the project. Through meticulous analysis and interpretation of user stories, architects can distill implicit quality concerns and translate them into actionable architectural decisions. In this paper, we proposed an approach for automated quality concerns extraction according to the ISO 25010 quality model from user stories and acceptance criteria. We fine-tuned Transformer-based pre-trained BERT variants to predict quality concerns present in user stories and acceptance criteria. Our findings indicate that the inclusion of acceptance criteria in user stories contributes to the identification of a greater number of quality concerns. Consequently, the results show that RoBERTa-large outperformed the other BERT variants, achieving 78.1% precision(Usability) and 83.3% accuracy(compatibility). In future research, we intend to extend our analysis to additional quality concerns sub-factors of the ISO-25010 quality model for fine-grained quality concerns extraction.

Data Availibility. The data that supports the findings of the current study, are available from the corresponding author upon reasonable request.

References

1. Pokharel, P., Vaidya, P.: A study of user story in practice. In: 2020 International Conference on Data Analytics for Business and Industry: Way Towards a Sustainable Economy (ICDABI), pp. 1–5 (2020)
2. Abrahamsson, P., Fronza, I., Moser, R., Vlasenko, J., Pedrycz, W.: Predicting development effort from user stories. In: 2011 International Symposium on Empirical Software Engineering and Measurement, pp. 400–403 (2011)
3. Popli, R., Chauhan, N., Sharma, H.: Prioritising user stories in agile environment. In: 2014 International Conference on Issues and Challenges in Intelligent Computing Techniques (ICICT), pp. 515–519 (2014)
4. Kustiawan, Y.A., Lim, T.Y.: User stories in requirements elicitation: a systematic literature review. In: 2023 IEEE 8th International Conference on Software Engineering and Computer Systems (ICSECS), pp. 211–216 (2023)
5. Siahaan, D., Raharjana, I.K., Fatichah, C.: User story extraction from natural language for requirements elicitation: identify software-related information from online news. Inf. Softw. Technol. **158**, 107195 (2023)
6. Tovar Onofre, M.Á., Camargo, J.E.: Automatic class extraction from Spanish text of user stories using natural language processing. In: Narváez, F.R., Urgilés, F., Bastos-Filho, T.F., Salgado-Guerrero, J.P. (eds.) SmartTech-IC 2022. CCIS, vol. 1705, pp. 33–47. Springer, Cham (2022). https://doi.org/10.1007/978-3-031-32213-6_3
7. Yuhana, U.L., Rochimah, S., et al.: Identification of conflicts in user story requirements using the clustering algorithm. In: 2022 International Conference on Computer Engineering, Network, and Intelligent Multimedia (CENIM), pp. 1–5. IEEE (2022)
8. Veitía, F.J.P., Roldán, L., Vegetti, M.: User stories identification in software's issues records using natural language processing. In: 2020 IEEE Congreso Bienal de Argentina (ARGENCON), pp. 1–7. IEEE (2020)
9. Ahmed, M., Khan, S.U.R., Alam, K.A.: An NLP-based quality attributes extraction and prioritization framework in agile-driven software development. Autom. Softw. Eng. **30**(1), 7 (2023)
10. Gilson, F., Galster, M., Georis, F.: Extracting quality attributes from user stories for early architecture decision making. In: 2019 IEEE International Conference on Software Architecture Companion (ICSA-C), pp. 129–136 (2019)

11. Devlin, J., Chang, M.-W., Lee, K., Toutanova, K.: BERT: pre-training of deep bidirectional transformers for language understanding. arXiv preprint arXiv:1810.04805 (2018)
12. Škorić, M., Utvić, M., Stanković, R.: Transformer-based composite language models for text evaluation and classification. Mathematics **11**(22) (2023)

Architecture Documentation

SCATS Framework for Software Integration in Software-Defined Vehicle with Cross-Organizational Agile Teams

Jasmin Jahić$^{(\boxtimes)}$ 🆔

University of Cambridge, Cambridge, UK
jj542@cam.ac.uk

Abstract. Embedded subsystems comprising software-defined vehicles have complex integration requirements regarding interfaces between them. While integration approaches today focus on functional alignment of interfaces, the alignment specific to triggering sequences of interfaces and quality (e.g., temporal) properties of interface signals is often overlooked. If such bugs remain undetected until the deployment of software to real hardware, they can introduce significant delivery delays.

To solve the problem of detecting the misalignment in interface and signals in early integration stages, we introduce a framework for Software embedded system integration in Cross-organizational Agile TeamS (SCATS) to transparently manage decisions about interfaces. By complementing Scrum, SCATS reduces the number of bugs that propagate to the later integration stages, and effectively reduce delivery delays.

Keywords: Agile · Scrum · software-defined vehicles · integration · delivery · testing · process · methodology

1 Introduction

Software-defined vehicles [1] are among the most complex software systems today [8]. They are jointly developed by numerous software system providers leading to integration challenges. The problem of designing interfaces and APIs to support software integration is a well-known software architecture topic [15]. While alignment of interface signatures is a relatively well-explored problem [10], the part of the problem that is not well addressed is related to i) managing a large number of interfaces in a system and ii) the quality properties of signals transferred through those interfaces. In systems with hundreds of interfaces, such as in software-defined vehicles, it is necessary to ensure that interfaces are properly implemented and also properly triggered (e.g., if an interface is not triggered, starting sequence of a vehicle might not be able to complete). Besides the functionality of the interfaces, it is necessary to ***test*** that the system parts behave with expected quality (e.g., signals at the interfaces appear within predicted deadlines). Bugs that originate due to misalignment of these requirements are often quite tricky to discover because they manifest only under very specific conditions (e.g., temporal) and tend to propagate into late integration stages. They are often discovered only once the software is deployed at the target hardware (real vehicle), at which point these bugs tend to cause significant delivery delays and are expensive to fix. Therefore, we focus on solving the following problem: *How to ensure alignment of interfaces and signals during software development in software-defined vehicles and effectively prevent propagation of the misalignment bugs to later integration stages?*

To solve this problem, we created a framework for Software embedded system integration in Cross-organizational Agile TeamS (SCATS framework). SCATS framework prescribes methodology that enforces systematic alignment and documentation of changes to interfaces and signals. To achieve this, SCATS introduces milestones and documentation artifacts within existing common development processes. Our framework complements Scrum methodologies and aims to reduce the number of interface misalignment bugs in development releases.

The rest of this paper is structured as follows. In Sect. 2, we discuss known challenges with software integration. In Sect. 3, we introduce our SCATS framework. In Sect. 4, we present the results of our evaluation of SCATS in an industrial environment, and conclude in Sect. 5.

2   Related Work

Van Moll et al. [5], Zafar et al. [13,14], and Ilyas et al. [3] highlighted that combination of stretched innovation goals, increased quality requirements, and shortened time-to-market causes integration challenges in the complex software systems. Herbsleb et al. [2] underlined that geographically distributed development teams face extraordinary communication and coordination problems. Miller [4] argues that Agile is challenged by geographically and temporally distributed teams, and that a team's heavy reliance on continuous integration further sharpens the impact of these challenges.

Mårtensson et al. [6] investigated continuous integration impediments in large-scale industry projects and argued that companies and teams might be practicing continuous integration at one level, but not at another, with different integration frequencies at different levels. Ståhl et al. [11] differentiated six levels of integration tiers and test scopes: 1) component; 2) subsystem; 3) partial product; 4) full product; 5) release; 6) customer. Additionally, they indicated that depending on these different levels, the length of the testing feedback loop varies significantly between i) immediate/minutes; ii) hours; iii) days; iv) weeks; v) once per release (where release is usually measured in months). Tufano et al. [12] acknowledged the problems caused by a separate integration stage in the software process: unpredictability and large integration effort. Shatil et al. [9] outlined the challenges that appear in the adoption process of agility into a system engineering project, and made preliminary suggestions for overcoming these challenges. Mårtensson et al. [7] highlighted that Agile development or continuous practices often implicitly assume a steady state where small features or partial features can be added one on top of the other in a continuous fashion. When that is not the case, however, they indicated it is less than obvious how to apply continuous practices "by the book". Zafar et al. [14] explored integration failure factors and devised a taxonomy of 40 such factors.

There is a good awareness of integration challenges in the community. However, what is missing are approaches to address these challenges.

3   SCATS: Framework for Software Embedded System Integration in Cross-Organizational Agile TeamS

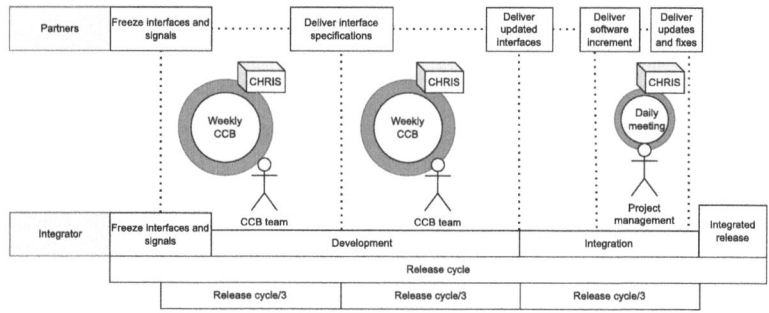

Fig. 1. SCATS Software Integration Process in Software-defined Vehicles

A common approach towards software integration in automotive industry is a collaboration between an integrator and partners to create regular software releases. All involved parties develop their software independently, and the integrator performs the integration based on commonly *aligned* interfaces and interface signals. The problem is that this *alignment* also occurs dynamically between individual developers and can lead to misalignments on the system level. To eliminate misalignment of interfaces and signals between different software systems, we have developed a framework for Software embedded system integration in Cross-organizational Agile TeamS (SCATS framework, Fig. 1). SCATS complements existing methodologies with new milestones and artifacts. These help engineers to explicitly prescribe interfaces between software systems (as architectural decisions). To transparently make changes in interfaces and enable discussion about the potential changes, we introduce a Change Request for Interfaces and Signals (CHRIS) template.

Let us start by discussing the complements that SCATS introduces to Agile. *Milestones* define deadlines when all involved parties have to provide certain artefacts. For example, align and provide information related to interface-related changes, in order to ensure that each party is enabled to proceed with its development, considering their respective development methodology. SCATS prescribes following milestones (Fig. 1): i) freeze interfaces and signals, ii) deliver interface specifications, iii) deliver updated interfaces, iv) deliver software increment, and v) deliver updates and fixes. At the beginning of a release cycle, SCATS prescribes that involved parties should agree on and freeze all interfaces and signals. After that, SCATS prescribes to put a deadline for delivery of interface specifications (the recommendation is to do it latest at 1/3 of the release cycle). At the end of the two thirds of a release cycle, SCATS prescribes to deliver final update to interfaces. Finally, the last third of the release cycle is divided into two parts. In the first third of it, partners need to deliver their software increments. After that, the integration (including testing) process takes place.

SCATS forces prescribing all interfaces and signals while planning a delivery cycle. To approve any request to change interfaces and signals once the delivery cycle commence, SCATS prescribes a Change Control Board meeting (CCB). To conduct a CCB, software project managers jointly send a meeting invitation. For changing any signal or interface, developers must now always follow a process. They first need to align about the change and then create a Change Request for Interfaces and Signals (CHRIS, Table 1) as a documentation *artifacts*.

For a change to take place it is necessary to do the following: i) developers create a CHRIS instance, ii) CCB participants agree that CHRIS conforms to prioritized targets in the backlog of the release cycle and to the the corresponding allocation of team capacities, and iii) there is no veto regarding non-functional requirements (product owners have a veto right to reject any change) and no consistency issues with the prioritized product backlog of the involved Scrum teams. CCB meeting is cancelled if there are no CHRISes reported before the meeting. During a CCB meeting, it can be decided to postpone a CHRIS (e.g., due to a request to provide more details in CHRIS, postpone it for the next release). CCB taking place between "Freeze interfaces and signals" and "Deliver updated interfaces" milestones happen weekly and last maximum one hour.

As the need for a Change Request for Interfaces and Signals (CHRIS) can possibly occur anytime, it is necessary to discuss different scenarios depending on the SCATS' milestone (Fig. 1). CCB taking place between "Freeze interfaces and signals" and "Deliver interface specifications" milestones has the following logic of deciding. If involved developers could implement all changes before the "Deliver interface specifications" milestone, then CCB decides about such CHRIS usually positively (based on its fields). If approved, these changes are included in the integrated release. If developers cannot guarantee to implement the changes before the "Deliver interface specifications" milestone, CCB can still decide to approve it (according to the priorities of the backlog) and include it in the integrated release. CCB can also decide to postpone the implementation of CHRIS to the next release cycle. Finally, a CCB can decide to reject a CHRIS. Regardless of the outcome, once CCB makes the decision it is the job of developers to update the CHRIS with that decision. In CCBs taking place between "deliver interface specifications" and "deliver interface updated interfaces", it is clear that any CHRIS originating then will not be a part of the "Deliver interface specifications" milestone. CCB can still decide to approve it (according to the priorities of the backlog) and include it in the integrated release. However, chances for this are slim and only high priority CHRISes are approved. If there is a CHRIS after the "deliver software increment" milestone, then such requests are handled in the following way. CCB is short and a part of a daily meeting (this is usually a chaotic period). Only software project managers, software integration experts, and affected developers attend the CCB meeting. The software project managers decide jointly if the CHRIS will be implemented in the current release cycle (usually reserved for the most critical fixes).

Table 1. Change Request for Interfaces and Signals Template

1	Organizational	
	Interface ID	a unique interface identifier
	Signal ID	a unique signal identifier
	CHRIS change counter	a subsequent change of one interface and one signal
	CHRIS ID	generated from the items above, traceability of changes
	Devs requesting the change	team name:developer name
	Current status	[created, approved, rejected, postponed]
2	Communication	
	Interface cluster	[HMI, sensors, actuators]
	Signal direction	[input, output]
	Connected modules	to the interface impacted by the CHRIS
	Deployment	processing element type (e.g., GPU)
3	Activity	
	Change type	[add; modify; remove]
	Implementation status	[not started, in progress, implemented]
	Implementation release cycle	in which the implementation is completed [R1; Rx...]
	Related feature	integrated functionality affected by the change
4	Signal	
	Signal description	
	Signal data type	
	Signal unit	(e.g., milliseconds)
	Initial value	
	Minimum value	
	Maximum value	
5	Temporal properties	
	Signal type	[periodic, event-triggered]
	Signal update period	time within which the signal must be updated

To use CHRIS, it is necessary to assign a unique ID to interfaces and signals. Changes to interfaces are centered around signals. No interface can exist without signals, and each signal is assigned to an interface. Each CHRIS is, therefore, associated with one singal. Developers that initiate a change use an instance of the CHRIS template (Table 1), and update it accordingly (before and after a CCB meeting, or once the implementation is completed). The template is organised around five categories. The organisational category captures information related to change identifiers for easier reference in the system and responsible developers (useful for the integration with other tools, such as Jira). The communication category captures information about type of interface, direction of a signal, modules involved in the communication over the interface, and deployment of a component containing the interfaces onto a computing device. The activity category captures data about implementation details and progress. The signal category describes the signal itself, with focus on data type and value range. The final category is reserved for temporal properties of the signal. All changes to interfaces and signals must be documented using this template.

4   Evaluation

4.1   Setup

Two companies (an integrator and a partner working together over two years) participated in applying SCATS, and provided an experience-based feedback. The integrator is an automotive manufacturer with over 100.000 employees. The partner is a hardware and software technology company with over 20.000 employees. For programming, they both use C++ programming language. The companies use Agile practices (based on Scrum), but with various different tweaks. Each company is planning separately their Scrum sprints based on their own priorities. Nevertheless, they align on functionalities they will develop and deliver within a release cycle. Duration of the release cycle is 5 weeks. Companies have noticed that alignment at the beginning of a delivery cycle and ad-hoc alignment between individual developers are challenging for their integration efforts (some interfaces are missing, some are not well-aligned, and for some there are issues regarding signal's quality properties). Before applying SCATS, they would regularly face *100 to 200 interface-related bugs per release cycle*.

We applied the SCATS framework in this setup over a **three-month period**. **Fifty experts** have been involved (software project managers: 2; software architects: 2; integration experts: 2; product owners: 3; scrum masters: 3; developers: 38). Regarding the **initial overhead**, we have measured and gathered feedback regarding the time to i) explain the SCATS approach to the experts and ii) create the CHRIS template in Jira. Regarding the **application overhead**, we have measured and gathered feedback regarding the i) time to align the timing of prescribed milestones (per a release cycle); ii) average time to discuss one instance of a CHRIS template in a CCB; iii) the number of CCB meetings that took place (over the three-month evaluation period); iv) average time to complete and align the fields of one CHRIS template instance; v) the number of CHRISes created in total (over the three-month evaluation period). Finally, we have measured **benefits** in terms of the reduction in the number of the interface-related bugs in the software integration.

4.2   Results

The **initial overhead**, when applying SCATS, sums up to approximately 12 h. Over the period of three months, the **application overhead** was mainly caused by 10 h to discuss the 120 CHRISes during CCB meetings, and 40 h invested to align and complete all fields of created CHRISes. The **application benefits** are up to 70% reduction in interface-related bugs, previously appearing in the integration stage. Interface-related bugs were previously in the range between 100 and 200, depending on the release cycle. The eliminated bugs were related to interfaces and signals issues known to at least one developer (e.g., a specific signal was needed). Therefore, we can conclude that the **overall overhead** of applying SCATS is minor compared to the benefits it brings (reducing the delays previously accounting up to dozens of days). However, we have also observed that **SCATS does not resolve all interface-related bugs**. These bugs are related to missing signals, which have not been identified as required by any developer until the software integration stage.

5   Conclusion

Companies that develop software systems and integrate them to create software-defined vehicles often face integration challenges. As a consequence, some bugs manifest in the later integration stages, introducing delays to the delivery.

To solve this problem, we have introduced our framework for Software embedded system integration in Cross-organizational Agile TeamS (SCATS framework). SCATS forces systematic synchronisation of parties involved in creating software systems for software-defined vehicles, in terms of their interfaces and signals. Consequently, SCATS exposes misalignment bugs during development and significantly reduces delivery delays. The evaluation confirms that SCATS solves the issue of misalignment of interfaces and signals with a minor effort.

The future work will focus on investigating how to resolve alignment issues that SCATS is currently not able to solve. These include situations when none of the developers is aware that there is a need to align certain signals and interfaces.

Data Availibility. The data that support the findings of this study (all data related to the evaluation, interviews, transcripts, etc.) are available on request from the corresponding author. The data are not publicly available due to containing information that could compromise the privacy of the industrial partner.

References

1. Aust, S.: Vehicle update management in software defined vehicles. In: 2022 IEEE 47th Conference on Local Computer Networks (LCN), pp. 261–263 (2022). https://doi.org/10.1109/LCN53696.2022.9843360
2. Herbsleb, J., Grinter, R.: Architectures, coordination, and distance: Conway's law and beyond. IEEE Softw. **16**(5), 63–70 (1999). https://doi.org/10.1109/52.795103

3. Ilyas, M., Khan, S.U.: Software integration model for global software development. In: 2012 15th International Multitopic Conference (INMIC), pp. 452–457 (2012). https://doi.org/10.1109/INMIC.2012.6511461
4. Miller, A.: A hundred days of continuous integration. In: Agile 2008 Conference, pp. 289–293 (2008). https://doi.org/10.1109/Agile.2008.8
5. van Moll, J., Ammerlaan, R.: Identifying pitfalls of system integration – an exploratory study. In: 2008 IEEE International Conference on Software Testing Verification and Validation Workshop, pp. 331–338 (2008). https://doi.org/10.1109/ICSTW.2008.22
6. Mårtensson, T., Ståhl, D., Bosch, J.: Continuous integration impediments in large-scale industry projects. In: 2017 IEEE International Conference on Software Architecture (ICSA), pp. 169–178 (2017). https://doi.org/10.1109/ICSA.2017.11
7. Mårtensson, T., Ståhl, D., Martini, A., Bosch, J.: Continuous architecture: towards the goldilocks zone and away from vicious circles. In: 2019 IEEE International Conference on Software Architecture (ICSA), pp. 131–140 (2019). https://doi.org/10.1109/ICSA.2019.00022
8. Charette, R.N.: How software is eating the car (2021). https://spectrum.ieee.org/software-eating-car
9. Shatil, A., Hazzan, O., Dubinsky, Y.: Agility in a large-scale system engineering project: a case-study of an advanced communication system project. In: 2010 IEEE International Conference on Software Science, Technology and Engineering, pp. 47–54 (2010). https://doi.org/10.1109/SwSTE.2010.18
10. Stocker, M., Zimmermann, O.: API refactoring to patterns: Catalog, template and tools for remote interface evolution. In: Proceedings of the 28th European Conference on Pattern Languages of Programs, EuroPLoP 2023. Association for Computing Machinery, New York (2024). https://doi.org/10.1145/3628034.3628073
11. Ståhl, D., Bosch, J.: Industry application of continuous integration modeling: a multiple-case study. In: 2016 IEEE/ACM 38th International Conference on Software Engineering Companion (ICSE-C), pp. 270–279 (2016)
12. Tufano, M., Sajnani, H., Herzig, K.: Towards predicting the impact of software changes on building activities. In: 2019 IEEE/ACM 41st International Conference on Software Engineering: New Ideas and Emerging Results (ICSE-NIER), pp. 49–52 (2019). https://doi.org/10.1109/ICSE-NIER.2019.00021
13. Zafar, A., Ali, S., Shahzad, R.K.: Investigating integration challenges and solutions in global software development. In: 2011 Frontiers of Information Technology, pp. 291–297 (2011). https://doi.org/10.1109/FIT.2011.61
14. Zafar, A.A., et al.: Taxonomy of factors causing integration failure during global software development. IEEE Access 6, 22228–22239 (2018). https://doi.org/10.1109/ACCESS.2017.2782843
15. Zimmermann, O., Stocker, M., Lübke, D., Zdun, U., Pautasso, C.: Patterns for API Design: Simplifying Integration with Loosely Coupled Message Exchanges (2022)

The Execution Perspective in Software Architecture Descriptions: A Systematic Mapping

Tales Viglioni[1,2,3](✉) , Thais Batista[2] , Everton Cavalcante[2,4] ,
and Flavio Oquendo[3]

[1] Federal Institute of Pernambuco, Barreiros, Brazil
talesviglioni@gmail.com
[2] Federal University of Rio Grande do Norte, Natal, Brazil
everton.cavalcante@ufrn.br
[3] IRISA-UMR CNRS/Université Bretagne Sud, Vannes, France
flavio.oquendo@irisa.fr
[4] Durham University, Durham, UK

Abstract. Executable software architecture descriptions enable architects to validate or verify the correctness of their designs early in the development process of software-intensive systems, reducing the risk of costly misconceptions. Executable architectures have been the subject of research for several decades, but there is no comprehensive panorama of the state of the art on this topic. To fill this gap, we conducted a systematic mapping study that explores activities and approaches related to executable architecture descriptions for designing software-intensive systems. We have selected and analyzed 45 studies from the last 30 years to understand the activities related to executing architecture descriptions and identify the existing approaches. We also devised a taxonomy with key concepts related to executable architecture descriptions for software-intensive systems. The resulting insights from our study can help researchers pinpoint areas for further investigation.

Keywords: Software architecture · Executable architecture ·
Systematic mapping study

1 Introduction

The representation of software architectures is one of the main activities of an architecture-driven software development process, as it allows for the anticipation of important decisions regarding system design. In this context, *architectural languages* have become well-accepted means for systematically representing and analyzing software architectures [5], producing *architecture descriptions* that can be used at design time or runtime.

Software architecture descriptions often cope with two major viewpoints, namely *structural* and *behavioral*. The former concerns the structure of the

M. Galster et al. (Eds.): ECSA 2024, LNCS 14889, pp. 379–395, 2024.
https://doi.org/10.1007/978-3-031-70797-1_26

software architecture, specifying its elements (components, ports, connectors) and how they are interconnected, while the latter is related to the behavior of the architecture and how its elements meet the established functional and extra-functional requirements. Nonetheless, an *execution* perspective (or even a viewpoint on its own) is relevant for software architects analyzing the architecture behavior. Executing a software architecture description, that is, having an *executable software architecture*, enables architects to verify and validate the architectural design against the desired requirements, as well as helps identify potential issues and bottlenecks early in the development life cycle, significantly reducing time and costs in the future.

Executing a software architecture description typically involves supporting parsing or transforming it into a second representation to be executed by an engine. Therefore, an executable software architecture description somehow combines an executable representation and an associated execution (analysis) engine [6]. Depending on the purpose of the architecture execution, this process can be complemented with the verification of properties of interest, which can also be specified in some notation.

The three decades of the Software Architecture discipline have witnessed not only ways of specifying architecture descriptions but also how to execute them. Several methods and tools have been proposed over this time, but, to the best of our knowledge, there is no comprehensive panorama of the state of the art on this topic. This work intends to fill this gap by investigating the literature on the execution perspective in software architecture descriptions for designing software-intensive systems. To achieve this aim, we have conducted a *systematic mapping study* by following consolidated guidelines for collecting, selecting, and analyzing primary studies while providing scientific value to the obtained findings. We have selected and analyzed 45 studies published in the last 30 years to understand the activities related to executing architecture descriptions and identify the existing approaches. We also devised a taxonomy with key concepts related to executable architecture descriptions for software-intensive systems.

The remainder of this paper is organized as follows. Section 2 describes the research methodology used in this study. Section 3 synthesizes the findings by answering the posed research questions. Section 4 presents a taxonomy for the execution perspective in software architecture descriptions. Section 5 discusses some possible threats to the study's validity under the lens of the classification schema proposed by Ampatzoglou et al. [1] for Software Engineering secondary studies. Section 6 brings the final remarks.

2   Research Methodology

The systematic mapping study reported in this paper followed consolidated guidelines available in the literature [8]. We have followed three basic steps: (i) *planning*, for outlining the protocol with the research questions to answer, the search strategy, the criteria to select primary studies, and the data extraction methods; (ii) *execution*, for identifying, selecting, and analyzing the studies in compliance with the protocol; and (iii) reporting, for aggregating information extracted from relevant studies considering the research questions and outlining conclusions. Due to space constraints, this section presents a concise view of the overall process. The detailed research protocol can be found in our replication package, available at https://doi.org/10.5281/zenodo.11508112.

Research Questions. Our study aimed to understand the primary purposes of executing software architecture descriptions and identify the existing approaches, methods, tools, and technologies. We defined two research questions (RQs) to refine this goal:

RQ1: What activities do the execution model or tool support?
RQ2: What approaches are employed in the execution of software architectures?

Search Strategy. We carried out an automated search process over five electronic publication databases to retrieve relevant studies to answer the posed RQs. Our sources were the ACM Digital Library, IEEE Xplore, ScienceDirect, Scopus, and Web of Science, which are significantly popular and have good coverage of the Software Engineering literature. We elected two main terms based on the RQs, namely *software architecture* and *execution*. After adding synonyms and related terms to increase the search coverage and precision, we reached the following search string: (``software architecture'' OR ``software modeling'') AND (executable OR executing OR execution OR run-time OR runtime OR simulating OR simulation) NOT hardware.

Selection Criteria. We adopted inclusion criteria to define circumstances that make a study relevant to our goal, and exclusion criteria helped us to exclude studies that would not help us to answer the RQs. We considered only primary studies presenting an approach or proposing a framework or tool for executing or simulating software architectures or those using some model to support these tasks. On the other hand, we have not considered primary studies related to the execution or simulation of software architectures or those associated with dynamic architectures or models@run.time, which are out of the scope of our study. We regarded a primary study as relevant if it did not meet any exclusion criterion and met at least one inclusion criterion. The protocol available in the replication package details all the inclusion and exclusion criteria.

Execution. Our search considered primary studies published until December 2023. Figure 1 depicts the steps to select the relevant primary studies. The automated search in the five databases using the defined search string retrieved 9,377 results. Removing duplicates resulted in 6,614 entries. After applying the selection criteria to their titles, abstracts, and keywords, we obtained 568 studies. Next, we analyzed the introduction and conclusion sections of the studies against the selection criteria and obtained 137 results. We read all of them in full and selected 45 studies for data extraction. In all these stages, we held consensus meetings to address divergent interpretations about the selection of studies. Appendix A lists the 45 selected primary studies, whose detailed references are presented in the replication package of our study.

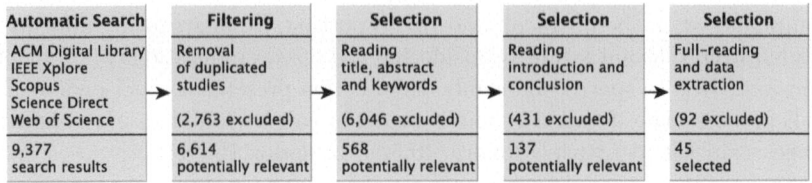

Automatic Search	Filtering	Selection	Selection	Selection
ACM Digital Library IEEE Xplore Scopus Science Direct Web of Science	Removal of duplicated studies	Reading title, abstract and keywords	Reading introduction and conclusion	Full–reading and data extraction
	(2,763 excluded)	(6,046 excluded)	(431 excluded)	(92 excluded)
9,377 search results	6,614 potentially relevant	568 potentially relevant	137 potentially relevant	45 selected

Fig. 1. Steps for selecting the relevant primary studies

Data Extraction and Synthesis. We organized a data extraction sheet with items related to the RQs and other relevant information from the selected primary studies. Besides metadata information (title, authors, citations, publication year, and venue), we extracted data related to activities supported in architecture executions and the approaches or technologies used. We also identified the type of validation of the studies and their challenges and limitations. Data extracted from the selected primary studies underwent an inductive open coding procedure [3] to help us answer the RQs while harmonizing our understanding to correctly categorize the studies. Our replication package contains the complete sheet with all the data extracted from the selected primary studies.

3   Results and Discussions

This section summarizes the results of the systematic mapping study, considering the RQs and data extracted from the analyzed studies. Section 3.1 presents an overview of the studies regarding temporal distribution, publication venues, and validation methods. Section 3.2 discusses the activities supported by the execution of software architectures (RQ1) and Sect. 3.3 describes the approaches we identified in the analyzed studies (RQ2). It is important to highlight that our research centers on the characteristics of executable software architectures, not other software artifacts.

3.1   Overview of the Selected Studies

Temporal Distribution. Figure 2-a depicts the evolution of the studies over time. Studies on executable software architectures have been continuously published since 1997, with an average of two to three papers per annum. This fact leads us to conclude that this research topic continues to attract the scientific community's interest. However, the absence of studies in the last two years deserves further investigation.

Publication Venues. Figure 2-b shows that 28 studies (62%) were published in conferences, workshops, or symposia, and 17 studies (38%) were published in journals. The studies were published in 36 different venues, thus indicating that there is no reference place for publications on executable software architectures. We noticed that the European Conference on Software Architecture (ECSA),

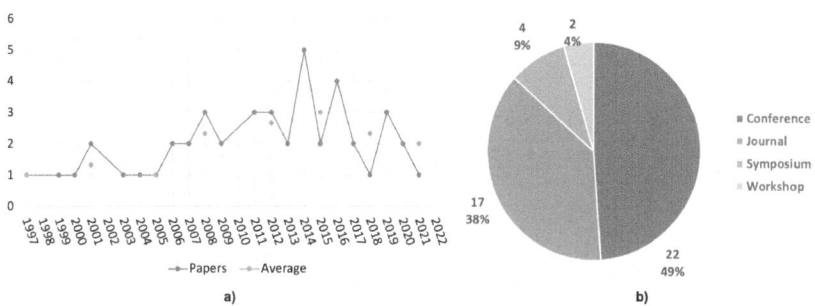

Fig. 2. Publications per year (a) and venue type (b)

one of the most prominent scientific international forums in the Software Architecture field, stood out among these venues with four studies.

Study Validation Methods. Although most of the studies (91%) include some form of validation, only 37% present empirical validations (15 case studies and two experiments), while the others offer only examples of potential scenarios of use. These data indicate that studies on executable software architectures need more solid empirical foundations regarding research validation.

3.2   Activities Supported by Executable Architecture (RQ1)

We analyzed the primary studies to understand the purpose of executing software architectures and to identify the different activities, techniques, and methods employed. Validation and verification are the primary purposes of executing software architecture to ensure its quality, mitigate risks, and enhance system reliability. As defined by the ISO/IEC/IEEE 24765:2017 International Standard [4], validation concerns "building the right product", i.e., ensuring that a system meets the requirements and expectations of stakeholders, while verification concerns "building the product right", i.e., the ability to determine whether a system has been correctly developed. We have identified 36 studies (80%) addressing validation and 20 studies (44.4%) addressing verification.

The software architecture execution is related to several activities. We have identified six activities as depicted in Fig. 3: simulation (34 studies), animation (eight studies), testing (five studies), model checking (nine studies), and transformation (four studies). Figure 3 also shows some intersections between different activities, which means that some studies have concerned multiple activities and purposes when executing software architectures. For instance, studies S10, S14, and S33 have addressed both verification and validation, while studies S14 and S44 employed animation and simulation for the same purposes.

Our analysis of the studies revealed the prevalence of the **simulation** activity, reported in 34 studies (75.5%), with 30 studies using it for validation and four for verification. Simulation-based techniques analyze a system at design time to determine, upon its execution, if it meets functional and performance criteria and

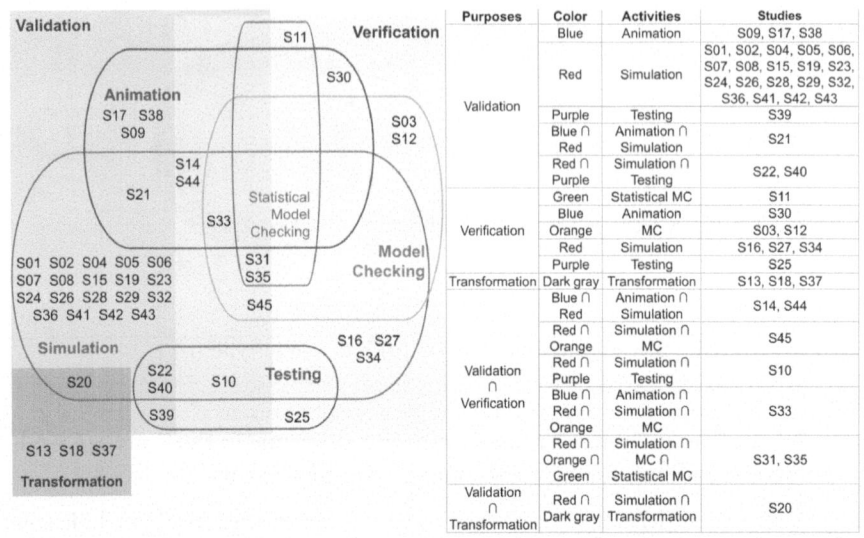

Fig. 3. Distribution of studies by purposes and activities

satisfies some constraints. Therefore, they align well with the purpose of validating architectural designs upon execution. Simulations can run either locally, in a single computing device within a closed environment, or distributed across multiple computing devices interconnected through a network. While local simulations offer simplicity and are suitable for small-scale applications, distributed simulations support larger, more complex systems, even though the simulation design is undoubtedly more complex. The choice between these two options depends on the size and complexity of the simulation model, the available computational resources, and the specific requirements of the simulation scenario.

Nine primary studies use **model checking** (MC) for verification purposes, which requires a behavioral model representing the software architecture execution. Well-studied in Computer Science for several decades, MC is one of the most used techniques for analyzing software architectures to verify if an architectural specification satisfies properties of interest [10]. MC takes as inputs a representation of the system (e.g., an architecture description) and property specifications expressed in some notation. The model checker returns true if the properties are satisfied or false in the case in which a given property is violated. Despite its wide and successful use and the number of supporting tools available, MC faces a critical challenge concerning scalability, impeding its use for some large-scale software systems. Statistical model checking (SMC) has arisen to overcome the limitations of traditional model checking, as reported in three studies (S11, S31, and S35). SMC is a probabilistic, simulation-based approach that consists of building a statistical model of finite executions of the system under verification and deducing the probability of satisfying a given property [7].

Animation is a technique of "walking through" a formal specification, enabling its dynamic analysis. We found eight studies related to animation, six studies using it for validation (S09, S17, S21, S33, S38, and S44) and two for verification (S14 and S30). Animation is usually coupled with a diagrammatic representation of the software architecture, which is transformed into executable code, and the results can be used to animate the original diagram. Therefore, animation enables architects to visualize the execution of the software architecture. This kind of activity seems interesting considering the progressive use of the diagrammatic notation of SysML and derived languages to describe software architectures due to its expressiveness and support for requirements documentation and traceability [9]. Colored Petri nets, used in study S44, also represent a feasible technique to animate software architecture models, ensuring that they follow an appropriate sequence of transitions associated with their execution.

We found **testing** activities in five studies (S10, S22, S25, S39, and S40) for validation and verification purposes. A common feature among these studies is their reliance on model transformations to automatically synthesize executable architectural elements for testing from high-level architectural specifications. The main concern is integrating the test activity in an early stage to improve the overall architecting process.

Four studies have addressed **transformations** (S13, S18, S20, and S37) in terms of generating a target model from a source model based on a set of rules that describe how to perform that transformation from one model to another. These models can be either graphical or textual and can be used as a foundation for architectural executions or as intermediary models for validation, verification, and other transformations. For instance, the methodology employed in study S37 involves using a parsing tool to analyze software architecture descriptions in the Acme ADL and automatically generate executable code by applying transformation rules. Moreover, multiple transformations can be employed in a given process, such as in the case of study S20. The reliance on transformations of software architecture descriptions into executable models is significantly influenced by model-driven engineering (MDE), which is well-acknowledged to effectively manage the inherent complexity of modern software systems.

Figure 3 also shows the intersections of studies within purposes and activities related to software architecture execution. We noticed a few studies spanned multiple purposes (e.g., validation and verification) and different techniques. This fact might point to a possible research avenue focusing on a holistic combination of approaches, languages, methods, and tools that are currently isolated or loosely integrated for supporting executable software architectures.

Highlights (RQ1). Executable software architectures have been primarily used to validate architectural designs. Simulation significantly prevails as a means of executing a software architecture to support validation. Animation in software architecture execution for verification and validation is also noteworthy due to the noticeable use of diagrammatic notations, such as UML, SysML, and derivatives, for architectural description. On the other hand, the reduced number of studies combining multiple purposes and activities might indicate a possible research gap in using executable software architectures.

3.3   Approaches for Software Architecture Execution (RQ2)

We analyzed the selected primary studies to identify the models, methods, tools, or technologies used to execute software architecture descriptions. We found two main approaches for executable software architectures (see Fig. 4). *Direct execution* implements the ability to directly execute from the architecture description, not requiring a transformation process. *Transform-to-execute* employs a transformation process to enable the software architecture execution. It is worthwhile to mention that multiple categories can apply to a given study.

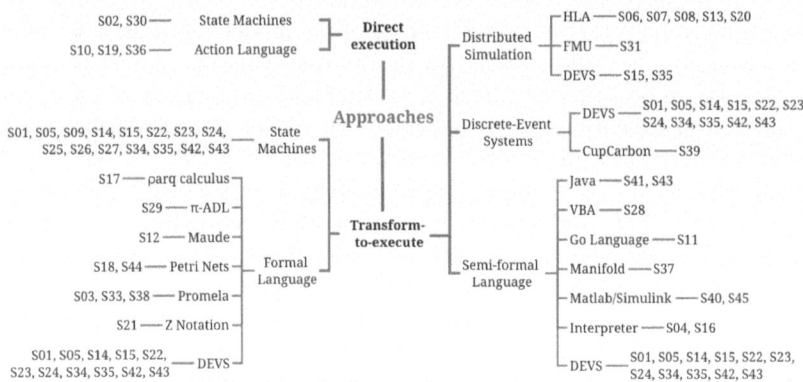

Fig. 4. Approaches for software architecture execution

Direct Execution. Studies relying on direct execution employ (i) *state machines* and (ii) *action languages* for execution. Both state machines and action languages can be used to model or specify the software architecture's behavior over time in terms of states or actions.

State Machines. In studies S02 and S30, state machines provide an internal perspective of the architecture's behavior, allowing for understanding the rationale of actions and controlling how an entity responds to events or actions, including state transitions. State machines can adequately represent the different states of the software architecture and track its execution upon transiting between them. While state machines are ideally suited for software systems with well-defined states, they can be combined with other approaches (e.g., event-driven paradigm) to cope with dynamic behavior.

Action Languages. An action language describes an aspect of a system's behavior in sufficient detail to allow it to be executed. When used with semi-formal notations like UML or SysML, an action language can provide an executable specification that can be run as a program independently from any specific platform or language. In study S36, the SysADL language, a specialization of SysML, is used to describe software architectures explicitly encompassing an executable viewpoint in addition to the traditional structural and behavioral ones. To enable execution, SysADL uses the OMG Action Language for Foundational UML (ALF) as the action language. Study S10 employs Executable and Translatable UML (xtUML), an enhanced version of UML that can be translated into code using model compilers. In this study, component, class, and state diagrams are used for structural and behavioral modeling, and actions are specified through an action language. Action languages play a relevant role because they complement existing modeling notations, with a clear separation between architectural modeling and the concretization of execution via actions. Even though model transformations can facilitate this process, action languages require dedicated engines and tool support to enable the architecture execution.

Transform-to-Execute. Studies relying on transform-to-execute employ a variety of approaches, including distributed simulation (8 studies), formal and semi-formal languages (40 studies), state machines (15 studies), and discrete event systems (12 studies). We describe these approaches in the following.

Distributed Simulation. In contrast to local simulation, which runs on a single server with all the resources, the distributed simulation uses multiple linked hosts distributed throughout a network architecture. DS is primarily employed to achieve scalability, aggregation, and reusability, as highlighted in study S06. This category includes well-known techniques such as High-Level Architecture (HLA), Functional Mock-up Unit (FMU), and Discrete Event System Specification (DEVS). Common concerns among these different techniques are facilitating the interoperability and reusing simulation components across different contexts.

Discrete Event Systems. Discrete event systems represent an approach relying on specifying the system's behavior as a series of events, each at a specific point in time, and includes transitions between the system's states. An executable architecture designed using this approach establishes sequences of events, triggering criteria for events, the production or consumption of information, and other resources involved in the system. DEVS is the most used technique in studies relying on discrete event systems (11 studies). This prevalence can be explained by some reasons, such as (i) the existence of tools supporting the DEVS metamodel representation and improving the generating simulation code and (ii) the integration of the DEVS metamodel with various simulators. Before using one of these simulators, software architecture models must be converted to DEVS through a transformation process. Some studies have used UML (S01 and S05),

UML-RT (S15), and SysML (S22, S23, S24, S42, and S43) as architectural languages, and the models are transformed into a DEVS specification to handle the simulation, analyze performance, and verify correctness.

State Machines. State machines focus on using state machine diagrams to express architectural behavior and transform them into an executable model. Each diagram depicts the different states of the architectural elements, the protocol for transitioning between states, and the corresponding effects of these transitions. Specifically, study S09 employs an approach that transforms these diagrams into source code in the Java programming language. In study S25, the operational semantics for state transitions, actions, and messages are given by the AsmL language, which is then transformed into executable code. The difference from using state machines in direct execution is that the state machine models expressing architectural behavior must be transformed into another model towards execution. We categorized the studies using DEVS as state machines because this class of formalism is founded on finite-state machines (FSMs).

Semi-formal Languages. Studies in this category transform semi-formal languages into executable or programming languages. This process usually involves mapping core concepts of the architectural language to the target language. Software architects often choose semi-formal notations, such as UML, SysML, or their derivatives, to describe the architecture's structure and behavior due to lower difficulty in learning, emphasis on visual elements, and general-purpose scope. Nevertheless, semi-formal notations are limited with respect to the automatic verification of architectural properties due to the lack of well-defined semantics and gaps in complex design decisions. Formal approaches address these deficiencies and provide enhanced assistance for architectural analysis.

Formal Languages. Studies in this category use a formalism with precise syntax and semantics to represent the software architecture, enabling automated tools to uncover flaws in architectural description. The main advantage of a formal approach is the ability to accurately assess whether a software system can satisfy the requirements and constraints and verify the precision and correctness of architectural designs [2]. However, while formal methods are imperative for safety-critical software systems, they are acknowledged for their complexity and require significant effort from software architects. Combining transformations for architecture execution with formal notations benefits from the automated analysis capabilities these approaches offer. Study S17 describes a visualizer of architectural execution flow for component-based software. The interpreter identifies the structural elements from a ρarc calculus formalism and transforms the architectural expressions into UML components based on pre-established transformation rules. The tool then shows the execution flow through ρarc calculus rewriting rules (operational semantics). Study S12 proposed an approach based on graph transformation to generate a visual modeling tool for translating UML models into equivalent Maude formal specifications for further analysis. The last step deals with analyzing and verifying the generated Maude specifications. We also identified the use of other formalisms, such as colored Petri nets (studies S18 and S44), Promela (studies S03, S33, and S38), and Z (study S21).

In summary, most of the selected primary studies (91%) rely on transformations of software architecture descriptions in a given architectural language to generate executable models. The main concern is minimizing the effort required to generate executable artifacts, which is undoubtedly facilitated by the many well-established MDE languages and tools available to support this process. Nevertheless, software architects and developers shall master these techniques. The direct execution of software architecture descriptions, albeit more complex, does not require software architects to know other languages and transformation processes to execute the architecture.

> **Highlights (RQ2).** Software architecture execution approaches are based either on the direct execution of models expressing the architecture or on transforming architecture descriptions into executable models. Most studies opted for a transform-to-execute approach due to the availability of language and tool support and the pursuit of reducing the effort required to generate executable software architecture models. However, relying on a transformation-based approach requires software architects to deal with the transformation process. Discrete-event systems, particularly DEVS, represent the most used transform-to-execute approach due to the wide availability and flexibility of tools complying with the DEVS metamodel.

4   A Taxonomy for the Execution Perspective of Software Architecture Descriptions

Taxonomies have been directly or indirectly used in the Software Architecture community to mature the knowledge field, mainly to understand and classify existing work. This work contributes with a taxonomy arising from the analysis of the studies selected in our systematic mapping study to capture important aspects related to the execution perspective of software architecture descriptions. The proposed taxonomy is not solely intended to summarize the current state of the art on this topic, but it also provides a basis to classify existing approaches and understand their characteristics. Furthermore, it can also serve as a reasoning framework for proposing novel or improved approaches in the field.

Figure 5 depicts our taxonomy for the executable software architectures to support software-intensive systems development. The taxonomy encompasses two main axes: (i) the *activities*, considering the purpose for execution and the employed techniques, and (ii) the software architecture execution *approaches*. The taxonomy categories are not disjointed, and others not currently covered can be included. Figure 5 also presents a classification of the selected primary studies according to the taxonomy to demonstrate its utility.

The initial level of the activities axis comprises the categories corresponding to the verification, validation, and transformation activities (see Sect. 3.2).

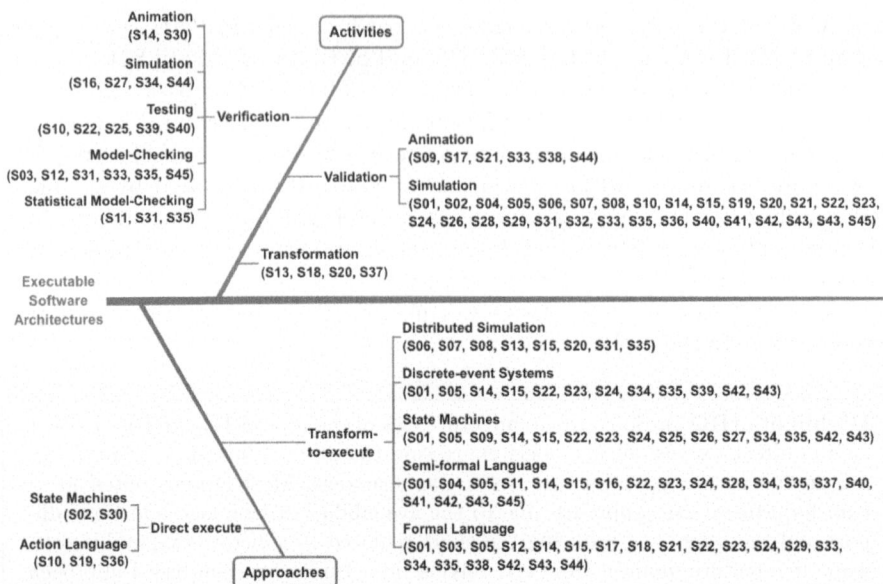

Fig. 5. A taxonomy for activities and approaches for executable software architectures

The second level considers each activity for a specific purpose. Some studies are classified into multiple categories, such as S45, which is categorized for validation through simulation and verification via model-checking. There are also cases where the same activity was considered for different purposes. In turn, the first level of the approaches axis refers to approaches that directly execute the architecture or require transformation to an executable model (see Sect. 3.3). The second level shows each approach employed to execute software architectures.

5   Threats to Validity

Threats to *research validity* are primarily related to the rigor of the study design and conduct. To mitigate this threat, we have followed well-established guidelines for conducting systematic mapping studies based on a precise protocol (available in the replication package) with goals, research questions, search procedures, and eligibility criteria for identifying relevant studies.

Threats to *study selection validity* are related to the study search and inclusion. A limitation often reported in secondary studies is the possibility of missing relevant studies due to inadequate selection of sources, deficiencies in the search string, study selection bias, etc., compromising completeness and coverage of the literature. Although we sought to augment the search coverage by considering five different publication databases, the automated search procedure might have missed relevant studies. To mitigate selection bias, consensus meetings enabled us to discuss the inclusion and exclusion of retrieved primary studies.

Threats to *data validity* are primarily related to data extraction and analysis processes, such as researcher bias, inaccuracies, and unverified extraction. Not all data extracted from the selected studies to answer the research questions were obvious, and some data had to be inferred. To minimize imprecision and bias in data collection and analysis, we reviewed the obtained data and consequent results, and disagreements have been solved to reach a consensus.

6   Final Remarks

This paper presented the results of a systematic mapping study aimed to provide an overview of the state of the art on the execution perspective of software architecture descriptions. Our selection and analysis of 45 primary studies enabled us to understand the activities related to executable software architectures and identify the existing approaches supporting them. Our work contributes to filling an existing gap in the literature related to the absence of studies offering a wide panorama of almost 30 years of research on this topic. Furthermore, we proposed a taxonomy representing the research landscape and that can be used to classify existing and future approaches.

Our results indicate that software architectures have been executed primarily for validation purposes, and simulation has been the prevalent activity. Animation plays a relevant role in software architecture execution due to the possibility of integrating it with visual, diagrammatic notations that have been often used in software architecture description, such as UML, SysML, and related architectural languages. The results also revealed that most existing approaches rely on transforming architectural models expressed in a given language into an executable model in another language. Besides taking advantage of MDE technologies and tools for reducing effort, model transformations promote a clear separation between architectural modeling and their concretization via execution. DEVS has been the most used transform-to-execute approach, possibly due to the wide availability and flexibility of tools complying with the DEVS metamodel.

Finally, the mapping results pointed out a limited number of studies integrating multiple purposes and activities. This may suggest a potential research direction in utilizing executable architectures as a combination of approaches, languages, and methods that are currently isolated or loosely integrated.

Acknowledgments. This study was financed in part by Coordenação de Aperfeiçoamento de Pessoal de Nível Superior (CAPES), Brazil - Finance Code 001, and INES (https://www.ines.org.br) - CNPq grant 465614/2014-0, CAPES grant 88887.136410/2017-00, and FACEPE grants APQ-0399-1.03/17 and PRONEX APQ/0388-1.03/14. T. Batista is a CNPq Research Fellow, grant 315963/2021-2. E. Cavalcante is supported by the Research Development Grant from the Durham University Research Staff Association via the Enhancing Research Culture Fund.

Data Availibility Statement. The study data can be found in the replication package available online at https://doi.org/10.5281/zenodo.11508112.

Disclosure of Interests. The authors have no competing interests to declare that are relevant to the content of this article.

A   Selected Primary Studies

S01 Alshareef, A. et al.: Activity-based DEVS modeling. Simulation Modelling Practice and Theory **82**, 116–131 (2018).

S02 Ammar, H.H. et al.: Modeling resources in a UML-based simulative environment. In: ACS/IEEE International Conference on Computer Systems and Applications, pp. 405–410. IEEE, USA (2001).

S03 Argote García, G. et al.: A formal approach for translating a SAM architecture to PROMELA. In: 20th International Conference on Software Engineering and Knowledge Engineering, pp. 440–447. Knowledge Systems Institute Graduate School, USA (2008).

S04 Balsamo, S., Marzolla, M.: A simulation-based approach to software performance modeling. In: 9th European Software Engineering Conference/11th ACM SIGSOFT International Symposium on Foundations of Software Engineering, pp. 363–366. ACM, USA (2003).

S05 Blas, M.J. et al.: Modeling and simulation framework for quality estimation of Web applications through architecture evaluation. SN Appl. Sci. **2**, 374 (2020).

S06 Bocciarelli, P. et al.: A model-driven approach to build HLA-based distributed simulations from SysML models. In: 2nd International Conference on Simulation and Modeling Methodologies, Technologies and Applications, pp. 49–60. SciTePress, Portugal (2012).

S07 Bocciarelli, P. et al.: A model-driven approach to enable the simulation of complex systems on distributed architectures. Simulation **95**, 12, 1185–1211 (2019).

S08 Bocciarelli, P. et al.: A model-driven framework for distributed simulation of autonomous systems. In: Symposium on Theory of Modeling & Simulation, pp. 213–220. Society for Computer Simulation International, USA (2015).

S09 Boger, M., Sturm, T.: Tool-support for model-driven Software Engineering. In: Workshop of the pUML-Group held together with the «UML»2001 on Practical UML-Based Rigorous Development Methods - Countering or Integrating the eXtremists, pp. 307–318. GI (2001).

S010 Burden, H. et al.: Executable and translatable UML – How difficult can it be? In: 2011 18th Asia-Pacific Software Engineering Conference, pp. 114–121. IEEE, USA (2011).

S011 Cavalcante, E. et al.: Statistical model checking of dynamic software architectures. In: Tekinerdogan, B. et al. (eds.) Software Architecture. LNCS, vol. 9839, pp. 185–200. Springer, Cham (2016).

S012 Chama, W. et al.: Using graph transformation and maude to simulate and verify UML models. In: 2013 The International Conference on Technological Advances in Electrical, Electronics and Computer Engineering, pp. 459–464. IEEE, USA (2013).

S013 D'Ambrogio, A. et al.: A model transformation approach for the development of HLA-based distributed simulation systems. In: 1st International Conference on Simulation and Modeling Methodologies, Technologies and Applications, pp. 155–160. SciTePress, Portugal (2011).

S014 Delecolle, A. et al.: Architectural strategy to enhance the availability quality attribute in system-of-systems architectures: a case study. In: 2020 IEEE 15th International Conference of System of Systems Engineering, pp. 93–98. IEEE, USA (2020).

S015 Dongping Huang, Sarjoughian, H.: Software and simulation modeling for real-time software-intensive systems. In: Eighth IEEE International Symposium on Distributed Simulation and Real-Time Applications, pp. 196–203. IEEE, USA (2004).

S016 Fuentes, L., Sánchez, P.: Execution of aspect oriented UML models. In: Akehurst, D.H. et al. (eds.) Model Driven Architecture- Foundations and Applications. LNCS, vol. 4530, pp. 83–98. Springer, Berlin, Heidelberg (2007).

S017 García, J.A.R., Diosa, H.A.: PintArq: A visualizer of architectural execution flow for component-based software architectures. In: Figueroa-García, J.C. et al. (eds.) Applied Computer Sciences in Engineering. CCIS, vol. 657. pp. 15–26. Springer, Cham (2016).

S018 Ge, B. et al.: A data-centric executable modeling approach for system-of-systems architecture. In: 2012 7th International Conference on System of Systems Engineering, pp. 368–373. IEEE, USA (2012).

S019 Gruhn, V., Schäfer, C.: Using mobile architecture modeling and simulation for enterprise applications. In: Draheim, D., Weber, G. (eds.) Trends in Enterprise Application Architecture. LNCS, vol. 4473, pp. 142–157. Springer, Berlin, Heidelberg (2007).

S020 Guiffard, E. et al.: CAPSULE: Application of the MDA methodology to the simulation domain. In: European Simulation Interoperability Workshop 2006, pp. 181–190. SISO (2006).

S021 Jia, X. et al.: Executable visual software modeling-the ZOOM approach. Software Qual J. **15**, 1, 27–51 (2007).

S022 Kapos, G.-D. et al.: A declarative approach for transforming SysML models to executable simulation models. IEEE Trans Sys Man Cyb, Sys. **51**,6, pp. 3330–3345 (2021).

S023 Kapos, G. et al.: An integrated framework for automated simulation of SysML models using DEVS. Simulation **90**, 6, pp. 717–744 (2014).

S024 Kapos, G.-D. et al.: Model-based System Engineering using SysML: Deriving executable simulation models with QVT. In: 2014 IEEE International Systems Conference Proceedings, pp. 531–538. IEEE, USA (2014).

S025 Leue, S., Wei, W.: An executable and extensible formal semantics for UML-RT. In: Giese, H. et al. (eds.) Model-Based Development of Embedded Systems 2009, pp. 182–188 (2009).

S026 Li, J.J. et al.: Automatic simulation to predict software architecture reliability. In: The Eighth International Symposium on Software Reliability Engineering, pp. 168–179. IEEE, USA (1997).

S027 Li, J.J., Horgan, J.R.: Applying formal description techniques to software architectural design. Computer Communications **23**, 12, 1169–1178 (2000).

S028 Liu, Y. et al.: SysML-based model driven discrete-event simulation. In: 21st ISPE Inc. International Conference on Concurrent Engineering, pp. 617–626. IOS Press (2014).

S029 López-Sanz, M. et al.: Representing service-oriented architectural models using π-ADL. In: Morrison, R. et al. (eds.) Software Architecture, LNCS 5292, pp. 273–280. Springer, Berlin, Heidelberg (2008).

S030 Lu, F. et al.: Application of executable UML technology for verification of armored vehicle information system model. Applied Mechanics and Materials **602-605**, pp. 1324–1328 (2014).

S031 Marinescu, R. et al.: Analyzing Industrial Architectural Models by Simulation and Model-Checking. In: Artho, C., Ölveczky, P.C. (eds.) Formal Techniques for Safety-Critical Systems. CCIS, vol 476, pp. 189–205. Springer, Cham (2015).

S032 Mokhtari, R., Chaoui, A.: Compiling, verifying and simulating dynamic software architectures using ANTLR and coloured-ADL. Int. J. of Communication Networks and Distributed Systems **19**, 4, pp. 406–433 (2017).

S033 Muccini, H., Pelliccione, P.: Simulating software architectures for functional analysis. In: Seventh Working IEEE/IFIP Conference on Software Architecture, pp. 289–292. IEEE, USA (2008).

S034 Graciano Neto, V.V. et al.: Modeling & simulation of software architectures of systems-of-systems: An industrial report on the Brazilian Space System. In: 2019 Spring Simulation Conference. IEEE, USA (2019).

S035 Oquendo, F.: Architecting exogenous software-intensive systems-of-systems on the internet-of-vehicles with SosADL. Systems Engineering **22**, 6, 502–518 (2019).

S036 Oquendo, F. et al.: Executing software architecture descriptions with SysADL. In: Tekinerdogan, B. et al. (eds.) Software Architecture. LNCS, vol. 9839, pp. 129–137. Springer, Cham (2016).

S037 Papadopoulos, G.A. et al.: An implementation framework for Software Architectures based on the coordination paradigm. Sci. of Comp. Prog. **60**, 1, 27–67 (2006).

S038 Pelliccione, P. et al.: CHARMY: A framework for designing and verifying architectural specifications. IEEE Trans. Software Eng. **35**, 3, 325–346 (2009).

S039 Sharaf, M. et al.: An architecture framework for modelling and simulation of situational-aware cyber-physical systems. In: Lopes, A., de Lemos, R. (eds.) Software Architecture. LNCS, vol. 10475, pp. 95–111 Springer, Cham (2017).

S040 Sindico, A. et al.: An industrial system engineering process integrating model driven architecture and model based design. In: France, R.B. et al. (eds.) Model Driven Engineering Languages and Systems. LNCS, vol. 7590, pp. 810–826. Springer, Berlin, Heidelberg (2012).

S041 Thomas, L. et al.: Architectural techniques for the description and validation of distributed real-time systems. In: Proceedings 2nd IEEE International Symposium on Object-Oriented Real-Time Distributed Computing, pp. 323–331. IEEE, USA (1999).

S042 Tsadimas, A. et al.: Integrating simulation capabilities into SysML for enterprise information system design. In: 2014 9th International Conference on System of Systems Engineering, pp. 272–277. IEEE, USA (2014).

S043 Tsadimas, A. et al.: Simulating simulation-agnostic SysML models for enterprise information systems via DEVS. Simulation Modelling Practice and Theory **66**, 243–259 (2016).

S044 Wang, R., Dagli, C.H.: Executable system architecting using systems modeling language in conjunction with colored Petri nets in a model-driven systems development process. Systems Engineering **14**, 4, 383–409 (2011).

S045 Yu, H. et al.: Polychronous modeling, analysis, verification and simulation for timed software architectures. J. of Systems Architecture **59**, 10, 1157–1170 (2013).

References

1. Ampatzoglou, A., Bibi, S., Avgeriou, P., Verbeek, M., Chatzigeorgiou, A.: Identifying, categorizing and mitigating threats to validity in software engineering secondary studies. Inf. Softw. Technol. **106**, 201–230 (2019)

2. Araujo, C., Cavalcante, E., Batista, T., Oliveira, M., Oquendo, F.: A research landscape on formal verification of software architecture description. IEEE Access **7**, 171752–171764 (2019)

3. Bandara, W., Furtmueller, E., Gorbacheva, E., Miskon, S., Beekhuyzen, J.: Achieving rigor in literature reviews: insights from qualitative data analysis and tool-support. Commun. Assoc. Inf. Syst. **37** (2015)

4. ISO/IEC/IEEE 24765:2017(E): Systems and software engineering – Vocabulary (2017)

5. ISO/IEC/IEEE 42010:2022(E): Software, systems and enterprise – Architecture description (2022)

6. Kim, E.S.: Executable architectures concept and methodology. In: 20th International Command and Control Research and Technology Symposium. IC2I (2015)

7. Legay, A., Delahaye, B., Bensalem, S.: Statistical model checking: an overview. In: Barringer, H., et al. (eds.) RV 2010. LNCS, vol. 6418, pp. 122–135. Springer, Heidelberg (2010). https://doi.org/10.1007/978-3-642-16612-9_11

8. Petersen, K., Vakkalanka, S., Kuzniarz, L.: Guidelines for conducting systematic mapping studies in software engineering: an update. Inf. Softw. Technol. **64**, 1–18 (2015)

9. Wolny, S., Mazak, A., Carpella, C., Geist, V., Wimmer, M.: Thirteen years of SysML: a systematic mapping study. Softw. Syst. Model. **19**(1), 111–169 (2020)

10. Zhang, P., Muccini, H., Li, B.: A classification and comparison of model checking software architecture techniques. J. Syst. Softw. **83**(5), 723–744 (2010)

Architectural Views: The State of Practice in Open-Source Software Projects

Sofia Migliorini[1], Roberto Verdecchia[1(✉)], Ivano Malavolta[2], Patricia Lago[2], and Enrico Vicario[1]

[1] University of Florence, Florence, Italy
sofia.migliorini@edu.unifi.it,
{roberto.verdecchia,enrico.vicario}@unifi.it
[2] Vrije Universiteit Amsterdam, Amsterdam, The Netherlands
{i.malavolta,p.lago}@vu.nl

Abstract. Context: Architectural views serve as fundamental artefacts for designing and communicating software architectures. In the context of collaborative software development, producing sound architectural documentation, where architectural views play a central role, is a crucial aspect for effective teamwork. Despite their importance, the use of architectural views in open-source projects to date remains only marginally explored.

Goal: We aim at conducting a comprehensive analysis on an extensive corpus of open-source architectural views. The goal is to understand (i) what the "history" of architectural views is, (ii) how architectural views are represented, and (iii) what architectural views are used for in the context of open-source projects.

Methods: We leverage a software repository mining process to systematically construct a dataset of 15k architectural views. Then, we perform (i) a quantitative analysis on the metadata of all 15k views and (ii) a qualitative analysis on a statistically-relevant sample of 373 views.

Results: Most projects rely on a single architectural view, which is often used to document a medium or high level description of the architecture. Views are usually created at either the beginning or at the end of a project, are rarely updated, and tend to be maintained by a single contributor. Views usually adopt an informal colored notation without a supporting legend and frequently report technologies used. Deployment and control flow are the most recurrent viewpoints, and commonly cover concerns related to software maintainability and functional suitability.

Conclusion: The state of the practice about architectural views in open-source software systems seems to favor informal descriptions. Despite this, the effort needed to create views might hinder keeping views up to date, and a common syntactic ground between viewpoints seems hard to find. To address current needs, we speculate that a solution could lie in defining and popularizing versionable, templateable views that can be integrated in collaborative programming environments.

M. Galster et al. (Eds.): ECSA 2024, LNCS 14889, pp. 396–415, 2024.
https://doi.org/10.1007/978-3-031-70797-1_27

Keywords: Architectural Views · Architectural Documentation · Repository Mining · Open Source Software

1 Introduction

In the vast landscape of software development, architecture plays a key role as a bridge between requirements and implementation [4]. A robust architecture has the potential to guarantee that a system will meet essential quality requirements in such areas as performance, reliability, portability, scalability, and interoperability [8]. One of the primary methodologies to design and communicate software architectures is through the use of *architecture viewpoints* and *architectural views* [6]. Different stakeholders generally have drastically differing mental models based on their experiences with a software-intensive system, and may only primarily focus on certain aspects of it [27]. By quoting Rozanski and Woods: *"A view is a representation of one or more structural aspects of an architecture that illustrates how the architecture addresses one or more concerns held by one or more of its stakeholders* [23]. A software architecture is a complex entity that defies a basic one-dimensional description. With the help of views, which enable the separation of concerns, we can break down the multidimensional structure into a number of, hopefully, engaging and comprehensible system representations [4].

In the domain of collaborative software development, GitHub is one of the leading platforms for open-source software projects [11]. With 420 million total projects at the end of 2023[1], GitHub actively promotes collaboration, enables transparent communication, and empowers developers to contribute to open-source projects globally. Although documentation is essential to the open-source ecosystem, as it helps developers learn about projects and, consequently, choose which ones to contribute to [28,31], open-source documentation is frequently regarded as inadequate, sparsely written, outdated, or nonexistent [9]. In the year 2000, IEEE introduced a standard for architecture documentation (the latest version updated in 2022 [2]) that advocates creating personalized views that best address the stakeholders and their concerns associated with the system to be represented. Although there are studies to identify different types of documentation provided in software repositories [12,21], none of them focuses on software architecture views. This work aims to fill this void by focusing on an extensive corpus of architectural views, intending to investigate their adoption within the context of open-source projects.

The main contributions of this study can be summarized as follows:

- A publicly-available dataset of 15k software architecture views extracted from 12.2k GitHub projects, supported by repository and view commit history metadata.
- A quantitative analysis of the collected data, casting light on the practice of software architecture views in open-source practice.

---

[1] https://github.blog/2023-11-08-the-state-of-open-source-and-ai.

- An in-depth manual categorization of a statistically representative sample of 373 views, conducted by studying both their syntactic and semantic properties.
- A comprehensive replication package of the study (see Sect. 8).

2   Related Work

To the best of our knowledge, this study represents the first analysis of an extensive corpus of software architecture views extracted from open-source projects.

The most similar large-scale study, conducted by Hebig et al. [12], focuses on the analysis of UML usage in open-source projects, involving a systematic mining of GitHub repositories. In addition, Buchgeher et al. [5] proposed a study on the adoption of architecture decision records (ADRs) in projects hosted on GitHub. Both studies aim at characterizing current practices of file creation and maintenance over time of either UML models or ADRs, without specifically focussing on software architecture views. Our study extends the research by also analyzing the visualization techniques and contents of software architecture views, without being confined to a specific language or format.

Expanding the perspective, Ding et al. [7] proposed a comprehensive investigation of 2k projects from four major open-source software sources, identifying 108 projects with documented software architecture. Their goal is to analyze what type of information is documented and how it is described. Similarly, Muszynski et al. [18] examine software architecture documentation practices in six large popular open-source software systems; the focus of this study is the classification of architectural views. Malavolta et al. [17] analyzed 335 open-source robotics projects and mined how roboticists document the software architecture of their system, which types of views are used and how they are represented (e.g., textually or visually). Unlike these studies, which examine a limited number of open-source projects, our investigation targets architectural views extracted from 12.2k repositories. Furthermore, instead of presenting a general overview, we focus on a particular type of software architecture documentation, namely software architecture views.

A classification of software architecture visualization techniques reported in the literature is presented by Shahin et al. [25]. A similar review of recent and key literature on software architecture visualization is performed by Ghanam and Carpendale [10]. Additionally, Alshuqayran et al. [3] presented a systematic mapping study on microservices architectures and their implementation, offering insights on architectural views/diagrams used. Rather than focusing on visualization techniques, our study aims at classifying software architecture views by analyzing both the languages syntax and views contents. In addition, we based our research on the practical use of software architecture views in open-source repositories, instead of relying on exsisting literature.

Several surveys have been conducted to understand the perspectives of practitioners on software architecture documentation and related tools. Rost et al. [22] investigated problems and wishes for the future with industrial participants.

Ozkaya [19] studied the level of knowledge and experience in software architectures among practitioners from both industry and academia. Malavolta *et al.* [16] examined the use of architectural languages in industry by interviewing practitioners from 40 different IT companies. Differently from such surveys, our research does not focus exclusively on industrial practices, but rather extracts data from GitHub repositories, characterizing the state of practice of software architecture views in the context of open-source software projects. Moreover, the main results of these surveys focus on architectural notations; our research also covers the contents of architecture documentation practices.

A more specific survey by Ozkaya and Erata [20] aims at understanding the usage of UML diagrams for modeling from different viewpoints. Additionally, a case study on the selection of viewpoints in projects from three different telecom-area software organizations was held by Smolander [26] to clarify how the conceptions of architects about architectural viewpoints differ, based on the prevalent situation and characteristics in an organization and in the project at hand. Unlike these studies that focus on a particular architectural language or a limited number of case studies, our analysis encompasses the usage of viewpoints across numerous software architecture views, without relying on any specific architectural language.

3 Research Methodology

3.1 Goal and Research Questions

The goal of this research is to characterize the state of practice of software architecture views in open-source projects. This study aims to answer, in the context of open-source software, the following research questions:

RQ1: *What is the history of architectural views?* This research question aims at understanding the timing of views introduction in repositories, how they evolve over time, and who contributes to this evolution.
RQ2: *How are architectural views represented?* With this question we aim at identifying the main characteristics of the notations used to describe architectural views.
RQ3: *What architectural information has been represented in the views?* This research question constitutes the core of the study. By answering it, we aim at characterizing our view dataset in terms of topics focused on, quality requirements considered, architectural styles, and technologies employed.

3.2 Research Process

Figure 1 shows an overview of the research process of this study. The details of each step are provided in the remainder of this section.

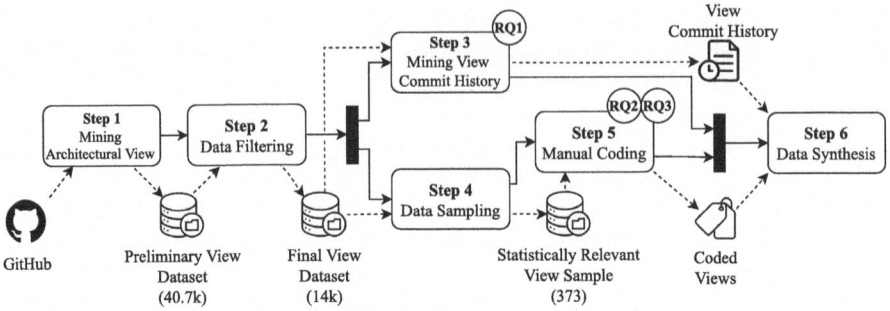

Fig. 1. Research Process Overview

Step 1: Mining Architectural Views. The objective of this mining step is to identify on GitHub images reporting an architectural view. In order to do so, we first automatically identify all files containing the substring "architect" in the filename that are linked in any of the README.md and CONTRIBUTING.md files hosted in a GitHub repositories (including the ones located in subfolders). This process is executed by sending requests to the Search Code endpoint of the GitHub REST API[2]. Below a list of the restrictions applied for this search:

- R1: Consider only the *default branch* of the repository,
- R2: Search for files smaller than 384 KB,
- R3: Search for repositories with fewer than 0.5M files,
- R4: Search for repositories that have had activity or have been returned in search results in the last year,
- R5: Do not include repository forks.

R1–R4 are forced by the Search code endpoint[3], whereas R5 is a choice that we make for our study in order to avoid duplication. Regarding R2, we observe that this limitation does not compromise our data collection as the file retrieval stops long before reaching the imposed limit. The mining phase was carried out between September and October 2023. For each extracted URL, we also gather relevant metadata about the repository that contains the file. At the end of this extraction phase, a duplicate removal is performed based on the image URL, and a dataset of 40.7k architectural views is produced.

Step 2: Data Filtering. After the first mining phase, a filtering process follows to ensure the quality of the mined views. The exclusion criteria applied are the following:

- E1: Views or repositories referring in their name to demos, examples, or exams,

---

[2] https://docs.github.com/en/rest. Accessed 9th April 2024.
[3] https://docs.github.com/en/rest/search/search?apiVersion=2022-11-28#search-code. Accessed 9th April 2024.

- E2: Views contained in a repository with less than 10 commits,
- E3: Views contained in a repository with less than 2 stars,
- E4: URLs linking to GitHub badges, such as "img.shields.io",
- E5: Non-downloadable images.

E1 serves to target real projects: We define a set of regular expressions and a view is discarded if either the repository name, the repository description, or the image URL matches any regular expression [17]. E2 and E3 are used to avoid inactive or non-maintained projects [15,17]. From a quick sample analysis of the dataset, we also identify a set of URLs that point to some GitHub badges, which are removed from the dataset (E4). Finally, we exclude views that produce an error in the downloading phase, pointing to views that are no longer available (E5). At the end of this filtering phase, we establish a dataset comprising 15k views. An example of view developed by the Google Cloud Platform[4] taken from the dataset is documented in Fig. 2.

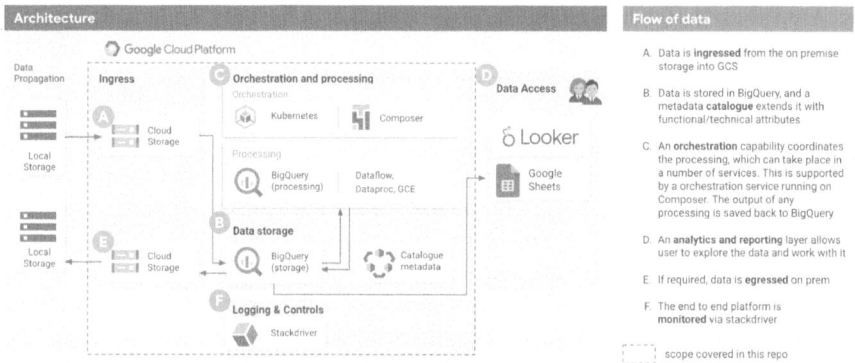

Fig. 2. Example of Mined Architectural View (See footnote 4)

Step 3: Mining View Commit History. For each view file we extract (i) the number of commits, (ii) the number of distinct contributors, (iii) the number of days between the first commit of the repository and the first commit adding the file representing the view, (iv) the total number of commits prior to the introduction of the view file, and (v) the number of commits between the first and last view update. The commit history is extracted for 14.2k views, as 6% of identified views resulted to be inaccessible (*e.g.,* images no longer in the repository, spelling errors, path errors). The data collected in this step is used to answer RQ1.

---

[4] https://github.com/GoogleCloudPlatform/reg-reporting-blueprint.

Step 4: Data Sampling. In order to answer RQ2 and RQ3 we need to extract a statistically significant sample from our dataset. For this purpose, an additional data filtering step is performed to consider exclusively one architectural view for each repository, leading to selecting 12.2k views. With the dataset of 12.2k views, to ensure 5% margin of error and 95% confidence level, we randomly sample 373 views. With a preliminary analysis of the first extracted sample, we can conclude that our dataset presents about 23% of false positives. In particular, the main categories of rejected views represent: the internal architecture of machine learning algorithms (*e.g.,* the layers of a deep neural network), with 59 views (15.8%), hardware architectures (9 views, 2.5%), network architectures (5 views, 1.3%), and logos (5 views, 1.3%). Views containing text in languages other than English are retained and automatically translated for subsequent analysis.

Step 5: Manual Coding. Three data exploration sessions follow, where researchers discuss the characteristics of the views that are relevant to answering RQ2 and RQ3, leading to the consolidation of the data extraction framework used for this study. The sample dataset is then thoroughly analyzed *via* iterative coding sessions. For scrutiny and replication purposes, detailed explanations of the codes used as coding guide are available in the replication package (see Sect. 8).

Given that RQ2 addresses the syntax of the visualization techniques employed, to answer it, we extract the following fields: (i) architectural notation employed, *i.e.,,* informal, semiformal or formal [6] (ii) shapes used, *e.g.,* rectangles or circles, (iii) use of color, (iv) presence of a legend, (v) presence of nested components, (vi) presence of explicit ports/interfaces between components, (vi) presence of explicit connectors, *i.e.,* lines connecting components, and (vii) connectors direction, *i.e.,* unidirectional, bidirectional, non-explicit or bus.

In order to answer RQ3, we take a step further and proceed to study the content of the views by considering the following fields: (i) architecture scope, *i.e.,* if the view represents a part of the system, the entire system, or the entire system plus how it interfaces with other systems, (ii) architectural style(s) used in the view, (iii) concern(s) addressed, *e.g.,,* deployment or connectivity, (iv) behavior, *i.e.,* if the view represents static, dynamic properties of the architecture, or both, (v) quality attribute(s) considered (first level of the *ISO/IEC 25010:2023* classification [13]), (vi) granularity of the system representation, *i.e.,* high, medium, or low, (vii) components nature, *e.g.,* servers or databases, (ix) connectors nature, (x) technologies reported (classification at the level of the software provider, *e.g.,* *AWS* and not *AWS Lambda*), and (xi) design overlays, *e.g.,* textual descriptions or code snippets [1].

Regarding the coding of *Architecture Scope* and *Granularity*, we use magnitude coding in order to capture the intensity of the feature considered. We use provisional coding [24] to classify the *Quality Attributes, Behavior, Connectors Direction,* and all the *yes/no* fields. For all other fields, we use open coding [14] to identify recurrent concepts. Two researchers independently analyzed 186/187 views, with weekly discussions used to revise, homogenize, and align the coding process. When necessary, doubts and disagreement points were resolved by involving a third researcher. The data gathered through this step is used to answer RQ2 and RQ3.

Step 6: Data Synthesis. To answer RQ1, which is purely quantitative, the extracted data are analyzed and interpreted by simple statistical means, such as data plotting and basic summary statistics. To answer RQ2 and RQ3, the codes obtained from the labeling process are first examined to see if any can be removed or merged with others due to a lack of representativeness. Then, the resulting ones are presented with different plots according to the information that we want to describe. For the discussion section, a phase of cross-matching results is added to extract potential interesting relationships among the various aspects analyzed.

4 Results

In this section, we present the results collected to answer our RQs (see Sect. 3.1).

4.1 Results RQ1: History of Architectural Views

Architectural Views Creation. Figure 3(a) illustrates the moment of the view introduction in the repository in terms of days passed since the start of the project, which is represented by the initial commit in the repository. Outliers are identified and excluded from the plot to enhance visualization. Our collected data reveals a tendency to introduce architectural views at the project start: the project's age when the view is introduced is on average 222.51 days, with a median of 39 days, while projects have an average duration of 1570.19 days, with a median of 1456 days.

The maximum value (9316 days) is recorded for the repository *PolarDB*[5], a cloud-native database developed by *Alibaba Cloud* active since 1996, which is characterized by 56 contributors, 2.6k stars and 427 forks. At the beginning, the repository contained the source code distribution of the *PostgreSQL* database management system. In 2021, it changes its subject to *PolarDB* and a complete new `README.md` file is introduced by a new contributor, including an architectural view.

However, the level of activities can highly vary throughout the lifespan of open-source projects. In order to place the view introduction relatively to other development activities of the project, Fig. 3(b) illustrates the distribution of views based on the percentage of repository commits done when the view is introduced. This figure presents a much more balanced perspective, showing that architectural views are introduced in repositories along all active phases of the projects, with peaks in the initial and final ones. The median and mean of the percentage of overall commits done when the view is introduced are 50% and 49.34%, respectively.

---

[5] https://github.com/ApsaraDB/PolarDB-for-PostgreSQL Accessed 14th April 2024.

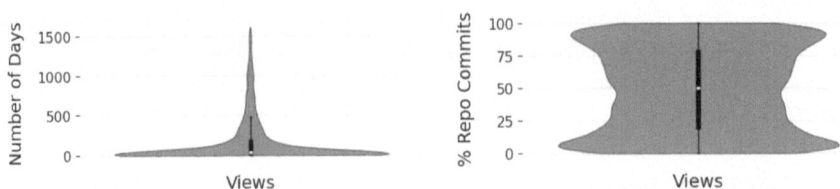

(a) *Project's age in days at the moment of the* (b) *Percentage of overall repository commits view introduction in the repository (without done when the view is introduced outliers)*

Fig. 3. Views Creation

Architectural Views Maintenance. Our results show that the vast majority of views with a retrieved commit history (10.6k out of 14.2k) remains unchanged. Yet, we found that around 25% of them (3.5k out of 14.2k) are updated one or more times. Figure 4 summarizes the distribution of views by number of commits. Outliers are excluded from the plot for a better visualization.

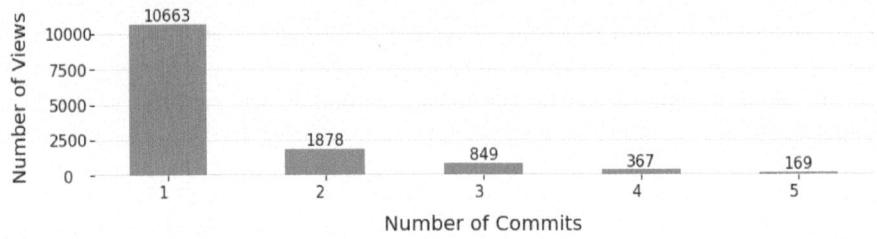

Fig. 4. Distribution of views by number of commits (without outliers)

When a view is updated, the number of commits is on average 3 (2 updates), with a median of 2 (1 update). The maximum number of commits on a view is 27 for a view that is updated over time by the same user[6], even though the repository has a total of seven contributors.

We have also conducted an analysis of the time span between the first and last commit targeting a view within a repository. This time span, shown in Fig. 5(a), is expressed as a percentage of overall commits of the repository. The maximum time span observed is 100% of repository commits, while the mean time span is 6.98%. The majority of projects seem to concentrate the view introduction and updating activities within a relatively short period, with a median of 33 days, without interspersing them with other substantial development tasks. The high frequency of 0% cases is due to the predominance of views with just 1

---

[6] https://github.com/digitaltwinconsortium/ManufacturingOntologies/blob/main/Docs/architecture.png. Accessed 15th April 2024.

commit. Updating activities extend beyond 10% of the project's commits only in a minority of repositories.

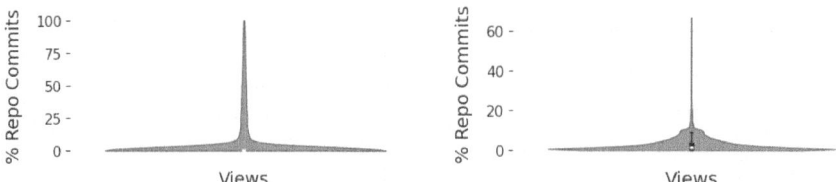

(a) *Time passed between first and last view commit expressed as percentage of overall repository commits* (b) *Percentage of overall repository commits dedicated to the view updates*

Fig. 5. Views Maintenance

Figure 5(b) shows a focus on the percentage of repository commits dedicated to the view creation and updating, with a mean of 2.91% and a median of 1.54%. The maximum of 66.67% is reached for a repository that has exclusively the purpose of describing the documentation of the project[7].

Contributors. More than 92% of views (13.2k out of 14.2k) have just one contributor, while the remaining 8% are updated by more than one user, with a maximum of 5. In particular, 72% of views that have at least one update are modified by the same contributor.

An additional analysis is conducted on the percentage of contributors editing the view. The median and mean are 50% and 57% of total contributors, respectively, that is due to the presence of many repositories with just 1 contributor. The minimum is 0.22% of repository contributors.

> **RQ1: Architectural Views History.** Architectural views tend to be introduced either at the beginning or at the very end of open-source projects. 75% of views are never updated. Activities on views are concentrated in short periods, without interspersing them with other significant development tasks. When updated, 71% of views are edited by the same user.

4.2   Results RQ2: Architecture Representation

Architectural Notation. Out of the statistically relevant sample considered to answer RQ2 (373 views), the vast majority of views use *informal* notations (96%), notably *boxes and lines*. Only a much smaller fraction (4%) employs a *semiformal* notation, in particular *UML*, mainly focussing on *low-level* aspects of the architecture. There is no evidence for the use of formal notations. Out of the views utilizing informal notations, seven resulted to be manual sketches.

---

[7] https://github.com/gridsuite/documentation. Accessed 15th April 2024.

Shapes. Figure 6 presents a distribution of views in terms of what kind of shapes are used in the views. The most popular shapes employed are *rectangles* (79%), and *icons* (46%), which mostly depict the icon of a technology. These *primary shapes* mainly appear as the only shape in the view, whereas all other shapes are always combined with each other or used as auxiliary shapes for specific elements, *e.g.*, cylinders for databases.

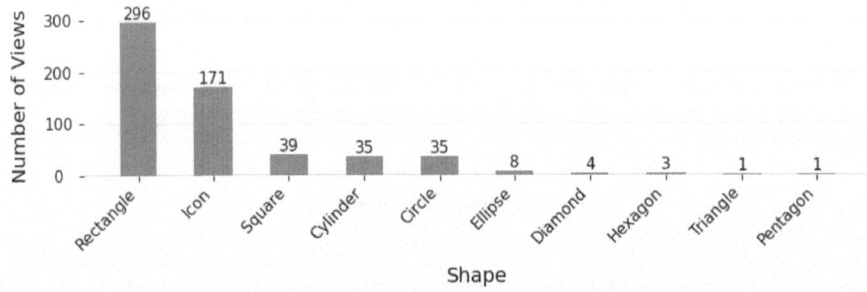

Fig. 6. Number of views *per* shape

Architectural views are mainly mapped to more than one shape. We observe the presence of a small group of 3-dimensional images (2%) generated by the Cloudcraft platform[8], which appears to be used also in a considerable number of bidimensional views.

Connectors. Most views use explicit connectors between architecture elements (92%). Different types of connectors may coexist in the same view, but we observe that when connectors are used, their direction is mostly *unidirectional* (74%). The presence of *bidirectional* connectors is detected in almost a third of the views (29%). We remark that sometimes bidirectional connections are shown as two unidirectional ones. Only three views present a bus (1%), while few more do not show explicit connector directions (5%).

Further Properties. Regarding further view properties, we noted that 81% of views are colored, even if the specific meaning of colors appears to be seldom defined, both for components and connectors. Only 8% of views include a legend to provide further explanations about the employed notations.

In terms of structural complexity, the use of nested components is slightly more frequent than unidimensional representations (56%), indicating a preference for hierarchical organizations within architectural views.

Regarding the presence of explicit ports or interfaces between architectural components, only a fraction of views show them (8%).

---

[8] https://www.cloudcraft.co. Accessed 15th April 2024.

RQ2: Architectural Views Syntax. Most views use an informal notation (96%), while only a minor portion (4%) a semiformal one (UML). Rectangles and icons are the most widely used shapes. Unidirectional connections are the most represented (74%). Views often use colors (81%), but only 7% of them include a legend documenting the color use. Explicit ports or interfaces between components are rarely documented (8%).

4.3   Results RQ3: Architectural Views Contents

Architecture Scope. Regarding the representation level of the architectural views, in most cases, the *entire* architecture of the system is represented (53%), instead of focusing only on a *part* of it (24%). A considerable fraction of the considered views (23%) also documents *interactions* between the system considered and other external systems.

Architectural Styles. The most recurrent architectural styles documented in the views are *client-server* (20%), *layered* (20%), *service-oriented* (15%), and *event-driven* (12%). Out of the views reporting service-oriented architectures, 35% of them focus specifically on *microservices*. Event-driven architectures instead often result to manage events with serverless functions, most notably AWS Lambda[9].

Concerns. Regarding the topics addressed by architectural views, we note various recurring subjects. Views can be mapped to more than one concern, with 20% of the total covering simultaneously more than one concern. A considerable portion of the views focus on *general* architectural documentation (30%), *i.e.,* rather than focusing on a specific topic, they adopt a broad and neutral technological viewpoint.

The second most recurrent concern covered by views is *deployment* (30%), with a recurrent focus on cloud and virtualization aspects. Following closely are *control flow* views (29%), used to document functional aspects of architectures. *connectivity* among architecture components and sub-systems is another recurrent concern (16%), followed by *data flow* (8%), and a smaller number of *security* views, considering mostly message cryptography and authentication mechanisms. *Performance* concerns are depicted only 3% of the views, and often illustrate architectural aspects related to load balancers and distributed systems. Finally, architectural viewpoints regarding *scheduling* are adopted in 1% of the views.

Granularity. In terms of granularity of the system representation, the majority of views exhibit either a *high* granularity one (53%), reporting a very coarse

---

[9] https://aws.amazon.com/lambda. Accessed 10th April 2024.

architectural representation, *e.g.*, just the architectural layers, or a *medium* granularity (37%), *e.g.*, considering the main sub-systems and key technologies used. A much smaller portion of views use a *low* level of granularity granularity (9%), and are mostly used to reason about source code implementation details, such as functions or attributes in addition to broader system components.

Behavior. Regarding the architecture behavior depicted, approximately half of the views consider *static* properties of architectures (49%). A smaller number instead considers *dynamic* properties (42%), such as the system *control flow*, are represented in a smaller portion of views. Finally, a subset of views considers both static and dynamic aspects (9%).

Quality Attributes. *Maintainability* is the most recurrent quality attribute considered in the views (68%), reflecting the high recurrence of the *general* concern of the views previously documented. *Functional suitability* is the second most recurrent concern (32%), and is mostly mapped to viewpoints detailing step-by-step use case executions through architectures. Least addressed QAs are instead *performance efficiency* (35%), *security* (9%), *flexibility* (8%), interaction capability (7%), compatibility (2%) and reliability (2%). As for the concerns category previously presented, a view may address more than one QA. Out of all ISO/IEC 25010 characteristics, *safety* is the only quality attribute that is not considered in the views.

Components Nature. Regarding the nature of components represented in the views, Fig. 7 documents the frequency of the different identified components across views. The most frequent components are *technologies* (present in 36% of the views). Other recurrent component types are, in order of frequency, *databases* (32%), *sub-systems* (24%), *clients* (24%). The *others* category reported in Fig. 7 includes components utilized only in one or two views, *e.g.*, virtual machines, firewalls, and coroutines.

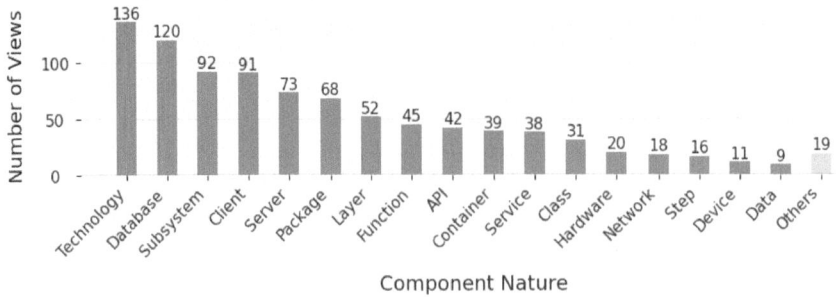

Fig. 7. Frequency of the different components across views

Connectors Nature. Regarding the interpretation of explicit connectors, it was possible to classify their nature in 91% of the dataset. In particular, 8% of views do not show explicit connectors, while for 1% of them the meaning of connectors does not emerge from the views' analysis. *Control flow* are the most used connectors type (28%), and are frequently used in combination with numerical indicators highlighting sequential operations transversing architectures at runtime. General *communication* among components, frequently used to highlight communication channels or protocols, are also recurrent (27%). Views with *data flow* connectors are a bit less frequent (25%), and commonly co-appear in views along database components. Least recurrent connector types are dependencies (12%), function calls (5%), inheritance dependencies (3%), API calls (2%) and composition relations (1.3%). As for other studied properties, a single view may incorporate multiple types of connectors, reflecting the complex interactions and relationships within the software system.

Technologies. Figure 8 documents the ten most frequently mentioned technologies in the architectural views[10].

The overall high recurrence of cloud-based and virtualization technologies, *e.g.,* Amazon Web Services (16%), Kubernetes (7%), and Docker (7%), may be attributed to the current growing adoption of cloud-native serverless applications.

Reporting at least a technology in the views results to be a rather widespread practice (35%). The median number of reported technologies per view is one and the mean is two, while the maximum number of technologies reported in a single view is ten.

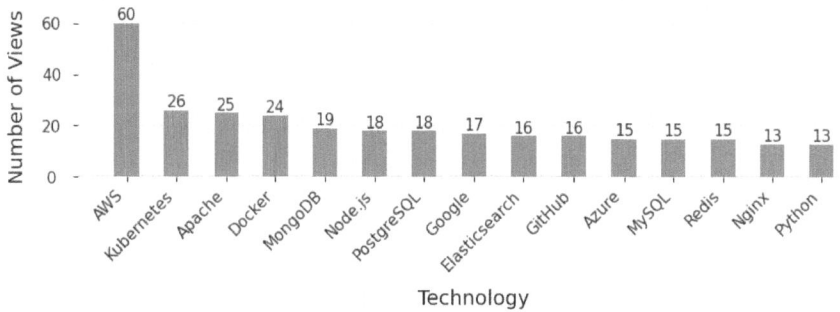

Fig. 8. Number of views per technology (top 10)

---

[10] The complete list of technologies and their recurrence is reported for completeness in the replication package of this study.

Design Overlays. Additional information to support the views is utilized in approximately a fifth of all views (18%). The majority of the overlays report complementary architectural information in form of *textual descriptions* (45% of views using overlays), which is primarily used to specify system or components functionalities. When *screenshots* are employed (17%), they mainly depict the user interface of an application or website. *miniviews* are instead less frequently used to zoom-in on particular aspects of the architecture (12%), *e.g.*, to document a data transformation process. Other information such as *URLs* (10%), *code snippets* (9%), and *configuration parameters* (7%) are also, but less frequently, used to highlight connection or low-level implementation details.

> **RQ3: Architectural Views Content.** 75% of views consider architectures in their entirety. Client-server is the most used architectural style (25%), followed by layered (20%), service-oriented (15%) and event-driven (12%) architectures. General architecture documentation, followed by deployment and control flow are the most recurrent topics covered. Views mostly consider a high or medium granularity level (91%). Static and dynamic aspect are equally represented. The most considered quality attribute is maintainability (68%), followed by functional suitability (33%). Around 65% of views explicitly report the technologies used, with AWS being the most recurrent one (25%).

5   Discussion

From the data collected a clear picture regarding the use of views in open-source practice emerges. *Most software projects rely on a single, seldom updated view, which mostly utilizes an informal notation to outline the high level architectural structure.* A common syntactic ground among viewpoints is very hard to be found.

Despite the importance of views for software documentation [23], their intuitive support for onboarding processes, and their potential to stimulate new contributions in collaborative environments such as GitHub, architectural view documentation in the open-source community does not seem to be a well established practice. As step forward, we ask ourselves: *What can we do to ease the adoption of architectural views in open-source practice, and how can we systematize the documentation of views?*

A potential answer to this question lies in *making views versionable*. In fact, to date, the process required to create and update architectural views on GitHub seems quite cumbersome. More specifically, editing an architectural view commonly relies on utilizing external tools which are not integratable with GitHub, and hence require first to update the view *via ad-hoc* standalone tools, and subsequently push an unmodifiable version of the view in shared repository. In addition to making views versionable, another aspect crucial for the adoption of view documentation in open-source could be to *allow views to be automatically rendered in the web interfaces of repositories*, in a similar fashion with which README.md files are currently visualized on GitHub websites. This would further support to move away from hard-coded immutable view image files in open-source repositories, while still providing a swift and intuitive graphical representation of views. Singular efforts striving to provide versionable web-friendly views, such as Mermaid[11], which allows to create various versionable diagrams on GitHub, *e.g.,* C4 and UML, should be therefore incentivized and given more visibility. Providing support and exposure to such projects would allow to move view documentation in open-source from a small community endeavors to mainstream, consolidated, and well integrated open-source practices. As related consideration, similar to current trends pushing for the standardization of architectural decision records in open-source[12], the community could spend efforts to *establish reusable templates to represent architectural views.*

We speculate that providing versionable, templateable views that can be integrated in collaborative programming environments would allow to further establish architectural view documentation practices in open-source. Through virtuous cycles, this could lead to further cascading effects, *e.g.,* the inclusion of GitHub Actions[13] dedicated to software architecture, and the rise of dedicated `ARCHITECTURE.md` documentation.

6   Threats to Validity

We document the study threats by following to the categorization proposed by Wholin *et al.* [30] and considering common pitfalls in assessing threats to validity [29].

Construct Validity. To ensure that the dataset of views addresses our research questions, we defined *a priori* the concept of *software architectural view* by relying on established definitions [2,6]. The entirety of the study is conducted by following such definition. *Via* a set of *a priori* defined inclusion criteria used during the mining phase, and a subsequent manual process based on a coding guide, views deviating from our construct definition were discarded to ensure adherence of the collected results to our research goal.

Internal Validity. A potential internal validity threat lies in the formulation of the search query used in the mining process (see Sect. 3, Step 1). To mitigate this threat, the query was refined over more iterations through a set of exploratory query trials. Another potential internal threat regards subjective biases that could have influenced the manual coding process (see Sect. 3, Step 6). To mitigate this threat, two annotators were involved in the process, weekly meetings were held to align the labeling process by discussing doubts and examples, and a third researcher was involved whenever necessary. The automated translation accuracy of non-English views could have also potentially influenced the internal validity

---

[11] https://github.com/mermaid-js/mermaid. Accessed 24th April 2024.

[12] https://adr.github.io. Accessed 24th April 2024.

[13] https://github.com/features/actions. Accessed 24th April 2024.

of the study. However, given their low recurrence (2%), we do not deem it could have majorly influenced our results. Exclusion criteria E1 (see Sect. 3.2) could had led to the inclusion of demo projects in our dataset. However, as no toy project was found in the manually scrutinized sample (see Sect. 3.2), we do not deem this threat majorly influenced our results.

External Validity. Albeit our best efforts, the automated mining process (see Sect. 3, Step 1) needed to rely on a keyword-based search of files linked in the `README.md`. While the search made it possible to identify a considerable number of views (14k), views that did not explicitly contain the substring "architect" in their filename could not be identified. In addition, albeit used as main source of GitHub repository documentation [21,28], files not linked to `README.md` or `CONTRIBUTING.md` files could not be identified. This choice constituted a trade-off between internal and external validity, which made it possible to automatically identify views while limiting the number of potential false positives. In future work, this threat could be addressed by using a more sophisticated view identification strategy, *e.g.*, by training an image classification model on the dataset constructed for this research, and automatically classify views from a more extensive pool of images.

Reliability. To ensure the reproducibility and verifiability of our results, we make a comprehensive replication package of the study available online (see Sect. 8).

7   Conclusions and Future Work

In this research we present an investigation on the architectural views state of practice in open-source projects. The study uses repository mining to build a dataset of 15k views, which are then analyzed both quantitatively and qualitatively.

From our study a clear picture on the use of architectural views in open-source project emerges. Views are used to intuitively convey in an informal manner key architectural concepts, are seldom updated, and are a single-contributor responsibility. A shared syntactic ground between views seems hard to be found, with *ad hoc* viewpoints being a widespread norm. Informally, the sole exception might be the general-purpose architectural set of icons and connectors supported by the AWS Architecture Center[14], that seems to be shared by a minority of the analyzed views. Overall, it appears as if the effort needed to create views, combined with the potential lack of intuitive yet structured set of viewpoints, hinders utilizing more than one view, and even keeping that single view up to date.

As future work, the topic of views in projects could be further investigated, *e.g.*, by conducting a survey with repository contributors. This would allow to further understand what influences view selection, how relying on a single view

---

[14] https://aws.amazon.com/architecture/icons. Accessed 17th April 2024.

affects system understanding and evolution, and what are the current requirements for new viewpoints. Building further on this study, we envision also to conduct research to automatically identify and classify architectural views, and characterize the contributors involved in editing architectural views. Paving the way for future research, our ultimate goal is to define and popularize versionable, templateable views that can be integrated in collaborative programming environments. Given its practical nature, such goal can be achieved solely through empirical evidence and close collaboration with open-source communities, in order to define views promoting simplicity, adoption, and consistency across projects.

Data Availability Statement. To support reproducibility and verifiability, we make all data used in this study, scripts, settings, coding guide, and results available in a replication package online (https://figshare.com/s/a796b8b414bbc7d09fe2. Accessed 24th April 2024). To encourage open science, the replication package is shared under open-source MIT license.

References

1. IEEE Standard for Information Technology-Systems Design-Software Design Descriptions: IEEE STD 1016-2009, pp. 1–35 (2009)
2. International Standard for Software, systems and enterprise Architecture description. ISO/IEC/IEEE 42010:2022(E), pp. 1–74 (2022)
3. Alshuqayran, N., Ali, N., Evans, R.: A systematic mapping study in microservice architecture. In: International Conference on Service-Oriented Computing and Applications, pp. 44–51. IEEE (2016)
4. Bass, L., Clements, P., Kazman, R.: Software Architecture in Practice, 4 edn. Addison-Wesley Professional (2021)
5. Buchgeher, G., Schöberl, S., Geist, V., Dorninger, B., Haindl, P., Weinreich, R.: Using architecture decision records in open source projects–an MSR study on GitHub. IEEE Access (2023)
6. Clements, P., et al.: Documenting Software Architectures: Views and Beyond. Addison-Wesley (2011)
7. Ding, W., Liang, P., Tang, A., Van Vliet, H., Shahin, M.: How do open source communities document software architecture: an exploratory survey. In: André, É., Zhang, L. (eds.) Proceedings - 19th International Conference on Engineering of Complex Computer Systems, ICECCS 2014, pp. 136–145. IEEE, Institute of Electrical and Electronics Engineers (2014)
8. Garlan, D.: Software architecture: a roadmap. In: ICSE 2000: Proceedings of the Conference on The Future of Software Engineering, pp. 91–101. Association for Computing Machinery, New York (2000)
9. Geiger, R., Varoquaux, N., Mazel-Cabasse, C., et al.: The types, roles, and practices of documentation in data analytics open source software libraries. Comput. Support. Coop. Work **27**, 767–802 (2018)
10. Ghanam, Y., Carpendale, S.: A survey paper on software architecture visualization. University of Calgary, Technical report, p. 17 (2008)
11. Gousios, G., Vasilescu, B., Serebrenik, A., Zaidman, A.: Lean GHTorrent: GitHub data on demand. In: Proceedings of the 11th Working Conference on Mining Software Repositories, MSR 2014, pp. 384–387. Association for Computing Machinery, New York (2014)

12. Hebig, R., Quang, T.H., Chaudron, M.R.V., Robles, G., Fernandez, M.A.: The quest for open source projects that use UML: mining GitHub. In: Proceedings of the ACM/IEEE 19th International Conference on Model Driven Engineering Languages and Systems, MODELS 2016, pp. 173–183. Association for Computing Machinery, New York (2016)

13. ISO/IEC 25010: Systems and software engineering - systems and software quality requirements and evaluation (square) - system and software quality models (2023)

14. Jenner, B., Flick, U., von Kardoff, E., Steinke, I.: A Companion to Qualitative Research. Sage (2021)

15. Kalliamvakou, E., Gousios, G., Blincoe, K., Singer, L., German, D., Damian, D.: An in-depth study of the promises and perils of mining GitHub. Empir. Softw. Eng. **21**(5), 2035–2071 (2016)

16. Malavolta, I., Lago, P., Muccini, H., Pelliccione, P., Tang, A.: What industry needs from architectural languages: a survey. IEEE Trans. Softw. Eng. **39**(6), 869–891 (2012)

17. Malavolta, I., Lewis, G.A., Schmerl, B., Lago, P., Garlan, D.: Mining guidelines for architecting robotics software. J. Syst. Softw. **178**, 110969 (2021)

18. Muszynski, M., Lugtigheid, S., Castor, F., Brinkkemper, S.: A study on the software architecture documentation practices and maturity in open-source software development. In: IEEE International Conference on Software Architecture, pp. 47–57 (2022)

19. Ozkaya, M.: What is software architecture to practitioners: a survey. In: International Conference on Model-Driven Engineering and Software Development (MODELSWARD) (2016)

20. Ozkaya, M., Erata, F.: A survey on the practical use of UML for different software architecture viewpoints. Inf. Softw. Technol. **121**, 106275 (2020)

21. Prana, G., Treude, C., Thung, F., et al.: Categorizing the content of GitHub README files. Empir. Softw. Eng. **24**, 1296–1327 (2019)

22. Rost, D., Naab, M., Lima, C., von Flach Garcia Chavez, C.: Software architecture documentation for developers: a survey. In: Drira, K. (ed.) ECSA 2013. LNCS, vol. 7957, pp. 72–88. Springer, Heidelberg (2013). https://doi.org/10.1007/978-3-642-39031-9_7

23. Rozanski, N., Woods, E.: Software Systems Architecture: Working With Stakeholders Using Viewpoints and Perspectives. Addison-Wesley Professional (2005)

24. Saldaña, J.: The Coding Manual for Qualitative Researchers. Sage (2021)

25. Shahin, M., Liang, P., Babar, M.A.: A systematic review of software architecture visualization techniques. J. Syst. Softw. **94**, 161–185 (2014)

26. Smolander, K.: What is included in software architecture? A case study in three software organizations. In: Proceedings Ninth Annual IEEE International Conference and Workshop on the Engineering of Computer-Based Systems, pp. 131–138 (2002)

27. Tu, Q., Godfrey, M.: The build-time software architecture view. In: Proceedings IEEE International Conference on Software Maintenance, ICSM 2001, pp. 398–407 (2001)

28. Venigalla, A.S.M., Chimalakonda, S.: What's in a GitHub repository?–a software documentation perspective. arXiv preprint arXiv:2102.12727 (2021)

29. Verdecchia, R., Engström, E., Lago, P., Runeson, P., Song, Q.: Threats to validity in software engineering research: a critical reflection. Inf. Softw. Technol. **164**, 107329 (2023)

30. Wohlin, C., Runeson, P., Höst, M., Ohlsson, M.C., Regnell, B., Wesslén, A.: Experimentation in Software Engineering. Springer, Heidelberg (2012). https://doi.org/10.1007/978-3-642-29044-2
31. Zagalsky, A., Feliciano, J., Storey, M.A., Zhao, Y., Wang, W.: The emergence of github as a collaborative platform for education. In: ACM Conference on Computer Supported Cooperative Work & Social Computing. Association for Computing Machinery (2015)

Author Index

A

Abdelfattah, Amr S. 20
Aguiar, Ademar 37
Ahmeti, Bardha 333
Alam, Khubaib Amjad 359
Amalfitano, Domenico 53
Amou Najafabadi, Faezeh 69
Arju, Muhammad Ashfakur Rahman 20
Asif, Hira 359
Astudillo, Hernán 350

B

Babar, Muhammad Ali 270
Bakhtin, Alexander 174
Batista, Thais 379
Becker, Steffen 121
Bogner, Justus 69
Boltz, Nicolas 253
Brogi, Antonio 350
Bzuneh, Teklie Belay 183

C

Cañas, Norberto 217
Capilla, Rafael 324
Cavalcante, Everton 379
Cerny, Tomas 20
Cetinkaya, Anil 183

D

De Angelis, Domenico Francesco 53
De Luca, Marco 53
Di Nitto, Elisabetta 288
Di Nucci, Dario 3
Díaz-Pace, J. Andrés 324
Dini, Vick 288

F

Fasolino, Anna Rita 53
Fatima, Iffat 233
Franz, Thomas 207

G

Gajewski, Darek 20
García, Javier 217
Genfer, Patric 157
Gerking, Christopher 253
Gerostathopoulos, Ilias 69
Groner, Raffaela 333
Guamán, Daniel 217

H

Harrison, Neil B. 37
Hebisch, Erik 207
Heinrich, Robert 121, 253

I

Inayat, Irum 359

J

Jahic, Jasmin 138
Jahić, Jasmin 371

K

Kaya, M. Cagri 183
Khan, Saif-Ur-Rehman 359
Krieger, Niklas 121
Kugele, Stefan 105
Kumara, Indika 3

L

Lago, Patricia 69, 233, 396
Lamprecht, Martin 105
Li, Xiaozhou 174
Linder, Maja 333

M

Malavolta, Ivano 396
Matar, Raghad 138
Mendonça Filho, Ricardo César 191
Mendonça, Nabor C. 191

M. Galster et al. (Eds.): ECSA 2024, LNCS 14889, pp. 417–418, 2024.
https://doi.org/10.1007/978-3-031-70797-1

Migliorini, Sofia   396
Mzid, Rania   86

O
Ochoa, Juan Sebastian   217
Oguztuzun, Halit   183
Oquendo, Flavio   379

P
Pallewatta, Samodha   270
Pérez, Jennifer   217
Ponce, Francisco   350

R
Rezgui, Ilyes   86
Rodriguez-Horcajo, Vanessa   217

S
Schmid, Larissa   253
Schoop, Sven   207
Schreyer, Lorenz   105
Soldani, Jacopo   350

Soliman, Mohamed   307
Speth, Sandro   121

T
Taghavi, Bahareh   253
Taibi, Davide   174
Tamburri, Damian Andrew   3, 288
Tommasel, Antonela   324

U
Ustunboyacioglu, Ipek   3

V
van den Heuvel, Willem-Jan   3
Verdecchia, Roberto   396
Vicario, Enrico   396
Viglioni, Tales   379

W
Wohlrab, Rebekka   333

Z
Zdun, Uwe   157
Ziadi, Tewfik   86

SPRINGER NATURE

GPSR Compliance

The European Union's (EU) General Product Safety Regulation (GPSR) is a set of rules that requires consumer products to be safe and our obligations to ensure this.

If you have any concerns about our products, you can contact us on ProductSafety@springernature.com

In case Publisher is established outside the EU, the EU authorized representative is:

Springer Nature Customer Service Center GmbH
Europaplatz 3
69115 Heidelberg, Germany

The manufacturer's authorised representative in the EU is Springer
Nature Customer Service Centre GmbH, Europaplatz 3, 69115 Heidelberg,
Germany. If you have any concerns regarding our products, please
contact ProductSafety@springernature.com

Printed and bound by CPI Group (UK) Ltd, Croydon, CR0 4YY
29/04/2026
02099532-0013